Pattern and Chaos in Art, Science and Everyday Life

Pattern and Chaos in Art, Science and Everyday Life

Critical Intersections and Creative Practice

EDITED BY

Sarah Horton and Victoria Mitchell

intellect

Bristol, UK / Chicago, USA

First published in the UK in 2023 by
Intellect, The Mill, Parnall Road, Fishponds, Bristol, BS16 3JG, UK

First published in the USA in 2023 by
Intellect, The University of Chicago Press, 1427 E. 60th Street,
Chicago, IL 60637, USA

A catalogue record for this book is available from
the British Library.

Copy editor: MPS Limited
Cover designer: Tanya Montefusco
Cover image: Lucy Ward, *You Guys Are So Stochastic*
(detail), 2019, ink and pencil on watercolour paper.
Production manager: Debora Nicosia
Typesetter: MPS Limited

Hardback ISBN 978-1-78938-871-8
Paperback ISBN 978-1-78938-780-3
ePDF ISBN 978-1-78938-781-0
ePUB ISBN 978-1-78938-782-7

Printed and bound by CPI Group (UK) Ltd, Croydon, CR0 4YY

This is a peer-reviewed publication.

Contents

Figures

Acknowledgements

The concept underlying the content of *Pattern and Chaos in Art, Science and Everyday Life: Critical Intersections and Creative Practice* was developed through discussions and debate with the Pattern and Chaos research group at Norwich University of the Arts (NUA). To the members of this inter-disciplinary group, some of whom have contributed to this volume and all of whom have supported us at every step of the way, we are indebted. The book would not have happened without them. We are grateful, especially, to Ben Salter, who participated in our initial discussions and selection of contributors, and to Krzysztof Fijalkowski, who generously read and commented upon drafts of the proposal and the introduction.

Beyond the Pattern and Chaos group, the wider research culture at NUA has similarly supported us, not only through a substantial financial contribution which has enabled colour printing throughout but also, especially through the attentions of Simon Willmoth and Marcia Pointon, through support in shaping the concept of the book in the initial stages of its development. Without the dogged support of a vibrant research culture, the relationship between ideas and outcomes is not inevitable, so we are especially grateful to have been beneficiaries of this enabling outlook and generosity. Beyond NUA, David Girling and John Mitchell have also been readers of drafts as well as keen supporters of the process leading to publication.

The contributors to this volume have, without exception, been a pleasure to work with, often in the face of finicky or last-minute demands. We are very grateful, not only for their positive commitment to and engagement with the underlying 'pattern-chaos' premise but also for their responsiveness to the shape and structure of the book. Furthermore, given that the visual aspect of this book has been at the core of the concept from its outset, we extend our gratitude also to those who have freely given copyright permission for images to be published.

The editors at Intellect Books, Mareike Wehner and (latterly) Jessica Lovett and Debora Nicosia, have been brilliantly supportive and committed throughout the process of steering the volume to publication. They have been utterly professional and always clear and generous in guiding the process. Similarly, the design team at Intellect have been consistently positive and skillful in accommodating our suggestions and needs.

Introduction

Sarah Horton and Victoria Mitchell

This is not a book about pattern, nor about chaos (though readers will encounter many readings or re-imaginings of these), but it is about the dynamics of their relationship. Although pattern and chaos are conventionally considered and structured as a duality (pattern as repetitive or rhythmic in contrast to chaos as random or muddled, for example), there is considerable overlap and intersection between them. While there are many books dealing with pattern (generally referencing nature or pattern design) and many books addressing chaos (mainly philosophical or scientific), we believe that by bringing pattern and chaos into conjunction with one another to examine their inter-relationship this anthology opens up a fresh constellation of ideas and examples for apposite academic inquiry. Pattern and chaos as discrete concepts are typically addressed through scholarship within disciplinary-specific contexts but rarely, if at all, do they appear together in such a way as to absorb a wide range of subject disciplines through a diverse range of processes.

This rich and complex territory is ripe for exploration and debate. What begins to emerge is a multifaceted but unified concept in which pattern and chaos are not only inseparable but are also, in combination, infinitely fascinating in the ways in which they can be exemplified through experimentation and probing research. Rather than existing in opposing spheres, the complex dynamic of pattern and chaos in interaction enhances and challenges critical frameworks and modes of understanding, perception and communication, often referencing the in-between territory of art and science. Although visual culture is a primary context for many contributors to this book, the research and outcomes are extensive in reach and myriad in the contexts to which they refer. They are also recognizable and relevant to the realities of the everyday, evident not only with reference to technologies but also through consideration of the way connections and disconnections between pattern and chaos reveal the personal and political instabilities of our time. As well as stimulating interdisciplinarity, the pattern and chaos combination sheds light on contemporary and future practices, thinking and behaviours, often with concomitant parallels drawn from historical precedent.

Pattern, pattern-thinking, pattern-breaking and pattern-recognition emerge not as the sole preserve of the visual, let alone of the arts, and chaos emerges not simply as the chaotic mess we make nor as the property of an unpredictable, complex system in physics such as may have existed before the creation of the universe. In their place, an as-yet uncharted territory of 'pattern–chaos' or 'chaos–pattern' begins to unfold, not only as a domain for critical, dialogic practice but also as a territory of subjective sensation, to be explored and experienced first-hand.

Developed through ongoing discussion in the Pattern and Chaos research group at Norwich University of the Arts,[1] the anthology aims to provide a valuable and versatile contribution to the way in which knowledge systems remain open to question and depend on innovative forms of communication. That many examples emphasize methods and processes relating to visual images may reflect not only the context within which the concept of the book developed but also the fact that interdisciplinary fields of knowledge production are intrinsic and conventional to practices and critical positions associated with visualization and the arts. Several contributions have involved bespoke drawing or writing made with the pages of this book in mind, often highlighting the intersection of text and image as a form of practice through which meaning is communicated. Speculative, playful and experimental approaches, however communicated, are viewed as central to the pattern–chaos concept as explored on these pages, including the structure of the book itself, which does not adhere to a formulaic model. Readers are encouraged to be experimental in their response, open to the possibility of rejecting the chronology and structure as presented in order to arrive at yet-to-be-determined conclusions or insights.

In selecting the contributions we were overwhelmed by the possibilities presented, as if the relationship between pattern and chaos was always there, waiting for an opportunity to be expressed or given focus. Contributors not only expressed enthusiasm and commitment to the field as outlined but were able to shape it to their varied practices with relative ease and with innovative examples based on original research. Although the methods employed are wide-ranging, they are drawn together coherently within the pattern–chaos nexus.

The remainder of this introduction is formed of three parts. Firstly, many of the ways in which pattern and chaos can be figured theoretically are considered, identifying why the recognition and mapping of the territory between pattern and chaos is both timely and worth investigation. Secondly, we consider the many ways in which the contributions in this anthology challenge and articulate pattern–chaos and chaos–pattern through processes of creative making, writing and thinking. Finally, we outline our approach to structuring the anthology and consider how the mapping of the spaces between the constructs of pattern and chaos within the six sections of the book represents a cartography of understanding for which the pattern–chaos nexus emerges as an active, facilitating process.

Mapping the territory

Considered from an anthropological perspective, pattern has long been understood as formative of semiotic coding and as a visualization of thought and behaviour, but has also demonstrated that a tendency to order is reversible (like a well-balanced figure–ground relationship), allowing chaos to appear within its midst as a not-so-distant relation. For Gregory Bateson (Bateson 1971: 148), all actions which are interlocking, directional, ever-changing, active, social and patterned, whether performative or material, are related to an 'almost universal linkage in aesthetics between skill and pattern', while for Suzanne Küchler patterns are 'good to think with' (Küchler and Were 2005: 172), both reflecting and formative of patterns of thought. Extending these perspectives, Alfred Gell (1998: 73–95) notes the animated effect of dense, complex patterns as they capture attention or convey enchantment as if referencing the confusion of a maze, or of chaos as a pattern that has not (yet) been discerned. Thus, rather than conforming to regular notions of repetition or recognizable shapes 'the *telos* of decorative art lies in the opposite direction, towards the complex, the ambiguous, and the multitudinous' (1998: 79). Patterns give order to the discordance of human experience, but they also contribute to the inevitable disarray which, perhaps increasingly, seems to dominate aspects of that experience. Pattern-making and pattern-perception are seen by Gell, as by many contributions here, as components of social technology, having an agency that is far wider reaching than the mere prettiness with which decorative patterns are familiarly associated.

Pattern takes many forms, but, as contributions to this volume attest, there is often an underlying or overt reference to geometry. For the architect and theorist Buckminster Fuller, pattern functions to translate geometry into material and *vice versa*, as in the 'World pattern inventory facility' which he envisaged as being able to track natural phenomena and cross-reference these to man-made satellite movements or trade routes using a 200 feet sphere onto which data would be projected from within and without to create 'an omnidirectional spherical picture' (Fuller in López-Pérez 2020: 127). This mass of data could then be used, for example, to track the movement of mined minerals as they moved around the globe, to capture 'the ecological pattern of man' [*sic*] (Fuller in López-Pérez 2020: 250).

The coincidence between conceptual and material found further resolution in Fuller's geodesic domes, through the use of tessellation. Familiar in the form of the honeycomb, and known to have been used by the Sumerians, circa 4000 BC, tessellation is also germane to Islamic patterning, both in design and as a means to access and celebrate the divine. Fusing geometry with aesthetically resonant mysticism, such patterning, when imagined as a harmony of the spheres, functions to ward off chaos and to cleanse the senses of disorder (Marks 2010: 162–63). Yet it is also very complex, with rhythms that shift as polygons and stars interlock and overlap in myriad ways, seeming to defy the reasoning that geometry purports to exemplify (Necipoğlu 1995: 231). Such complex geometric patterning may engage with the mystical, but it can also evoke chaos in the mind's eye.

Within patterned matter, it is not just enchanting entanglement or maze-like confusion that gives rise to chaos. Contagion can also lurk, as in the potential of synthetic biology to take a wrong turn, ethically speaking, or, as identified in contributions to this volume, surfacing as virus or glitch. Artificial Intelligence, computer-based pattern recognition and artificial neural networks continue to expand the complexity and range of pattern's identity, sometimes provoking the consternation of those who identify the way in which such systems harbour 'mis-recognitions' and can negatively impact justice and truth. As Hito Steyerl argued in her seminal essay 'A sea of data: Apophenia and pattern (mis-recognition)' (2017), a tendency to perceive meaningful (patterned) connections between unrelated phenomena (apophenia) has long formed and informed the imagination, enabling the unimaginable and improbable to be articulated, but as they impact contemporary life their unreliability indicates the urgency for bringing pattern–chaos intersections to the fore. The dynamics of infinitesimal uncertainty are inherently unstable but also ever-present and potentially uncomfortable in their effect on reality. For Quentin Meillassoux (2009), this state of unreasoned inseparability is an absolute or limitless 'hyper-chaos', which equates with reality. Although order and chaos co-exist, Melliassoux's understanding of reality indicates that they can never be fully disentangled from one another.

The spaces in which we live and work are permeated by conjunctions of pattern and chaos. For Henri Lefebvre ([1992] 2013), social rhythms, along with the cycles and habits of the body, are in a state of perpetual transformation in our everyday lives, effecting states of conflict as different rhythms jostle against one another. 'Habit is the way we make patterns in the world' (Highmore 2016) and whilst there is productivity in habit-learning, and repetition can lead to accomplishment (William James 1890), Ben Highmore identifies habit as both possibility and problem. Excessive repetition of language and other habitual behaviour can prove problematic, not only in the Freudian (1920) or Lacanian (1978) sense of compulsive repetition but also in stultification. Too much of the same pattern ceases to be comforting or pleasurable and becomes torturous or disconcerting: we lose our ability to differentiate between edges, between days and weeks. Habits become part of our habitat, our habitus even (Bourdieu 1984), but can also be problematic if we become stuck in our ways. James considers that measuring the world based on what we already know and avoiding that which disturbs our sense of ease is the very thing that prevents revolution. Habit leads to routine leads to conservatism. Questioning the 'pattern maintenance' that moderates the values that are institutionalized in society (Talcott Parsons 1951) creates space for change.

From the perspective of the social imaginary, patterns of the projection within which social life is framed and understood can be conceived as layers of dominant ideologies or of groupings such as the familial, local, national and global. While the structures of the human psyche can't be reduced to a social world, they are implicated in the imaginary formation and complexity of its patterning. The imaginary and the real intersect, as do pattern and chaos, in a manner which is essentially undetermined and thus open to invention and variation. The imaginary has been described as 'the unceasing and essentially undetermined (social-historical and psychical) creation of figures/forms/images' (Castoriadis 1975: 3) and in this sense, it indicates an apt identification of visual practices through which irregularity and fixed schemata are constantly in negotiation, in ways that mirror the shifting but symbiotic relationship between pattern and chaos and inform the research basis of many of the contributions included in this volume.

In combination, pattern and chaos attract a search for origins and aetiological explanations which serve to make sense of totalities and worldviews. For biologist and theorist Donna Harraway (2016), the chthonic (that which inhabits the underworld) and the tentacular are necessary to one another. She speaks, for example, of 'material-semiotic composting, as theory in the mud, as muddle' and of 'the chthonic powers of Terra' infusing 'its tissues everywhere' in a tentacular fashion. The chthonic entanglements of living and being are implicated in catastrophic extinctions but as 'everything is connected to something' and 'nothing is connected to everything', intricate networks and pathways also form myriad patterns of potential for sustained life. In *Entangled Life: How Fungi Make Our Worlds* (2020), Merlin Sheldrake articulates a similar terrain of microbial connectivity. For contributors to *Pattern and Chaos* tentacular patterns of actions and chthonic forces are perceived in combination as intra-actively and materially embedded and as possessing agency.

Such imagining and scrutiny of matter, rooted in challenges to ecosystems and (especially now) viruses, is not without reference to the classical notion of Chaos

Theory as pertaining to the way in which a minute change in one context can rise exponentially to initiate a tsunami of change in another context. Infinitesimal detail structures both pattern and matter, thus the patterned minutiae of organic life as formed within the orderly expansion of galactic space and time attest to processes which are everywhere active, emergent, materially self-replicating and self-organizing. Whether organic or inorganic, matter is inherently creative, such that 'even the humblest forms of matter and energy have the potential for *self-organization* beyond the relatively simple type involved in the creation of crystals' (De Landa 1997: 16, original emphasis).

Our thinking, feeling and believing are not excluded from self-organization (Coveney and Highfield 1995: 279–327). The neo-cortical columns of the brain process information for consciousness and sensory perception through responsive, intricate, dense neural networks, the inner workings of which are 'chaotically patterned' (Korn and Faure 2003). With over twenty billion neurons in the neocortex, each of which forming as many as 10,000 electrical or chemical synaptic connections, it is perhaps not surprising that scientists have yet to fathom their complexity. This does not deter the imagination from imagining, the mind from thinking or the hands from exploring materials through touch or sight, as contributors in this volume, some of whom have engaged specifically with neuroscience, can attest. The relationship between excitation and inhibition, which is central to the orderly or disorderly manner in which neurons interact with one another, is arguably the biochemical equivalent of the zeros and ones of artificial or digital processes. Sir Charles Sherrington's inspired description of the brain (1942: 176), awakening 'as if the Milky Way entered upon some cosmic dance', and swiftly becoming 'an enchanted loom, where millions of flashing shuttles weave a dissolving pattern, always a meaningful pattern, though never an abiding one: a shifting harmony of sub-patterns' might also describe the agency of the Jacquard loom, or the computer, or other man-made devices through which the intersection between pattern and chaos comes to be known.

Dissolving patterns are also evoked as a form of chaos by Gilles Deleuze and Félix Guattari in their consideration of the way in which the mind attains a degree of consistency, as in rhythmically holding heterogeneous elements together. They characterize chaos 'not so much by its disorder as by the infinite speed with which every form taking shape in it vanishes' or similarly, as a void 'containing all possible particles and drawing out all possible forms, which spring up only to disappear immediately, without consistency or reference, without consequence' (1994: 118). This ceaseless movement, which is not so much a chance mixture as an active undoing of 'every consistency in the infinite', is also a source of energy, impulse and strength which can be slowed down and made coherent through *sensation*, thereby enabling consistency 'without losing the infinite into which thought plunges' (Deleuze and Guattari 1994: 42; see also Grosz 2008: 8–9). Especially, sensations of rhythm and vibration function as a filtering, connective tissue and as a form of shielding from the void of chaos, marking constant fluctuation as waves, pacing, measure or refrain (Deleuze and Guattari 1994: 168; Grosz 2008: 19). In the present context we identify glimmers of pattern–chaos and chaos–pattern in this process. Although Deleuze and Guattari consider science, art and philosophy as tending towards different ways of processing the relationship between chaos and pattern (with science more likely to develop formulae rather than the representations associated with art, for example), a number of contributors here have reached solutions which draw on combined approaches, often through collaboration.

Struggling sublimely against the forces of chaos has been characterized through myth as futile or maddening, figured as a forcefield of darkness that cannot be overcome and that cannot be reasoned. This infinite unknowability and formlessness of chaos, beyond the bounds of reason and as pertaining to the unconscious, draws critical and creative attention towards it, functioning as a source of (in)formation, however random, blurred, difficult or undifferentiated this might be. Chaos may be un-knowable, but it has agency and is active in its continuous presence and emergence, experienced in and through the dimension of time. In Greek mythology the limitless abyss or primal, dark void of Chaos (χάος), as in Hesiod's poetic *Theogony* (1953), is the initial state out of which the universe arises, an undifferentiated, unstable, shapeless cosmogonic imagining. This Chaos is drawn upon in Plato's reflections on *chôra* (χώρα) (*Timaeus* 1971) as a space of ambiguity within which a disorderly motion of the elements forms a tentative something, from which material becoming is conceived as emergent. For philosopher and psychoanalyst Julia Kristeva, this *chôra* shapes the relationship between the semiotic and the Symbolic which precedes and conditions that which can be named – prior to the subject and to meaning but nevertheless active in the formation of that which, as such, is coming into being (Kristeva 1984). By this account, the coding of differentiation and hence of pattern would not exist without the unknowable chaos that precedes it. This coming into being of language can also, as for many contributors here, be figured as a process of 'return', whereby a dissolving of language enables a regaining of the undifferentiated or amorphous knowing and becoming that has been lost or forgotten over time.

Making meaning

Each contribution in this volume is led by new research and scholarship, whether through visual practice, text or scientific enquiry. Although approaches are varied, they all capture aspects of the interstices between pattern and chaos, often as a dynamic exchange or dialogue within and between ideas and/or as interchange resulting from processes, whether 'natural' or occurring in the context of a studio or laboratory. Research can be conveyed visually but is also reflected upon, through written exegesis and critical scrutiny, engaging different disciplines, contexts and styles, including poetic, scientific, historical, theoretical, practical and speculative modes of expression.

The theoretical context for the variety of approaches is not easily pigeon-holed but arguably this is a strength rather than an impediment. In 'Tactics for not knowing: Preparing for the unexpected' (2013) the writer and artist Emma Cocker considers the unknown as 'an anomalous breach or gap in existing thought that must be [...]. bridged by the production of new knowledge (2013: n.pag.)'. She notes how zones of indeterminacy are often actively sought out within the context of art practice and that disorientation of that which can be named is a valued strategy. In the context of the interstices of pattern and chaos, knowing and not knowing are necessarily interdependent, thus that which is not yet known can engage bafflement as a welcome state or stage of coming-into-being, exemplified by delight in a process of experimentation which avoids the restrictions of accepted conceptual or procedural frameworks. This is not to say that art is less ordered or less patterned than science, rather that both are underpinned by that which is not yet known and therefore open to question. Sarat Maharaj (2009) notes that 'what we lump together as "science" is often a congeries of quite divergent activities, disciplines and domains, each with its own kit of objectives and logical procedures' and that in considering the scope of knowledge production 'we might do better to keep matters open, perhaps with a feel for the hodgepodge of methods, even muddle, that attends the lab workbench (2009: n.pag.)'. The 'craziness' of collage is also seen to be useful, for example, in challenging accepted forms of representation. As a preface to their comparison between the 'smooth' but entangled, rhizomic complexity of felt and the 'striated' regularity of warp and weft weaving, including the ideological implications that arise from each, Deleuze and Guattari show the image of a 'crazy' quilt ([1988] 2020: 551). The quilt is 'an amorphous collection of juxtaposed pieces that can be joined together in an infinite number of ways' (554) and as such is symbolic of the anti-war machine which is a key focus of their account of smooth space. Among the contributions in this volume, many incorporate aspects of collage, of fragments re-cast to form alternative, rhizomic notions of states of knowing and being in the world.

The theoretical often registers as a discursive method, whereby discourse describes formal ways of thinking expressed through language, but as Maharaj argues (2009), the non-discursive is equally significant 'in its penumbra of the non-verbal, its somatic scope, its smoky atmospherics, its performative range', and as he notes elsewhere (2002: 71), 'the kinds of think-feel-know it throws up are quite different from discursive modalities'. In avoiding binary discursive/non-discursive approaches, contributors to this anthology not only reflect the indeterminate connective tissue between chaos and pattern but also the oscillating 'whatever' between word and image. Thought is a performative practice, shaped through materials, just as practice is thought forming. Breakthroughs into the not-yet-known are likely to be unscripted in origin which for Maharaj (2002: 71) can give rise to an alternative, contaminated geometric syntax.

Diagrams, tessellations, dust, knots, mazes, folds, creases, flux, virus, fire and flow are just some of the phenomena indicative of processes through which issues of pattern–chaos and chaos–pattern are addressed. These, and related terms, often evoke transition or passage, or movement between one state and another. Together with the specific techniques used in the research, such as photography, cyanotype, video, ceramics, typography, graphic design, textiles, collage, drawing, coding and digital media, there is a clear sense that attention to the relationships between pattern and chaos enables a rich seam of visual ideas to be brought to the surface. Making and meaning are in dialogue in articulating the intricacy of their intersection.

Drawing upon fields such as visual culture, sociology, physics, neurobiology, linguistics or critical theory, for example, contributors have experimented with forms, processes, materials, sounds and language or have reflected on the work of other artists, scientists and scholars. How things are made, how forms take shape and how matter is perceived are issues addressed through experimentation, chance and inventive strategies. A tendency of pattern-like order to break down 'chaotically', as in entropy, is encouraged and/or noted by artists and scientists alike. Many of the contributors take risks, beginning with a process that they deliberately break from. This is a purposeful interruption or derailment, as if the pattern is not enough. Although not commonly used as such, pattern can function as a verb – thus 'to pattern' as a shortened version of 'making a pattern for (something)'. But there is no chaos equivalent, no active 'to chaos' or 'chaosing' as a shortened form of making chaos, which in a sense is being enacted here: there is a curiosity about making that pushes the artist to explore the 'what ifs' that deliberately invoke chaos. The artist might add into their process a catalyst or an additional factor that problematizes the equation – a type of 'bug', not unlike the virus that is able to gain a hold or footing within its host, thus building traction. Concepts are inert until the artist/writer/theorist brings out the latent potential within them. The etymological sense of the word pattern as 'taking as a model' is overturned in various instances where the makers here overturn or question a prescribed model or their own template. For the visual arts in the twenty-first century, such challenging of measured forms not only bears comparison to the Romanticism which once challenged the classical but also to the provocations of Dada's response to the atrocities of the First World War. There is a sense of disruption and chaos as characteristic of that which is not yet known, of processes which reach beyond status quo stability or pattern as a fundamental principle of logic.

When a 'troubling' of material or process is undertaken through a predominantly visual approach, the research and its effects can bypass or usurp the power that language is often afforded (Barad 2007). For example, the eighteenth-century naturalist Carl Linnaeus created calligrams to describe botanical forms, using paragraphs and typographical distinctions to re-visualize the material he was describing (Foucault [1966] 2002: 147). Our investigation of pattern and chaos has resulted in many contributions which address how language can be re-envisaged, ordered or disordered through the use of image, composition and typography as well as through grammar and syntax, articulating that which is sometimes not yet fully resolved through visual-material means. Visual and written forms are frequently fused, articulating the hinterland of pattern and chaos in such a way as to provoke a reconsideration of syntax and understanding. The complex, experimental processes employed in visual and material practices can be especially resonant and direct in terms of accessing and engaging with pattern–chaos, and where signs and structures of language are referenced they are often untethered from their recognized meaning to prompt surprising and atypical implications.

In this pattern–chaos universe, new meanings emerge and old meanings are challenged. Always manifold and multifarious, such meanings are a site of continuous construction and demolition, of coming into being or melting into thin air. Whether the agency of becoming meaningful is perceived as human-centred or multispecies in origin, pattern–chaos is intrinsic to the (bio)morphology of connectivity as it infiltrates tissues of knowing and making.

Order, order

Arriving at a structure for this book was itself a creative endeavour. Michel Foucault (1966) describes the drive to classify and bring together accepted knowledge with the yet-to-be-known, redefining anachronistic topologies in light of newly emerging knowledge. These are patterns of thought that are disrupted and fragmented, made chaotic even, by the imposition of new and updated patterns. In shaping the order of this book there were any number of ways in which we, as editors, could have structured this group of contributions, for example by word length, in which case dust butts up against water movement in the Norfolk Broads, or by alphabetical order of the title, by the number of images contained, and so on, many of which contain a logic of sorts, but are still largely arbitrary, and not especially helpful in uncovering clusters of ideas. Why has this structuring been important? Because although pre-existing categories make things nameable, writers and makers cast into doubt and re-position these assumptions. By re-ordering, re-categorizing and re-configuring the old, new patterns of understanding can be brought to light.

Patterns and categories can also obfuscate, narrowing valences or possibilities, the danger being that by naming something we capture it, pin it down, and 'know it' in a way that reassures us, but is then able to prejudice our thinking. To avoid this, the headings we have used are deliberately poetic as well as descriptive, intentionally permeable, and hopefully leave room for the miscellaneous and anomalous that do not neatly fit, that could indicate a category not previously imagined. It is therefore an important feature of our structure that nothing is too tightly fixed. It is open to new additions or configurations by readers and future researchers. The section groupings also enable the structure to be consistent in highlighting interdisciplinarity, rather than (say) putting all art–science collaborations or all collage-related applications in the same 'box'.

The contributions have been organized into clusters of subjects which reflect a robust, yet also experimental, arrangement. Each section begins with a short introduction which explains this clustering and further embeds the contributions into the overall pattern–chaos nexus. The overall structure is organized with a leaning towards pattern at the outset (Part 1, 'Pattern Dynamics') and a tendency to the chaos of disintegration and decomposition at the end (Part 6, 'Nothings in Particular'). 'Pattern Dynamics' brings together contributions, across a range of disciplines, which address pattern quite directly but in ways which challenge notions of process and/or reorganize forms or materials according to less predictable outcomes or metaphorical resonance. In Part 2, 'Morph, Flux, Mutate', the contributions emphasize the passage between pattern and chaos and its creation of evolving and mutating form. For example, pattern painted or drawn across a surface, and the aberrations of that pattern, might become continuous mutations – patterns that morph towards the unpredictable and ambiguous. Generative systems spawn multiple and iterative forms. Part 3, 'Decompose–Recompose', includes practice-led contributions which use disruption and instability as creative strategies and an imposition of chaos upon known and controllable variables. Products deriving from chance processes such as the glitch, deterioration and shattering demonstrate the potential for collage and related practices to challenge traditionally held patterns of thought and form.

Extending the realm of the proliferous, and especially given the impact of COVID-19, we felt that no current publication on pattern and chaos could avoid mention of viruses (Part 4). Here the viral is explored as 'lively matter', examining the virus' existence as between the living and nonliving, relying for survival on a host upon which it imposes its own patterns. Pattern in the artworks featured in this part employs the metaphor of virus or the DNA of a virus itself as generative code and/or as insidiously adaptable.

The impact of COVID-19 on social life resulted in new patterns of experience. Although not specifically focused on the virus, Part 5, 'Social Imaginary', nevertheless considers pattern and chaos in socially sensitive, visually attuned and psychologically poignant arenas. The notion of matter out of place, whereby a conventional (recognizably patterned) situation is subject to interference or noted for the disruption that an outsider can bring, is a particular focus for many contributors in this section. There is frequently a breaking down or collapse of 'acceptable' codes or norms of behaviour, whether through impact on the human body, perception, the psyche or the social.

Finally, 'Nothings in Particular' brings together contributions which address the almost invisible, the barely noticed and the inconsequential. Entropy, disintegration, collapse, intense observation and mind-altering activities – states or processes which shape subjectivity, making practices and critical thinking, for example – are explored.

Conclusion

Deleuze and Guattari have been influential and prescient guides in mapping the unfolding of this territory, not only in respect of the rhizomic pathways which counteract the tap-root formation of pattern and chaos as separate constructs but also for their understanding that pure chaos can be forestalled through sensing its vibrations and rhythms. A boundary between non-sense (chaos) and sense (rhythm, pattern) is crossed, as likewise in Guattari's *Chaosmosis* (1995), whereby chaos crosses over into cosmos and 'all the disciplines will have to combine their creativity to ward off the ordeals of barbarism, the mental implosion and chaosmic spasm looming on the horizon' (1995: 135). For Mark Featherstone, this chaosmic spasm is 'a transitional moment where nonsensical chaos folds over into meaningful complexity' and suggests parallels with the way in which the post-modern subject comes to terms with the vast territories of contemporary technology, ecological crisis and global complexity (Featherstone 2016: 244).

For Deleuze, the not yet known might also be figured as 'perplication', the state of 'Problems-Ideas' (Deleuze 2014: 367) within which both unknown and known, question and potential answer might exist together, and so we frame the contributions presented here using the terms pattern–chaos and chaos–pattern, to perplex the gaps in existing thought and uncover the fruitful dialogues that writing and making can uncover. Although this anthology could never be an exhaustive survey of pattern and chaos, so endemic and far-reaching are their implications, we present these concepts and the way they speak to each other as ever-changing, expanding, re-positioning and re-forming. In conclusion, therefore, we invite other 'makers of meaning' to activate our conjoining of pattern–chaos or chaos–pattern as a singular 'doing' word, and so as to continue this dialogue.

NOTE

1. Previous outcomes from the group include a symposium, 'Beyond Pattern and Chaos' (Norwich University of the Arts, 2018) and 'Drawing Pattern and Chaos', a single-issue volume of *Drawing Research Theory and Practice* (Horton 2019).

PART 1

PATTERN DYNAMICS

Introduction

Pattern presents a maze of possibilities. Before it came to be popularly understood as repeated decorative motifs, pattern (which is etymologically linked to the Latin *pater*, meaning 'father') more commonly referred to notions of model or template from which further examples could be made. In the context of chaos, this latter sense of exemplar invites additional connotations of order and certainty against which uncertainty and variety take their bearings. In the contributions to Pattern Dynamics, interactions between precision and disparity result in the opening up of notions and characteristics of pattern to experiment and scrutiny across a range of disciplines.

These include 'the dynamic patterns of living processes' which Gemma Anderson-Tempini, Jonathan 'J. J.' Phillips and John Dupré address in the context of protein science and for which Anderson-Tempini's drawings articulate a protein folding energy landscape as an elaborate maze. In a similar but reverse vein, the visual complexity of neurons as patterns is explored by Gill Brown through graphic representations of chemical neurotransmission. Distinguishing the technical meaning of chaos from 'random' effects and further resonating across the multidimensional network of energies connecting art and science, Brian and Harry Whalley explore how the dynamics of pattern-breaking are expressed in a number of different contexts, including swarm behaviour and sonification.

Many pattern motifs are familiar, likewise the oft-repeated actions of making them, but how a piece of paper is folded or crumpled, how a knot is tied or how a canvas is woven can be nevertheless open to investigation, technical invention, imagination and reflection. Together with the weaver Angharad McLaren, the artist Geoff Diego Litherland grew flax from which McLaren wove a patterned canvas onto which and in response to which Litherland painted speculative geometries, the whole process tracing a nuanced intermeshing of pattern in action. In Alun Kirby's 'metamorphograms', cyanotype and origami combine to fold memory into light while for the collaborative explorations of CRIMP, examples of chaotic and organized crumpling convey their belief in the fold as a reflection of a fundamental process of nature. Like the fold, the behaviour of knots and tangles exhibits opposing tendencies, towards orderly pattern or chaotic snarling, for example. Eleanor Morgan's 'attempts to draw a single knot' are precarious, but nevertheless parallel the embodied rhythm of weaver birds and a knot-making orangutan.

For folds and folding or knots and knotting, the patterns and the action of forming them work together, but not only in material, physical union. They are also formative of patterns of thought, embedded in our patterns of communication and intrinsic to being and to life, as evidenced scientifically in the workings of DNA. In Krzysztof Fijalkowski's essay on the spiral, which begins this volume, it is not only the perpetual motion of its form that shapes its workings but also the way in which it evokes spirit or spins out of control through force of imagination. While science and art might be bridged in myriad ways through pattern and chaos, this spiral also alerts us to a 'coiled articulation between disparate systems' that is never, ultimately, resolved.

1

The Anxious Spiral

Krzysztof Fijalkowski

The spiral is a vacuum [...]. It represents something [...] the void, the anxiety void, the void of anxiety (Bourgeois 2018: 64).

Rather than being mutually exclusive opposites, are pattern and chaos really two poles of a continuum, each implying and anticipating its other, in perpetual flux? This dynamic motivates the figure of the spiral, linked to both pattern and chaos while expressing their reciprocity in its movement between them. In play here is a loosening of the spiral's emotional or ethical appeal: where many approaches emphasize its value as a universal model for growth and structure in nature, and by extension for spiritual order, we might also read it as an unstable, disruptive sign of anxiety and disorder, a coiled articulation between disparate systems. In this way, it traces a path into divergent models of thinking, whose family connections might include Umberto Eco's characterization of the Baroque as a kinetic, 'progressively dilated' space in which works must be seen in perpetual transformation (1989: 7), or Roger Caillois's proposition for a 'diagonal science' bringing together incompatible systems of knowledge through common patterns in scientific and poetic disciplines, with the figure of the spiral singled out for special investigation (2003: 247). Among other spiral models for knowledge we might also consider Walter Benjamin's adoption of the 'convolute' (*konvolut*, an idea borrowed from Adorno) as a structuring principle for the sprawling research archive that could never become the book of *The Arcades Project* – meaning literally a bundle of printed papers, but with echoes of rotated and rolled form, of the textual scroll as an echo of a volute shell (see Benjamin 1999: xiv). The ancestry of such proposals includes the work of Goethe, one of the first scholars to examine the spiral as a principle of natural growth, for whom the 'spiral tendency' – in reciprocity with verticality – promises both a dynamic harmony resolving opposing forces and a model for thought capable of exceeding restrictions and conventions (Mainberger 2010: 214).

A universal figure in the cosmos – Goethe calls it a 'primeval phenomenon' (Mainberger 2010: 213) – the spiral is found everywhere in the natural world. Particularly striking is its occurrence at the furthest ends of the spectrum of scale, from sub-atomic particle events caught in bubble chamber experiments to vast spiral galaxy formations (Figure 1.1). Between these extremes are myriad instances of simple or complex natural spiral structures, from the coiled and radiating forms of shells and plants to bones, horns and other anatomical features (in the human body, for instance, the cochlea of our ear, muscle structures around the torso or the whorls of hair on our scalp).[1] What connects them all is growth, making them central to the study of natural morphology. A key part in D'Arcy Wentworth Thompson's

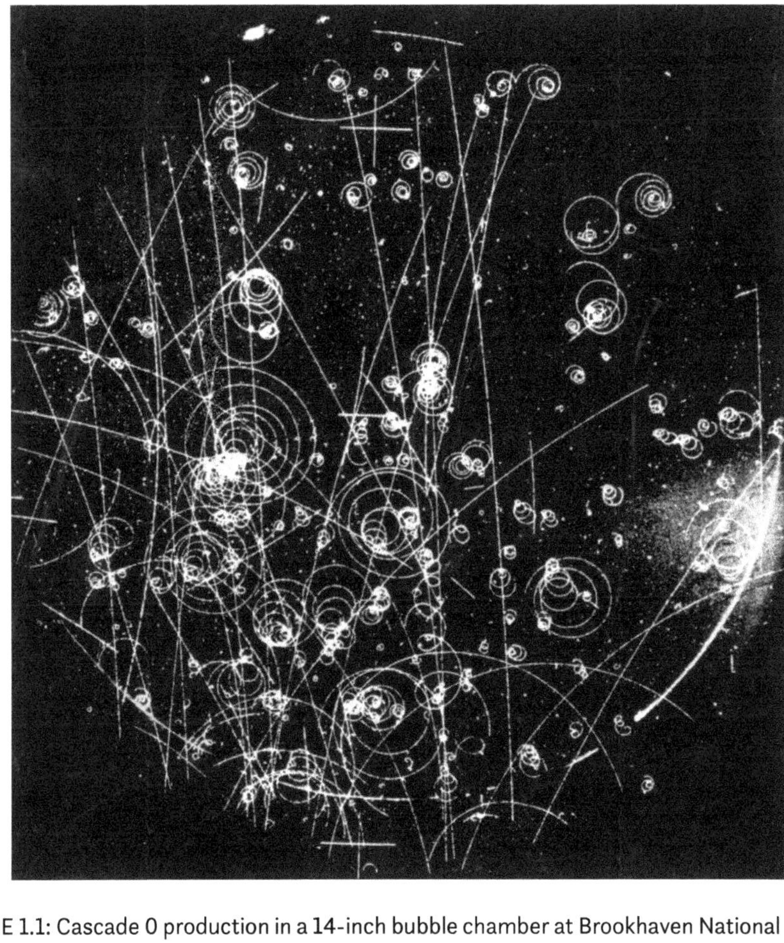

FIGURE 1.1: Cascade 0 production in a 14-inch bubble chamber at Brookhaven National Laboratory. Photograph taken on 12 March 1964. Bubble Chamber-1366. Photo courtesy of Berkeley Lab. © The Regents of the University of California, Lawrence Berkeley National Laboratory.

On Growth and Form of 1917 (2014: 172–220), the spiral is also the subject of Theodore Andrea Cook's (1979) slightly earlier *The Curves of Life*, advancing parallels between nature, architecture and art, including the relationship between logarithmic spirals and the golden section. Other research incorporating spirals within perspectives on structure and change include Édouard Monod-Herzen's *Principes de morphologie générale* (1927), or Theodor Schwenk's *Sensitive Chaos* (1965, first published 1962), on patterns of eddies, vortices and currents in water (Figure 1.2). All these works notably straddle scientific and creative approaches, their rich iconography making them particularly attractive to visual practitioners (Monod-Herzen's work was taken up eagerly by surrealists in Paris before the war, for instance,

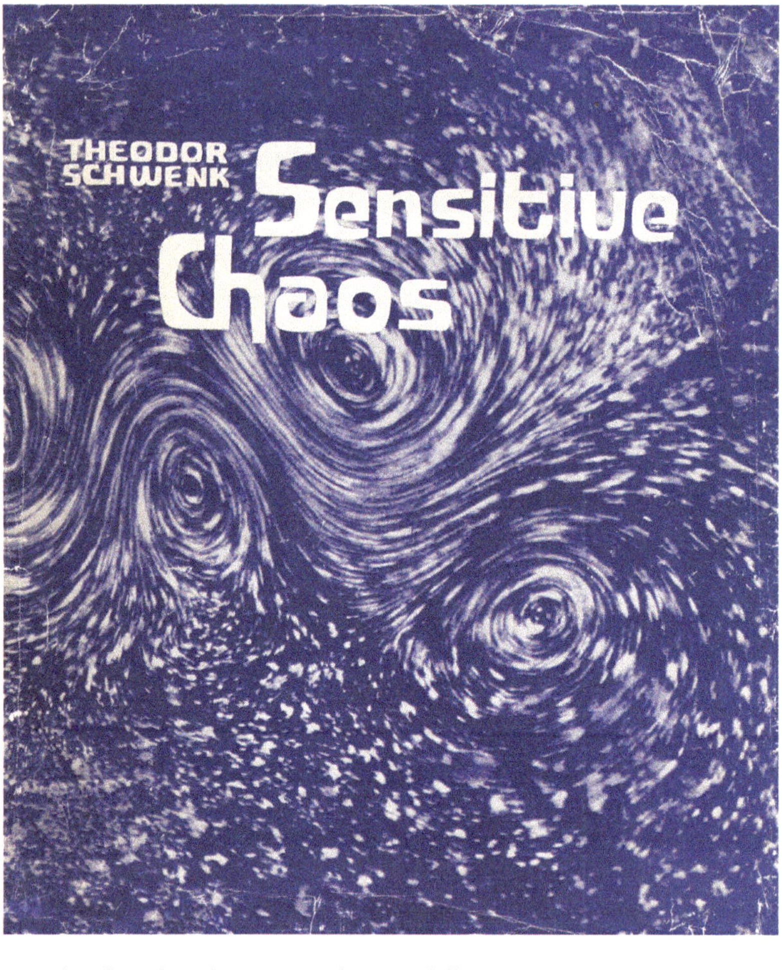

FIGURE 1.2: Theodor Schwenk, *Sensitive Chaos*, Rudolf Steiner Press, 1965. With permission.

Thompson's by the Independent Group in London after it), and despite their age, the majority of these popular science books remain in print to this day.

These universalizing tendencies, and the recurrence of spiral patterns across worldwide cultures, including the most ancient human mark-making, also make the spiral available for spiritual interpretations. Here it can function as an archaic symbol of eternity, cyclical processes, journeys or the divine; emphasizing wholeness and resolution, the spiral's ability to link inside and outside, self and other, implies that part of life contingent on time, becoming and realization (Purce 1974: 7–32). The spiral's combination of particular but indeterminate forms and meanings lends it a powerful charge as a sign of change and aspiration, as found for example in the abundant use of the motif in Hilma af Klint's abstract spiritualist paintings and drawings from 1906 onwards. Occult and alchemical thinking is part of this perspective, and indeed Goethe's conceptual framework for the spiral tendency drew upon esoteric as well as scientific models (Klausmeyer 2021: n.pag.). What emerges is the spiral's attraction for speculative thinking, its ability to remain open or reversible in its affordances, just as a spiral can be seen alternately as questing outwards from the centre or tracking inexorably inwards to a point of origin. Geometrical instability and tensions between centrifugal and centripetal forces make it a potent symbol of persuasion and alterity, from visual systems such as Celtic ornamentation to representations of hypnotic suggestibility. In its propulsion between inner and outer realms, beyond optical dimensions towards the shapes of thought, the spiral acts as a membrane or mutable zone pulsing between and across other systems, a model of universal structuring principles that might be traced, for instance, back to the ideas of the Ancient Greek philosopher Anaxagoras (fifth century BCE), who thought of the cosmos as in perpetual motion, constantly rotating and expanding, gradually separating all things as a rearrangement of the primary unity of matter (Curd 2019: n.pag.).

This dual nature – dynamic growth/tensile umbilical – characterizes many constructive spiral structures of the early twentieth century. Visionary designer and architect Frederick Kiesler's unbuilt Parisian department store project of 1925 imagined a helicoidal ramp clad in glass aiming to maximize customers' flow through sales settings, encouraging shoppers to traipse down the gentle incline from floor to floor without even realizing it (Kiesler 1996: 51; Phillips 2017: 80–85). This dream of unlimited contact between bodies and environments spurred Kiesler's proposals for spiral theatres around 1924–26, exploding the static stage and mingling actors and audience. Other ideas for theatres and architectural spaces featuring spiral configurations were in the air at this time, notably in contemporary projects by Andreas Weininger and Gert

Caden (see Phillips 2017: 29–38 and 57–67), encapsulating a post-war vision of the spiral as a utopian principle linking spaces or ideas, thrusting tendrils beyond conventional form into a radical tomorrow. Behind all of these lies Vladimir Tatlin's unrealized *Monument to the Third International* (1919–20), a giant rotating spiral scaffold intended to host multi-purpose communication functions, twisting upwards as though the marriage of progress and socialism had re-thought the Eiffel Tower as a technological ammonite. Here, perhaps, one might discern a dialectical building principle that underlies many natural spiral structures, where each new section improves on the previous one, resolving as it revolves (see Dawkins 1996: 183). Looking back three decades later, Kiesler reviewed the promise of this 'spiral-theme' (later realized in Frank Lloyd-Wright's Guggenheim Museum) in terms of the forces connecting experimental architecture, emerging science and the demands of a mechanized society driven to destruction: 'Gravitation (attraction) and radiation. Inward drive: (centripetal) equal to relaxation, exhaustion, shelter, peace. Outward-drive: ambition to expansion (centrifugal), acceleration distribution, contact, conquest' (1996: 48). These themes already informed the mythos of the tower in mediaeval or Renaissance architecture (see Cook 1979), where spiral staircase technologies, resonating with natural structures such as shells, are both defensive and colonizing devices (from this perspective, a tower without a spiral is simply a prison: the magic illogic of Rapunzel's doom).

But if spiral construction principles promise new rhythms and connections, might they also promote uncertainty, the unsettling loss of order? This slippage is found in several of Marcel Duchamp's kinetic works of this era: the mock-scientific apparatus *Rotary Demisphere (Precision Optics)* of 1925, in which a revolving half-sphere bearing a black and white whorl creates a pulsing space, and the set of six cardboard *Rotoreliefs* (1935) whose arrangements of circles give the illusion of receding or protruding once set in motion on the turntable of a phonograph, linking this visual play to the mechanics of acoustics where record grooves, and later magnetic tapes, store sound as a spiral (Figure 1.3). The *Rotoreliefs* developed similar rotating spirals used in Duchamp's trance-inducing film made with Man Ray, *Anemic Cinema* (1926). The subject would have chimed well for Man Ray, since spirals also populate his own images and objects of this period: 'When I was a young man', he recalled, 'I was already obsessed with this form; when working as a draughtsman I was fascinated by curves, spirals, parabolas, hyperbolas' (Grossman 2014: 195). Spiral forms recur in a number of other avant-garde abstract films of this era, not to mention in camera-less photograms of the early 1920s by Man Ray and László Moholy-Nagy. For Duchamp this concern also generated one of the more enigmatic elements of the *Large Glass* (1915–23), the curled-up chute along

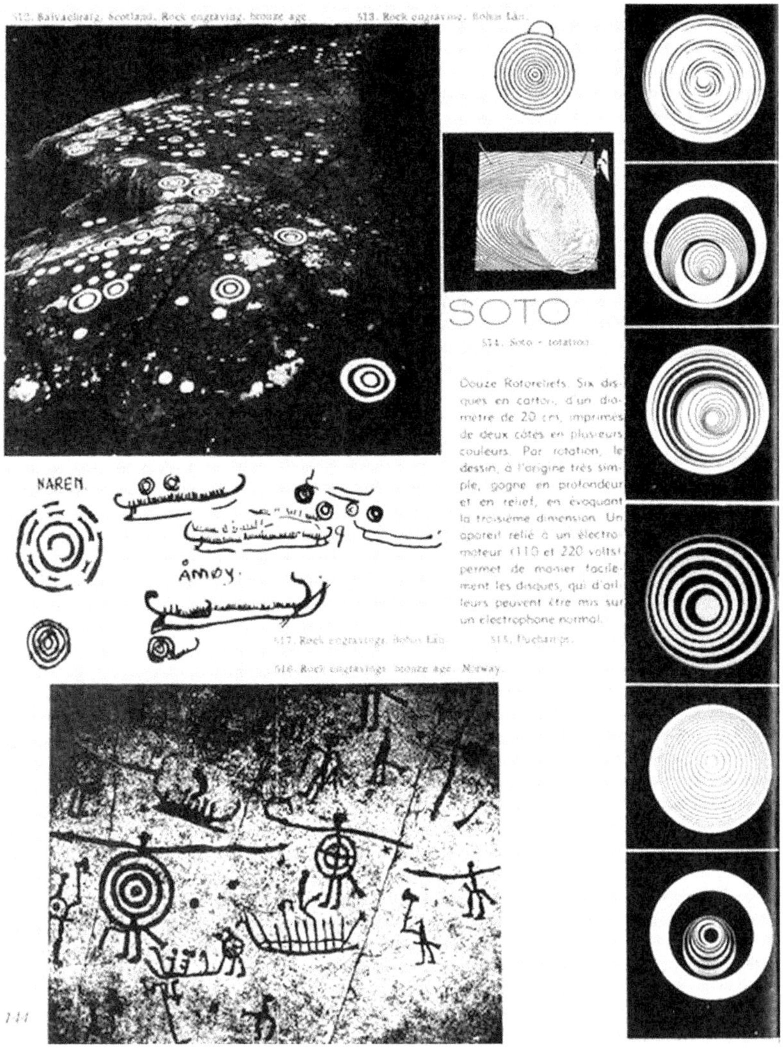

FIGURE 1.3: *The Situationist Times*, no. 5, 1964, p. 144. Image courtesy of the Institute for Computational Vandalism.

which desire liquid flows and splashes, known as the 'Toboggan, or Corkscrew' – a component that was never completed. Arguments have been advanced that spiral forms in the *Large Glass* relate to Duchamp's interest in Henri Poincaré's theory of multi-dimensionality, 'unstable equilibrium' and chance (itself one of the sources of modern chaos theory), with Poincaré's mathematical diagrams repeatedly featuring spirals (Shearer n.d.: n.pag.). Discussion of Duchamp's spirals tends to focus on optics or kinetics, but what might be emphasized here is their uncertainty and propensity towards obsessive, hypnotic qualities – themes that also characterized later mass-market eruptions of the spiral.

The presence of the spiral as a cross-cultural symbol of mythical forces would be taken up at the turn of the 1950s by the Cobra movement – among whose constituents were the Danish group Spiralen, which included painter and writer Asger Jorn. Issue 7 of the movement's eclectic journal *Cobra* was particularly rich in representations, from the Babylonian coiled serpent symbol for heaven to a magical carved stone from the Congo. Extending the spiral across space and time, Christian Dotremont considered the now forgotten author Stefane Christesco, a former military engineer whose refutation of Einstein's theories included an elaborate cosmogony dividing the universe into atoms, ether and 'ultra-ether', patterned with spirals at every level of being, every stage of evolution. Linking natural and mental energy 'from one whorl to the next', in Christesco's vision 'the very thought processes of the cosmogenist are prey to spirals, like the tentacles of knowledge' (Dotremont 1950: 21). *Cobra*'s interest in the spiral planted one foot in myth, one in speculative science, just as Kiesler knew of the cosmological theories and intense coloured diagrams of Walter Russell, reconciling scientific and spiritual visions of the universe through 'cosmic illumination': 'all direction is curved – all motion is spiral' (Phillips 2017: 126–28; Binder 1995).

Jorn's theoretical research into the meaning of form is a link to sustained attention to the spiral in *The Situationist Times* (1962–67). The journal's diverse interests recalled earlier *Cobra* publications, presenting typologies of form and pattern across history and cultures with essays and documentation of knots, mazes and spiral geometries, especially in issues 3 and 4 (January/October 1963) (Scandinavian Institute for Computational Vandalism 2019). Issue 4 concentrated on the labyrinth as an emblem condensing geometry, topology and psychology: while Max Bucaille's article 'Spiral and Labyrinth' (2019: 146–47) argues that the former's predicable regularity cannot be conflated with mazes, where all is 'difficulty, hesitation, impediment, errancy, anguish', copious images evidenced a porous exchange between the two figures, constantly threatening to lead eye and body astray towards some fateful encounter. Issue 5 found evidence of spirals as mental maps – in Larry Busbea's (2019) phrase 'geometries of thinking' – with a discussion of children's hopscotch games (Schade 1964), while Jorn's essay 'Mind and Sense' (Jorn 1964) considers Nordic archaic and folk art through ambivalent structures of knotting and weaving, illustrated by an ancient spiral labyrinth from Gotland.[2]

If the spiral is a model of thought, Goethe's admission of feeling caught 'in the vortex of the spiral tendency' (Mainberger 2010: 203) suggests it also encapsulates the vagaries of the intellect, a cipher for confusion and delusion. The emergence of the rotating spiral to evoke mind control or loss of agency – presented for example by Melvin Powers's mass-market paperback *Advanced Techniques of Hypnosis* (1953) describing a 'Powers Hypnodisc' the same size as Duchamp's *Rotoreliefs* and similarly spun on a record turntable (Powers 1956: 8 and 76; see Figure 1.4) – is typified by the opening credits of Alfred Hitchcock's *Vertigo* (1958). Designed by Saul Bass, in this graphic sequence sinister geometries twist from the pupil of Kim Novak's eye, announcing that protagonists and spectators alike are about to succumb to the crisis in the cross-currents of obsession, enigma and the emerging science of the mind. The elision of vortex and vision returns in the iconic shower scene of *Psycho* (1960) that ends with a dissolve from the blood-stained water swirling down the plug hole to Marion Crane's dead eye.[3] From its second season onwards, the opening sequence of TV fantasy series *The Twilight Zone* (1960–64) featured a mesmeric graphic spiral, signalling not only the blurring of time and space, but a heady mix of Cold War sci-fi and conspiracy theories, and the hint that whether human or alien, the controlling powers also pull the strings of the psychic manipulation of mass media and advertising. More recently another film's opening sequence features a spiral figuring moral ambiguity and the failure of free will: the Coen Brothers's *The Man Who Wasn't There* (2001) begins with a slow vertical pan down a barber shop pole and into the premises of Ed Crane, whose unwitting descent towards murder and punishment the movie recounts. A sign dating back to the twelfth century announcing that barbers were also employed for minor surgery, the twisting red and white strands suggestive of blood and bandages, here the pole's endless motion discloses what the audience will have to wait to discover: that good and evil are contingent, inextricable, the slide from one to another as inevitable and banal as the spinning of a shop sign.[4]

In this context, the anxious spiral portends at the same time a renunciation of selfhood to unfathomable laws (the ubiquitous situation 'spiralling out of control', the stricken aircraft in a tailspin) and shadowy agencies, or a portal to the chthonic forces of nature, internalized as the labyrinth of the unconscious. Vortices, vacuums, whirlwinds, tornados, cyclones, waterspouts, whirlpools; Charybdis, cursed to imprisonment beneath the sea in monstrous form, gulping and spewing out brine.[5] As David Lomas establishes in considering the significance of vertigo and labyrinth for artists and writers like André Masson and Georges Bataille in the aftermath of the First World War, these natural or mythical – albeit perceived as aberrant – motifs announce trauma and the pathological foundering of space, proclaiming flux and instability in the cosmic order, the lure of ecstatic collapse. For Bataille, a galaxy becomes a Catherine wheel, while in Masson's 1930s artworks, gyrations evoke cyclical returns but also dizziness and rupture (Lomas 2006: 83–85 and 96–104). This visionary perception of the spiral as a motive force of potentially overwhelming universal

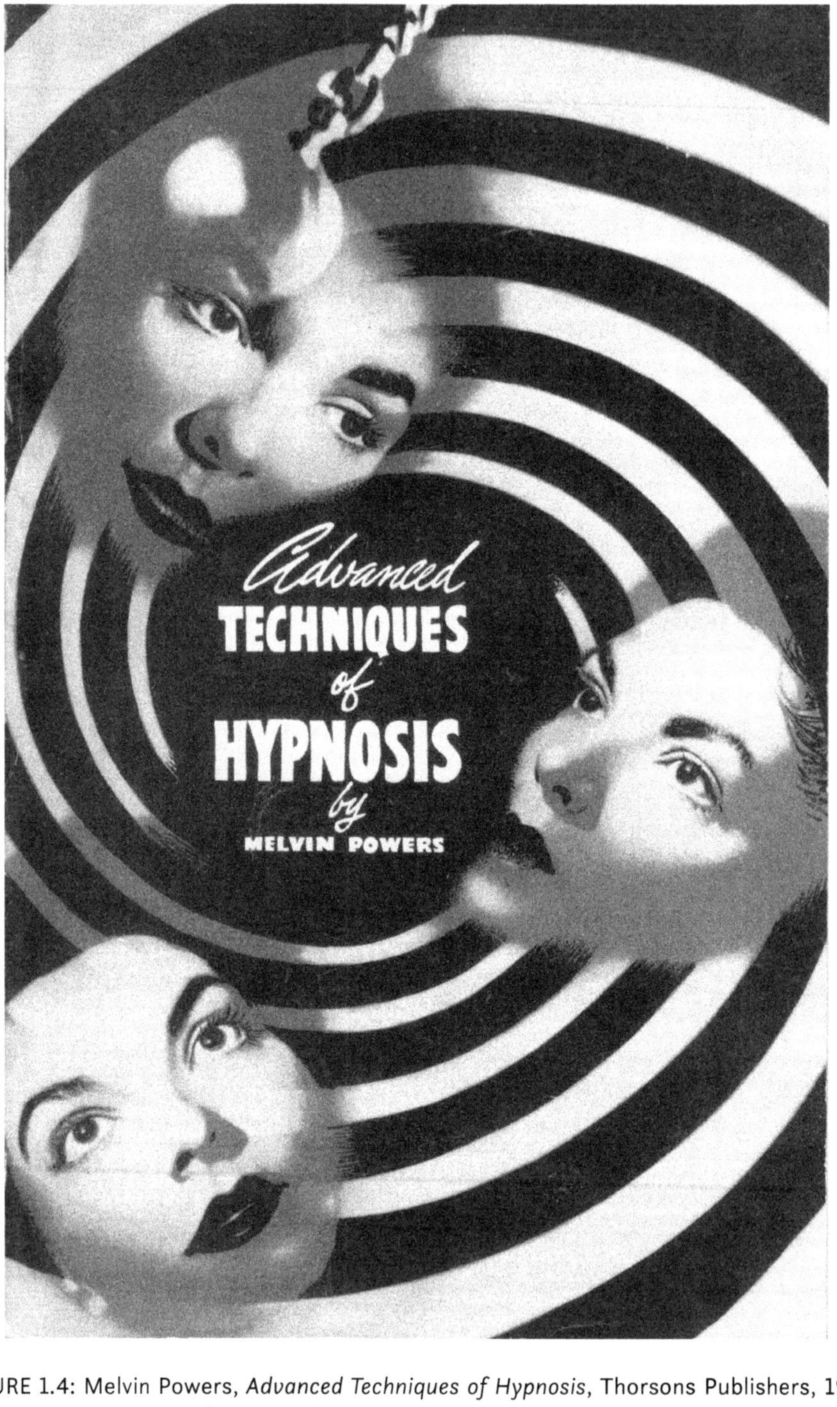

FIGURE 1.4: Melvin Powers, *Advanced Techniques of Hypnosis*, Thorsons Publishers, 1953. Reprinted by permission of HarperCollins Publishers Ltd. © 1953 Melvin Powers.

principles can be found more recently in the intense scraper-board drawings of Cathy Ward, with their whorls and oscillations of hair-like forms figuring, in Doug Harvey's words (2020: 12) 'cosmic psychic spaces animated by sublime tidal paroxysms and cataclysmic shudders' (Figure 1.5). A chilling eruption of this disarray presides over Junji Ito's horror manga *Uzumaki* (2013), first published in Japan as a serial in 1998–99. In a sequence of linked and increasingly gruesome stories, the inhabitants of Kurouzu-cho succumb to spiral phenomena that obsess and devastate the population, mutating every form, every human body, into a tortured vortex in thrall to a spiral curse. When a series of hurricanes eventually destroys the town, it is rebuilt as a circular labyrinth, as though at the same time to shield against and embrace the terror.[6]

One figure above all proclaims the spiral as an emblem of the slide into chaos: Alfred Jarry's turbulent character Pere Ubu (from the play *Ubu Roi* of 1896 and other works), whose grotesque stomach bears a giant graphic *gidouille* – Jarry's neologism for the logarithmic spiral stretching from navel to pocket (Figure 1.6).

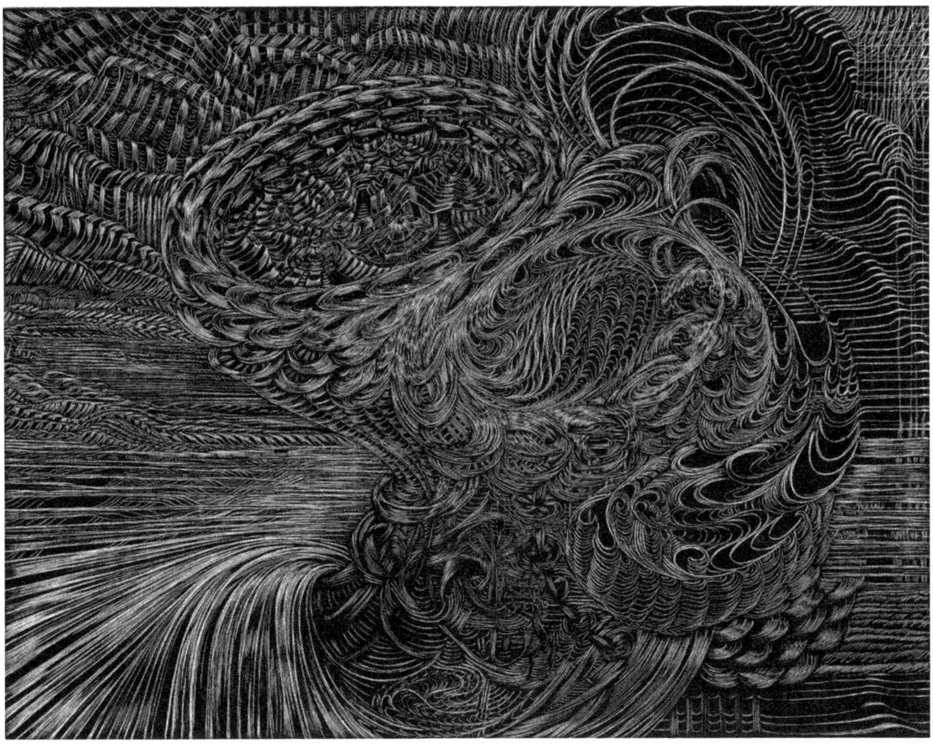

FIGURE 1.5: Cathy Ward, *Transcension to Liberty Realm*, incised ink on clay, 2012. Courtesy of the artist.

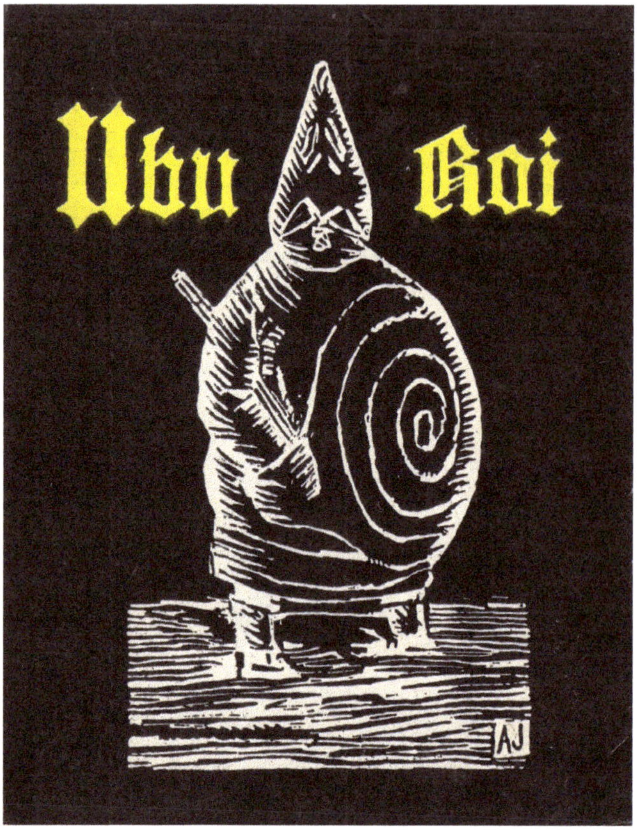

FIGURE 1.6: Alfred Jarry, *Ubu Roi*, Gabberbocchus Press, London, 1961. © Themerson Estate.

An absurd but all too familiar tyrant, Ubu combines the wilful, irrational and carnivalesque in his insatiable ambition. On his abdomen, the spiral affirms and dissolves binaries: the *gidouille* both crowns its wearer as he tramples upon structural oppositions, and mocks our obligation to return endlessly to them in deciding whether to acknowledge that morsel of Ubu in us all. In this giddy movement in and out of form, between conscious and unconscious forces, the resplendent helter-skelter on Ubu's belly is simultaneously aggressive geometry and the queasy return of the despot's guts, the scatological plumbing on display for all to see (see Tremblay 2004: 29–46 and Lowe 2006). Four decades later physician and surrealist theorist Pierre Mabille's book on human anatomy *The Construction of Man* (1936) partitions the body's relationships into a peripheral zone (the limbs and skeleton, ordered and symmetrical), and an interior zone (organs and viscera, formless and chaotic). This applies across all functions, including psychological frameworks in which organized and structured conscious processes interact with shapeless, unknowable unconscious ones. Yet this maze of divergences and lostness is somehow formative: 'The tendencies of the unconscious, being unable to express themselves directly, adopt artifices and tortuous paths. A whole mechanism of sublimations, repressions and replacements emerges, making us the individuals we were observing' (Mabille 1936: 21–36 and 120–21).

In this schema, the spiral swerves from state to state, from the chaotic to the structured, from loss to power – and back again, always as exchange, flux and reflux. In the words of another profligate spiral-maker, Louise Bourgeois,

> The spiral is an attempt at controlling the chaos. It has two directions. Where do you place yourself, at the periphery or the vortex? Beginning at the outside is the fear of losing control; the winding in is a tightening, a retreating, a compacting to the point of disappearance. Beginning at the center is affirmation, the move outward is a representation of giving, and giving up control; of trust, positive energy, of life itself (Bourgeois 2018: 7).

As a principle of incessant growth, as a model of divergent thought, the spiral winds and pulses at the heart of a cosmic and personal order, undoing certainty, hastening anxiety, in order that we may always start again.

NOTES

1. Space here precludes a detailed examination of the spiral's varieties, though I should note that in this discussion these include both flat spirals and helix shapes that extend spiral rotation from two to three dimensions.
2. The use of the spiral as resonating between mathematics and myth had already been found in another adjacent context, on the front cover of the final issue of painter and theorist Wolfgang Paalen's Mexican cross-disciplinary journal *Dyn* (1944). Both Paalen's paintings and ethnographic material from *Dyn*'s pages occasionally featured spiral themes.
3. In Hitchcock's two sequences the connection between the iris of the pupil and the mechanical iris of the lens, opening and closing as a flattened helix, is also a tempting connection. On the psychopathology of the sensation of vertigo as a loss of orientation and a collapse of boundaries of space in the context of inter-war surrealism, see Lomas (2006).
4. One further example of film and television adaptations of this theme is the long-running French police procedural and legal series *Spiral*. The English title is in fact a fertile mistranslation of the French *Engrenages*, meaning the interlocking of gears, but again the implication is that no moral distinction can be made between 'right' and 'wrong' sides of the law, 'good' and 'bad' characters: all is contingency, compromise and interlocked slippage (*Engrenages* [*Spiral*] 2005–present).

5. Were it not for the spiral's characteristic as either connecting or dissolving binaries, it would be tempting (in line with Goethe's schema of spiral and vertical) to see in the myth of Scylla and Charybdis – the rock and the whirlpool, both personified as cursed monsters – the complementary forces of a universal structuring principle for which the spiral is one essential half: object and void, enduring geology versus liquid turbulence; timeless feature versus unpredictable event; the thrust skywards or the plunge to the depths; the geometry of the crystal and the radiation of the sunflower.
6. Junji has affirmed that the impulse for his story is partly a deliberate deflection of the affirmative use of spirals in Japanese graphics towards horror: see Winsby (2006). An extraordinary real-life tale of spiral obsession turned sour is El Helicoide, a vast, never completed and now partially ruined modernist shopping mall in Caracas coiled into a hillside, hailed as a masterpiece on its instigation in the 1950s but on falling into disrepair taken over by the Venezuelan military to become a political prison: see Olalquiaga and Blackmore (2018), among other presentations by Olalquiaga.

2

Representing Kinematics and Dynamics by Pattern Breaking in Nature, Art and Music

W. Brian Whalley and J. Harry Whalley

Introduction

The two-dimensional (2D) plane exists 'out of time' whereas, in contrast, sound and music exist in the one-dimensional domain of time. In both cases, patterns can be established and broken, and dynamics and kinematics can be represented. Here we explore some approaches that represent movement and motion through 'pattern breaking'.

There are many ways of generating 'random' numbers, that is, statistical randomness. Randomness is not the same as the unpredictability of events for example. Chaotic behaviour suggests unpredictable movement. Here, chaos has a specific meaning, of a complex system appearing to be random as a result of sensitivity to tiny changes in initial conditions. Albert Einstein's *annus mirabilis* of 1909 included a paper on 'Brownian motion' where a small particle, such as a pollen grain on a liquid surface, is jostled by the random movement of molecules. (For related animations, see Brownian motion, online n.d.) This activity, movement over time, can be modelled. The movement of pollen, unpredictable or not, comes from observations over some time period. However, as a snapshot, a single image will not indicate movement.

The question we pose is, can motion be represented statically and in two dimensions? What are some of the processes of translation between dimensions? We link static patterns, which can be used in various ways, to help represent movement in some manner. We consider that this may be an experiential, psychological effect as well as relating to a physical phenomenon. In particular, we suggest that knowing or recognizing how something behaves or is *about* to happen is significant. Cognitively, what might we expect to happen from memory or from previous behaviour, experience or knowledge?

Some basics

Considering random behaviour, say a gas held in a sealed container, we know that there is random movement of molecules. If the container is heated, the pressure increases, even though the number of molecules remains the same. This statistical behaviour is described by the various 'gas laws' and summarized by the 'ideal' gas law. Although this equation (the 'General Gas Equation', $pV = nRT$) implies – that is represents – chaotic or random behaviour, it does not itself show movement. Heating some water in a can with a lid may pop the lid off if heating is continued, or if cooled, make a thin metal can collapse. Some 'laws' (or law-like statements) can imply movement or reaction. This might be even used to teach, even explain, some basic science, or better say something about how the world 'works'. Pattern and chaos appear to be opposites but can give way to each other. Condensation of vapour or freezing of liquid water produces more order in moving to another state, such as ice crystals – thermodynamic changes can produce visibly ordered or disordered 'patterns'. Theodore Schwenk (1996) gives numerous examples of flow transitions and 'critical' behaviour – related to sensitivity and chaos – associated with fluids and mobile solids such as sand. Here we unpack some ideas relating to the recognition of (regular or nearly regular) patterns. Two dimensions can indicate extension(s) into a third (cartesian) dimension and in time. We also impose 'boundary conditions' on problems. Can we represent movement without recourse to animals – a prancing horse, a pole vaulter – and without associated machines, words or phrases (e.g. Roy Lichtenstein's *Whaam!*)?

Stoppard and physics

A first example is the well-known 'Newton Cradle'; a predictable effect being recognized from a static starting point (Figure 2.1). Both configurations (or patterns) are simple but in B the mind's eye suggests what will happen next. Here we coin the phrase 'pattern breaking' to suggest that something unexpected happens (or might happen) from a static starting point. It is cognitively related to 'symmetry breaking' in physical sciences. Both initial conditions (A and B) are static, and experience shows that we 'know' what will happen next in B. We build this into experience. This is 'prior knowledge', and the ability for us to form a heuristic is important in pattern recognition algorithms. Experimental conditions might also include the properties of the spheres,

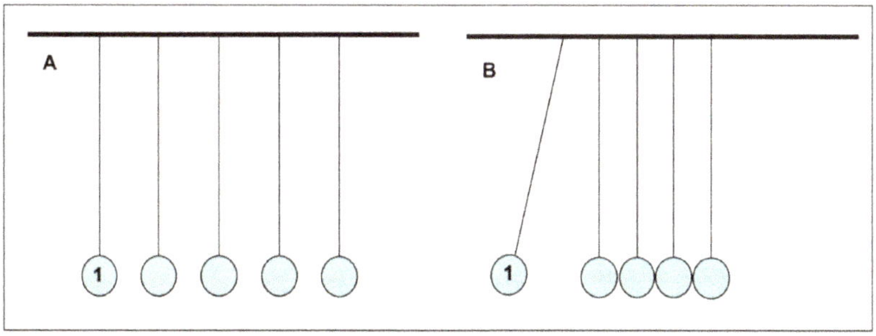

FIGURE 2.1: Suspended metallic spheres in profile. 'A' suggests stasis but experimental experience of 'Newton's Cradle'. 'B' suggests a dynamic response when the spheres are touching. Different patterns emerge, depending on how far back sphere A1 is if released. Image: W. B. Whalley CC BY-SA 4.0, 2023.

whether solid steel, glass spheres or rubber balls. A literal 'pattern breaking' occurs in the screen version of Tom Stoppard's *Rosencrantz and Guildenstern are Dead* (1990) (Player: 'There's a design at work in all art').

Designed symmetry and pattern breaking

In Gillian Boyd's design (Figure 2.2), the flying bird image might be seen as infringing a design parameter, but it is the symmetry breaking that actually gives the impression of movement. A stylized shape would do something similar. Note that the pattern also suggests a volume in three dimensions. It might be considered to have a fractal dimension of between two and three. There is a conflict between the gestalt notation of the whole (through grouping), and how our attention is drawn to the 'parts' through the breaking of symmetry. This is also seen in some pictures by Bridget Riley and Victor Vasarely, although generally, volume, rather than movement, tends to be projected.

Further examples of pattern breaking, or what happens next, that suggest incipient movement are visible in Figure 2.3; in some cases (bottom right, producing chaos – or at least 'randomness') after a critical event of a moving circle/sphere acting on the regular pattern. This effect can also be seen in Per Bak's (1996) 'sand pile' avalanching models. More fully, these are Bak-Tang-Wiesenfeld or Abelian models and are produced using rule-based cellular automata (CA). Examples can be seen, with one changing dynamically, in the Abelian sandpile model (n.d.). CA models are examples of 'agent-based models' and can be used to explore social interactions (Flache n.d.) and as pattern-related simulations of epidemics modelling on Grant Sanderson's (2020) '3blue1brown' YouTube channel.

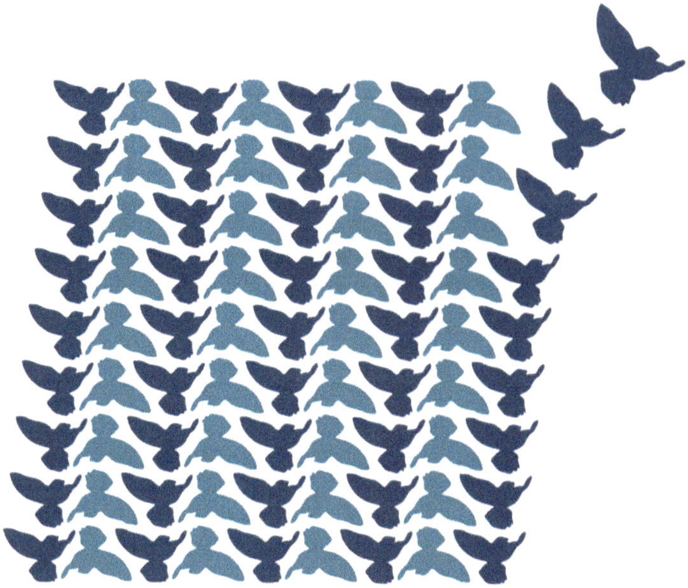

FIGURE 2.2: Gillian Boyd, *Flight*, 2021. 'Symmetry breaking' of a basic rhomboid envelope. The effect is present even without the bird symbol. Screen-print. © Gillian Boyd, 2023.

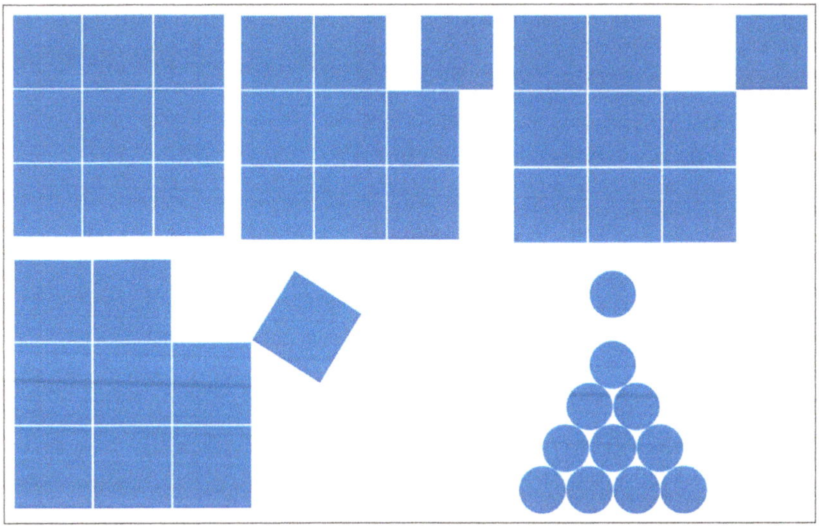

FIGURE 2.3: Exploration of pattern changes to invoke volume, movement and anticipated events via pattern breaking. The effects of change may vary according to the view taken. Image: W. B. Whalley, CC BY-SA 4.0, 2023.

Visualization of the top three sets is likely to be in plan view (static) but change to vertical section ('dynamic') when a slight rotation is added to the moving square. The round objects may be subjected to symmetry breaking by the arrival of a billiard ball (in plan) or descending grain on a pile (side view) at a 'critical state'. The movement response of the grains would differ according to the plan/section view. These are examples related to the gestalt and 'flipping' between dimensions, a form of pattern breaking.

Networks in nature

Branching patterns or networks are common in nature and examples can be found in the significant books by Kemp (2000) and Ball (2016), for example. Stevens (1974) gives a particularly rich array of 'patterns in nature'. Rivers developing in a landscape are akin to trees, 'space filling' branching structures or 'channel avoiding' networks and can be extended into three dimensions. Although these systems might characterize movement, they do not *show* movement.

Trees are 3D networks, often distinctive in 2D in profile. Symmetry is roughly axially symmetrical, what we expect of a tree. If the profile is pushed by a strong prevailing wind, then the static image gives the impression of movement (Figure 2.4). It is an example of translation of a relatively 'regular' pattern. From this static 2D image the

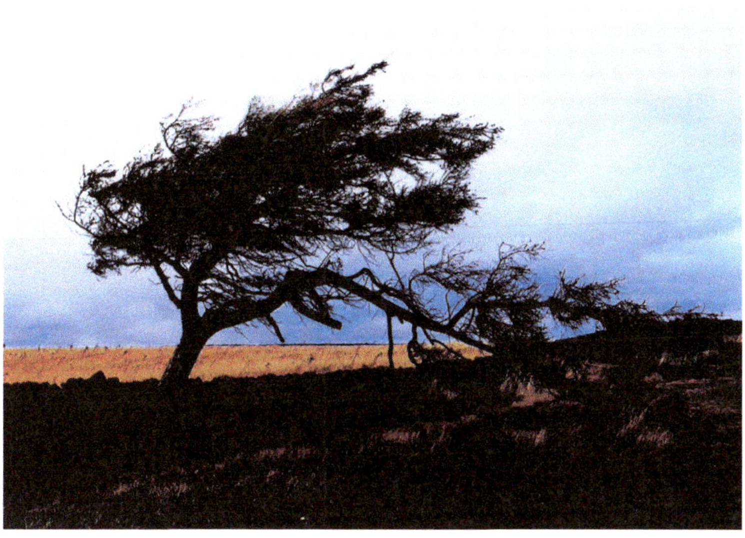

FIGURE 2.4: Windswept tree after westerly gales of many years in upland Derbyshire with the prevailing wind from the left (west). Image: W. B. Whalley, CC BY-SA 4.0, 2023.

direction of the prevailing wind is evident, and there is a translation of historical growth. We understand this through the same cognitive heuristics as with a Newton Cradle.

Chaos and patterns, especially in music

Joseph Haydn in his *The Creation* (*Die Schöpfung*, 1798) starts with a 'Representation of Chaos' ('*Die Vorstellung des Chaos*'). But this is far from presenting random notes; it *represents* chaos rather than *is* chaotic. As previously noted, 'chaos' has a technical meaning and is distinct from 'random'. However, in signal processing, the trick of information extraction may be to show a signal in another form, and this may accentuate the pattern of a complex system.

The balance between order and chaos is also typically at play in the mind of the composer when building melodic ideas; between monotone and random is a continuum in which melodies can fall. In a recent instance, Scott Rickard (2011) expanded on the relationship between randomness and beauty with a counterexample, *the world's ugliest melody*. In this example, the composition avoids repetition, and in so doing demonstrates how repetition (at all scales) is an integral and intuitive aspect of melody and music creation. Musical forms such as theme and variation, sonata, rondo, blues and two-sixteens all provide a template or, in a visual language, a repeat structure, in which a composition is provided with a structurally defined proportion of repetition.

The concept of sonification of patterns is explored in a composition (Whalley 2016) that reflects the principles of play in the board game Go. On a Go board, players alternate in placing pieces called stones on the intersections of a rectangular grid. Movement is implied by stone positions on the board, their connections to similar and opposing pieces and the 'direction' of play. Therefore, any musical interpretation of this game into music must translate the implied movement in the game through musical tension and other devices. These patterns and rules are understood by each player, and like 'gliders' in John Conway's *Game of Life* (Adamatzky 2010), initial conditions of pieces on the board will imply movement or progression.

Visualization and numerical series

Pattern visualizations can be formed or implied in various ways. Sound and visual images may combine to produce a pattern. A plate with sand on the surface and subjected to a constant frequency will give a characteristic 'Chladni figure' (after the German physicist and musician, Ernst Chladni, 1756–1837). This is unexpected and a good illustration of pattern formation from a random scattering of sand. (For an example of the effect of pitch on a Chladni plate, see St. Mary's University, online 2009.) Numerical links can be made between Chladni figures and the (cathode ray) 'TV screen' experiments of Nam June Paik in the 1960s.

In physics labs 'Lissajous figures' on oscilloscope screens lead to polynomials, Fourier analysis and knot theory. Lissajous (or Bowditch) figures can be simply investigated by a pendulum streaming sand onto a flat surface and relate to the chaotic motion of a 'three-body' problem represented as a 2D trace. Deterministic patterns can be produced by linked pendulums: a 'Blackburn pendulum' or 'harmonograph'.

The realization of the oscilloscope trace into recognizable shapes and patterns (Figure 2.5) takes the concept that Chladni described with the extra dimension of time. In this example, the aim of the visualization is artistic, both regarding the visual produced and also in respect of the sound and composition.

Philip Ball (2016) provides many other examples of changing forms and issues of complexity and 'emergence'. A good example of showing the effects of simple computer-based rules to produce changing patterns is Craig Reynold's simulation of bird flocking (Eversham and Ruiz 2011). Wildlife films also show us that a static image of murmurating birds provides a feel of movement, even if the direction of the flock is unknown. Three simple 'steering' rules applied to each 'boid' in the simulation produced a realistic effect, emergent behaviour, of a murmuration with a random starting situation.

Moving fluids and chaotic behaviour

A jug of water and a closed rubber balloon contains air or water molecules in random motion. A balloon with a needle just below implies movement if the needle were to pierce the rubber. This is our cognitive experiential heuristic. A thin stream of water from a jug can be seen as poured into a cup. What happens if the 'pouring angle' is decreased slowly? Depending on the spout's design it may be that the flow no longer projects out but starts to 'stick' to the side of the jug (the Coanda effect). This flip, or collapse, is a change in the pattern of the 'what happens next?' kind. This is an example of a catastrophe (as in 'Catastrophe Theory' of Thom and Zeeman) or a sudden phase transition from one form to another. Allan McRobie's *The Seduction of Curves* (2017) provides examples and explorations of curves, caustics and catastrophes and links to a wide variety of phenomena and aesthetics. This can be explored further, as in Hans Jenny's (2001) *Cymatics* which provides many examples to explore design possibilities of interactions of natural phenomena and vibrations. A repeating pattern

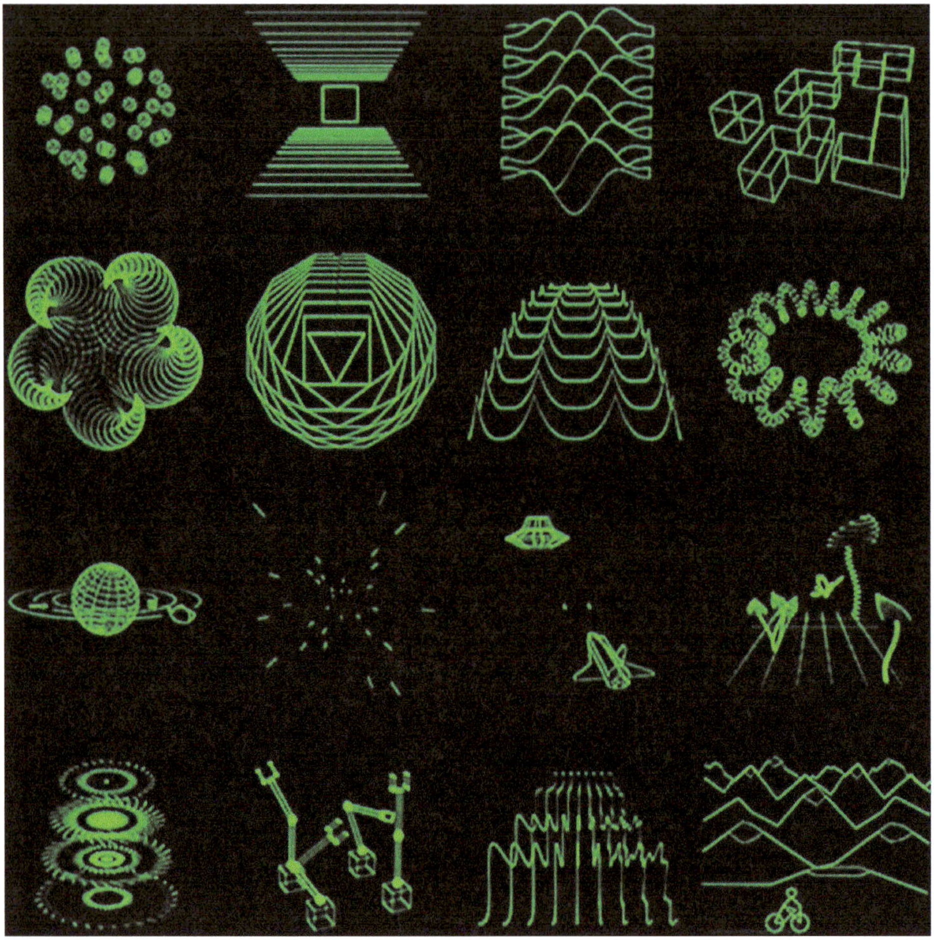

FIGURE 2.5: Linking the visual with the audio. The cover of the audio–visual album *Oscilloscope Music* (2016): https://oscilloscopemusic.com. This allows uncompressed audio files to be played through an oscilloscope. © Jerobeam Fenderson.

in sand produced by a moving fluid (air or water) produces the well-known phenomenon of sand ripples. They may be produced on a beach and some of these effects are shown in Schwenk (1996). The transition from laminar to turbulent (chaotic) flow can be seen when a candle is extinguished and the vertical plume of smoke ultimately diffuses, an effect seen with Schlieren imagery. Again, visualization and the organization of the 'random' into patterns are important. Philip Ball's *Patterns in Nature* gives more examples and also mentions ideas of symmetry breaking (2016: 16–17). A rounded landscape skyline, representing a gestalt landscape, can be described mathematically by the Fourier series. Where it is jagged, or 'spiky', such as with alpine peaks, a fractal description can be incorporated. This 'jarring' of visual perception is part of the observational process (Whalley 2019).

Discussion

The terms random, chaos and complex systems have technical meanings, although they may not always be obvious in the natural world. Our examples seek to demonstrate translations from one dimension to another in a process that can be intuited. This intuition comes from a set of heuristics developed to understand movement and from our ability to find patterns and find breaks in those patterns. Particularly remarkable is how these translations happen effortlessly, even though they are complex and mathematical. A rhythmic pattern, when broken, creates instability, much in the same way a disturbance in a visual pattern does. Understanding how our gestalt experience from one domain can be understood in another is an area from which we might learn much about how our minds operate. The 'new eye' on a phenomenon can come from many directions: art and music, engineering or mathematics.

Conjectural 'conclusions'

'Scientific' concepts such as the Fourier series, Chladni figures, phase changes and catastrophes produce creative possibilities, not just for research in science and technology but with implications for cognition, design and graphics. The interplay between human heuristics of our physical environment, the gestalt and the underlying mathematical principles is multi-faceted. Yet the boundaries of pattern and chaos lend themselves to evocative and striking artworks that imply deep mathematical concepts that can be intuitively understood. Such ideas play off against each other and can inject humanities back into the world that restrictive educational views, presented solely by 'Science-Technology-Engineering-Mathematics', seem to lack.

ACKNOWLEDGEMENTS
We thank Gillian Boyd for the use of her image *Flight* and the editors for their assistance.

3

Drawing Dynamic Patterns: The Protein Maze

Gemma Anderson-Tempini, Jonathan 'J. J.' Phillips and John Dupré

Introduction

Any visual representation is an abstraction from the complexity of living process and requires selecting some point of view. We are an interdisciplinary team of artist, molecular scientist and philosopher of biology, interested in drawing as a way of investigating the dynamic patterns of living processes. This short chapter presents a fragment from our project 'Representing Biology as Process' (2017–21) in which we developed new methods for drawing processes at the molecular, cellular and organismal scale. Here we focus on a few maze drawings we created as a metaphor and a model for the protein folding energy landscape. We detail how the maze drawing acts as a template in which to draw the dynamic pattern of the processes of protein folding, binding and more.

Proteins are a fundamental and highly diverse unit of organization common to all living organisms. An overwhelming majority of biological processes at the molecular level depend on the function of protein molecules and this is central to current efforts in biological and medical research. The creation, functional states and eventual release or recycling of a protein are spatiotemporally dynamic processes that have unique rhythms and patterns to observe. Here we are interested in the problem of representing protein dynamics and their multidimensional reality.

The current conventional image for representing protein folding (thermo)dynamics in biology is the 'folding funnel'. This is broadly speaking a downward-pointing cone, adapted with additional geometric features to represent attributes of the energy landscape (Hartl and Hayer-Hartl 2009). In this model, some features are represented well, others are not. For example, the cone suggests that proteins fold to the lowest energy state whereas, in fact, they may simply fold to a stable state (not necessarily the global energy minimum) that is relatively thermodynamically stable, given the physiological constraints. In its current state, the 'folding funnel' image does not function well as an intuitive image or teaching aid, for example because the landscape fails to represent the undirected exploratory stochasticity of the trajectory experienced by an individual protein molecule. Here we conceptualize the protein folding landscape as an *n*-dimensional manifold, as it is commonly perceived. The images we create through drawing are a flat projection of that manifold.[1]

It is common to reduce these high-dimensional descriptions in order to create a more tangible representation – the process of creating a 2D or 3D view of high-dimensional objects is termed 'projection'. In this collaboration, we explore how representing protein dynamics can be addressed by participatory one-to-one drawing sessions and group 'drawing labs' where both artistic and scientific inputs interact with and affect the image to enhance iteratively the quality of connection with and observation of the phenomena of investigation.

The protein maze

After a few months of exploratory collaboration, we had created a portfolio of drawings and notes. All of these were laid out on the floor of one large room for a meeting between the team members. Anderson-Tempini suggested that a circular maze could be an effective visual tool for imagining the protein energy landscape if the centre of the maze is considered equal to the bottom of the 'folding funnel' (the 2D circular maze is an unfolded 3D funnel/cone). For example, through our drawings we found that the embodied process of drawing the path/trajectory through the maze structure (a maze is a complicated irregular network of passages or paths in which it is difficult to find one's way; it has many paths, whereas the labyrinth only has one) is comparable to the way the protein finds its way through the landscape (Figure 3.1). This was followed up by discussions about the nature and composition of mazes (dynamic mazes, fluidity, interconnection, layering, maze vs. labyrinth, etc.). In doing so, we explored in what ways these features might be used as a metaphor for the dynamic pattern of a protein. One example of this is the dynamic pattern of branching and fusion of paths through the maze which may depict the potential trajectories of a protein travelling through a protein folding landscape: this could begin to add features, such as stochasticity and thermodynamic equivalence of microstates.

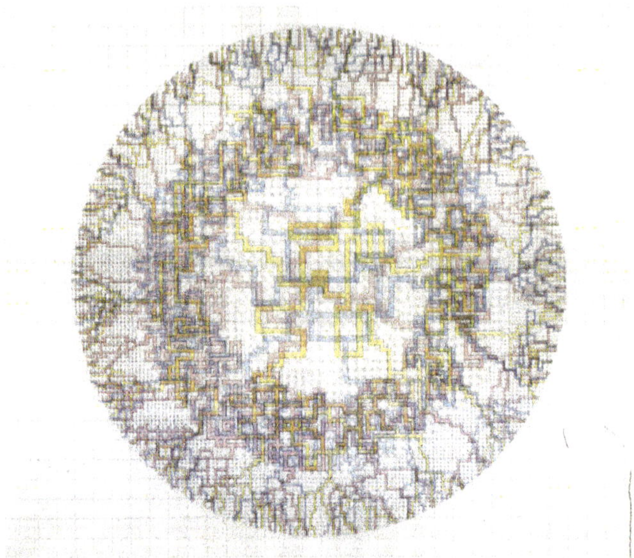

FIGURE 3.1: Gemma Anderson-Tempini, 2018. *Protein Maze Experiments*. Pencil and colour pencil on graph paper. All images © Gemma Anderson-Tempini.

The concept of interactively exploring the maze was agreed upon; Anderson-Tempini, began to experiment with ways to draw a maze and then together with Phillips devised ways to adjust the drawing of the maze to parallel and depict dynamic patterns of the protein folding landscape. In doing so, we utilize a latent feature of the maze that was not there before: granularity (i.e. the size of the grid, Figure 3.2). This feature defines the proximity of available changes in direction, which may be altered throughout the maze image to reflect physical concepts for the protein, such as the steepness of the energy landscape, the degree of parallelization of the pathways and the density of thermodynamically equivalent conformations. The maze now provided us with a metaphor for the protein landscape and an interactive drawing method to visually create variations of structures and features that pattern the landscape. Drawing the dynamic nature of the protein in the maze enabled us to go beyond illustration and standard representations and to enact features of the protein itself.

The circular maze

The circular maze is comprised of three areas: the high potential energy entry stage on the outside, the intermediate potential energy exploratory stage in the middle and the low potential energy stage in the centre (Figure 3.3).

FIGURE 3.2: Gemma Anderson-Tempini, 2018. *Circular Maze with granularity*. Pencil on graph paper.

The outer ring has a fine-grained structure. This supports features of a protein folding landscape, such as multiple entry points, steep descent (intentional and noticeable avoidance of turns), fusion and branching. The middle ring has intermediate granularity and two interfaces (one with the inner ring and one with the outer ring). This supports features of partially folded immature protein, such as exploration and meandering through turns, uphill and backwards travel (i.e. 'local minima and maxima') and fewer entry points from the outer ring to the middle ring than into the outer ring itself. The inner circle has a coarse-grained (sparse) structure with a single interface that connects it to the middle ring. This supports features of the end of protein folding, such as the global minimum (i.e. the termination of the maze), landscape roughness, yet an intuitive reduction in choice and fewer entry points from the middle ring than in the upper levels. The granularity and features (e.g. colour can represent a further dimension) sufficiently represent many of the important features of the landscape: relative stability/instability (centre:stable/outer:unstable), stochasticity (backwards and off-pathway randomized turns), nonlinearity (zig-zag trajectory), vibrational energy and entropy (line as a high-frequency scribble), potential energy (e.g. colour gradation), postural and morphological change, etc. These features can be changed for different protein energy landscapes and so have relevance to multiple protein processes.

The maze structure allows for countless possible paths of development and multiple 'solutions' to the maze. Aside from the 'ideal' path, there are likely to be several

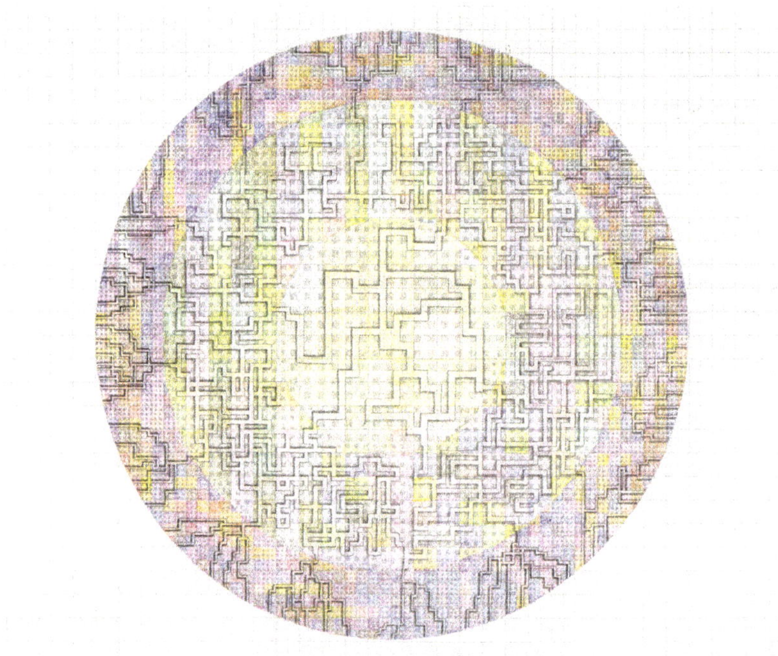

FIGURE 3.3: Gemma Anderson-Tempini, 2019. *Protein Maze no. 2.* Pencil and colour pencil on graph paper.

ways to reach the point of lowest energy. In fact, although there may be one single grouping of paths that are thermodynamically or kinetically equivalent to each other and these collectively are the most efficient pathway (somewhat representative of 'foldons'), there is not one single optimal route. How the maze is drawn is a personal choice – it could be a single trajectory through the maze that represents an efficient thermodynamic pathway or multiple trajectories with varying levels of efficiency. The image becomes interactive through the drawing process. The drawing is informed by the lab's knowledge of the protein and created by enlivening this knowledge and visually imagining the dynamic nature of the protein. Together, these parameters reflect the complex interplay of features in the energy landscape of the protein. The grid is like a game and the drawer is the player. There are many ways to play it and many ways to resemble the protein. In the imaginative sense, the maze drawings allow for a becoming of the protein landscape. Importantly, these last attributes of the maze are afforded by its rich ability to incorporate features that represent attributes of the protein and the folding pathway which are missing from one-dimensional energy landscapes and to do so trivially and interactively, thus potentially facilitating an interactive mode, as distinct from 'off-line' engagement with a pre-prepared image.

In form, the maze drawing resembles a mandala and in function to some extent relates to the mandala as an image for meditation and learning. We can therefore look at the maze cones as intuitive images for learning and molecular meditation. The maze is an image to think with, and a base for intuitive drawing exercises while the analogue and low-tech nature of the tool ensures wide reach and practicality.

Qualities of the maze drawings:

The form of the maze is composed of a grid structure; the Maze path is created by hand drawing possible paths through the grid architecture. The grid structure acts as an analogue for the multi-dimensional developmental landscape. The multi-dimensionality of the landscape here is not represented spatially as 3D or 4D; instead, the many dimensions are represented qualitatively. In drawing, the parameters that can be changed are related to granularity: shape of the grid unit (i.e. squares/triangles); size of the grid unit; definition of the grid unit (hard edge, soft edge); opacity of the grid unit (darkness, lightness); and colour combinations layered on top of the grid structure. The use of colour is intuitive, constructed as a play of colour, each addition in response to the last. Colours can be interpreted as dimensions; the multi-colour landscape is analogous to a multi-dimensional landscape. There is no observable colour in proteins, only their energy gradient so we can assign hot and cold colours to this; but colour is arbitrary and therefore without bounds, rules or particular meaning. Colours are like attractor points as the path is drawn towards different colours as it moves and so the pathway and the landscape are co-created and in continuous relationship.

Drawing dynamic patterns: Productive engagement with the unknown

Much of the visual material that we encounter in biology fails to capture the dynamic nature of biological processes in a 2D image. Here, however, we propose that a collaborative approach that shares methods of art and science can be used to produce novel biological images that represent dimensionality and dynamic patterns. Drawing helps to consolidate ideas, clarify concepts and bring visibility to thought.

Increasingly, the problem of how to move from the representation of the object towards the representation of the process is now also recognized by science. The images created here can also assist the general philosophical struggle to find good metaphors and practices for representing, discussing and thinking about the process. Art and science therefore can be understood to have a shared goal to discover visual ways of exploring dynamic patterns, processes and complex systems and concepts. Documenting these extra-scientific forms of knowing is crucial in order to open up rather than foreclose what is possible to see, say, feel, imagine and know in today's life sciences.

ACKNOWLEDGEMENTS

We thank the Arts and Humanities Research Council for funding our project 'Representing Biology as Process', 2017–21 (AH/P007457/1) and we thank members of the Phillips lab for participating and enthusiastically engaging with the project, especially the Drawing Labs.

NOTE

1. If the maze circle was a grid, we could glue the edges together into the shape of a toroid (doughnut) surface.

4

The Metamorphogram: Pattern as Memory of Experience

Alun Kirby

'Metamorphogram', a splicing of metamorphosis and photogram, describes an image created through a combination of origami and cyanotype. Thus, an initial three-dimensional structure is exposed before being returned to a flat sheet for development and display. The metamorphogram goes beyond traditional cameraless photography by removing the external object typical of that process; the prepared paper is given an origami form which 'takes a photo' of itself. Examined in terms of philosophy of memory, this process records direct experience, analogous to the creation of subjective autobiographical memory. This novel process facilitates the creation and visualization of patterns invisible in the original form.

For example, Figure 4.1 shows a *tato* (an origami 'purse') folded from tissue paper coated with cyanotype chemicals and exposed to sunlight. Its external

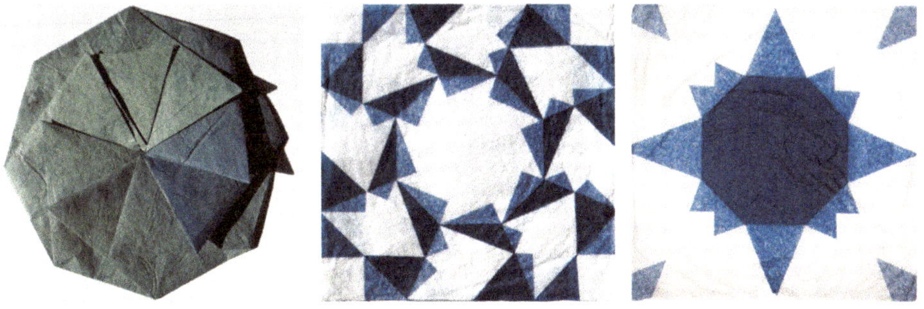

FIGURE 4.1: Coated tissue paper folded into a *tato* form (left). A metamorphogram resulting from a *tato* exposed folds up (centre) or folds down (right). All images © Alun Kirby.

form had a visible geometric pattern of folded flaps. Unfolding and development of the image reveal a new geometric pattern related to, but different from, both its crease pattern and origami form. This pattern goes beyond the folded surface; three layers within the form are combined into the final image. Dark blue areas indicate paper from the surface of the form, mid blue the layer below and white areas at the bottom. The experience-dependent nature of the pattern is clearly shown in another *tato*, identically treated except for being exposed face-down.

The metamorphogram's process produces patterns which can be interpreted in terms of direct experience. In *Memory Boxes* (Figure 4.2), cyanotype-coated origami boxes were placed in the homes of people living with dementia, as part of a project on memory. Two identical boxes were given to different households, each having a unique experience, with the patterns produced relating directly to the context of this experience. The experience of the first box (left) lasted about three weeks, in a fairly bright place, with light coming mostly from the right. It also recorded a splash where some tea was spilt nearby. The square in the centre was the bottom of the box, and so remains white. The second box (right), in place for three months, exhibits more even and intense colour, showing it experienced more light, and from all sides. The central square is blue, as the bottom of the box was exposed to light. The participant moved house during the project, and the box recorded that experience, including being turned upside-down. Thus, we see again that pattern is experience-dependent, and also how the metamorphogram allows us to visualize aspects of the 'biography'

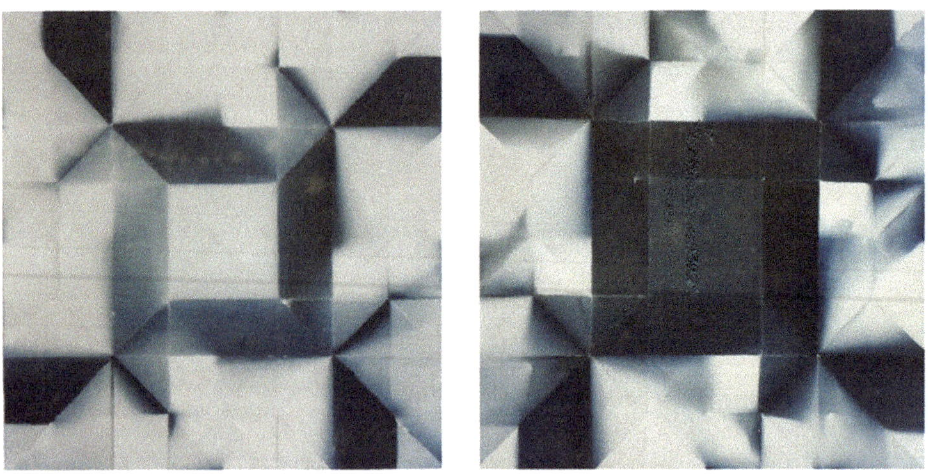

FIGURE 4.2: Metamorphogram images from *Memory Boxes*, 2018. All images © Alun Kirby.

of a form in the pattern it expresses. We may say that each object records its experiences subjectively.

In another project, 'One Thousandth of a Wish', the principle that each pattern image can be considered philosophically as an individual memory is further developed, and ways in which social and cultural memory are formed from experiences common to many individuals are explored. *Senbazuru*, the act of folding 1000 classical origami cranes (*orizuru*), is said to grant the maker a wish. Folded from cyanotype-coated tissue paper and left to expose randomly, each crane creates a unique variation of the same basic pattern (Figure 4.3).

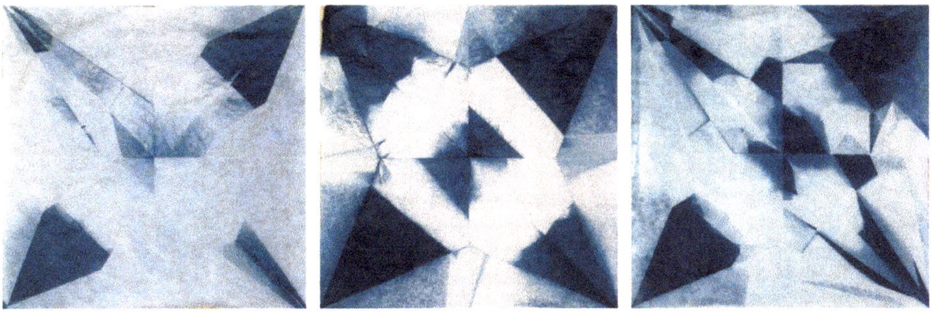

FIGURE 4.3: Examples of *orizuru* metamorphograms produced as part of the 'One Thousandth of a Wish' project, 2017–ongoing. All images © Alun Kirby.

A collage of 81 of these metamorphogram cranes, *Cliques and Networks* (Figure 4.4), reveals new relationships formed entirely from the 'memories' of the process. The patterns recorded while in crane form allow them to create connections between themselves, forming yet more patterns. Circular forms – the 'cliques' – appear where several individuals connect directly. An extended diagonal grid – the 'network' – is formed, connecting individuals beyond their immediate contacts. Based on these explorations of pattern using an artistic process, a new form of cultural memory is proposed, reflecting a sociobiographical memory in which individuals have agency both to alter an existing socially shared narrative and to value that narrative as part of their autobiographical memory – their 'self'.

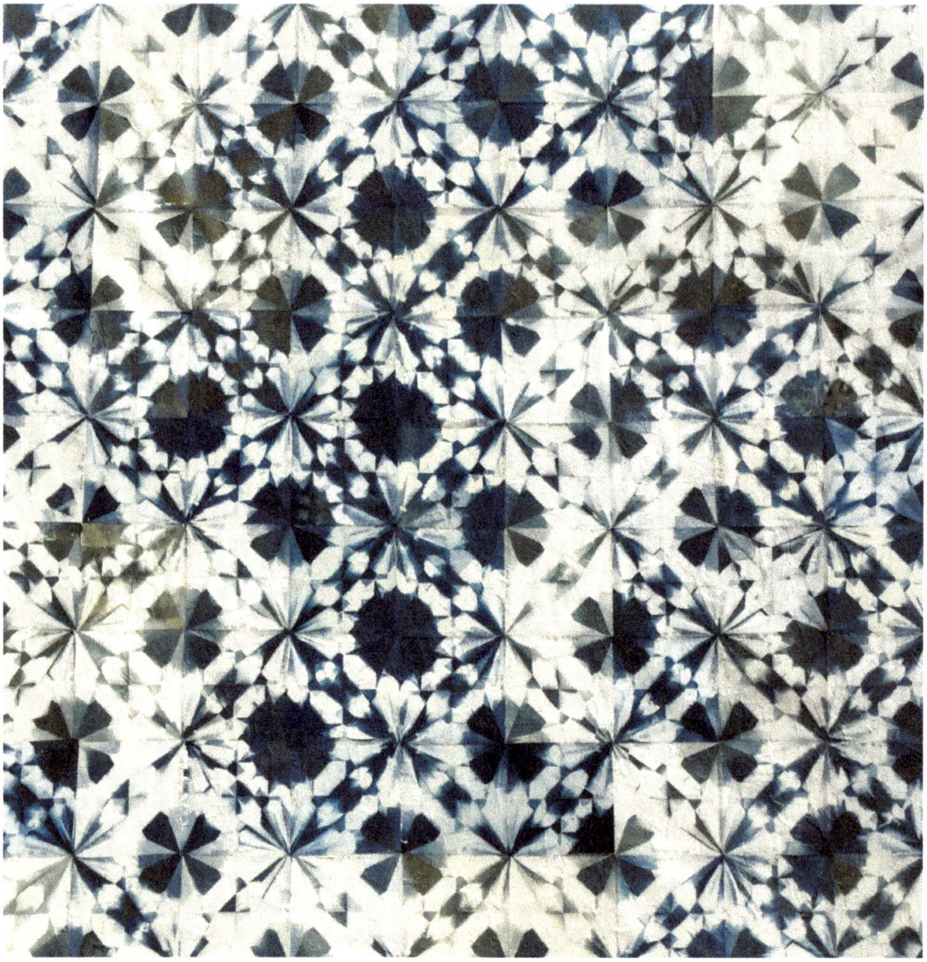

FIGURE 4.4: *Cliques and Networks*, 2019. © Alun Kirby.

5

Crumpling: An Exploration of Nature

Dewi Brunet and Gwenaël Prost, for the CRIMP Collective

For twenty years, the CRIMP Collective (International Research Centre for Folded Creation/*Centre de Recherche International en Modélisation par le Pli*) has been developing crumpling techniques through an applied and delicate approach. Whether infinitely small or extremely large, the fold is a universal process governing the structure of the natural world, including mountains, flowers and DNA, for example. Drawing inspiration from these, the CRIMP Collective uses a unique folding technique combining chaotic and organized crumpling to create original artwork.

Crumpling compounds intricate patterns and randomness. A mechanical and repetitive process allows perpetual shifting, bringing together the folded and unfolded state of the material that is used. From the early 1960s, in his 'folding as method', the artist Simon Hantaï crumpled and knotted canvases before painting over them. These were then spread out, revealing the structure of the crumples as a kind of randomized geometry; what appears to be a geometric and structured crumpled shape reveals instead an entire network of microfibers and randomized folds. This duality, whereby pattern structures while chaos creates, is, in our view, mesmerizingly aesthetic.

Structured crumpled patterns do not appear by mere chance however. Beyond the technique, folding turns out to be a comprehensive tool for exploring our relationship with the world. The seemingly chaotic aspect of crumpled structures (Figure 5.1), be it sea waves or the brain in our head, is in fact organized by the language of the natural world: mathematics. In other words, such structures obey a principle of logical construction, evolution and development over time, through trial and error, alongside, and in response to physical constraints. Found in numerous typologies,

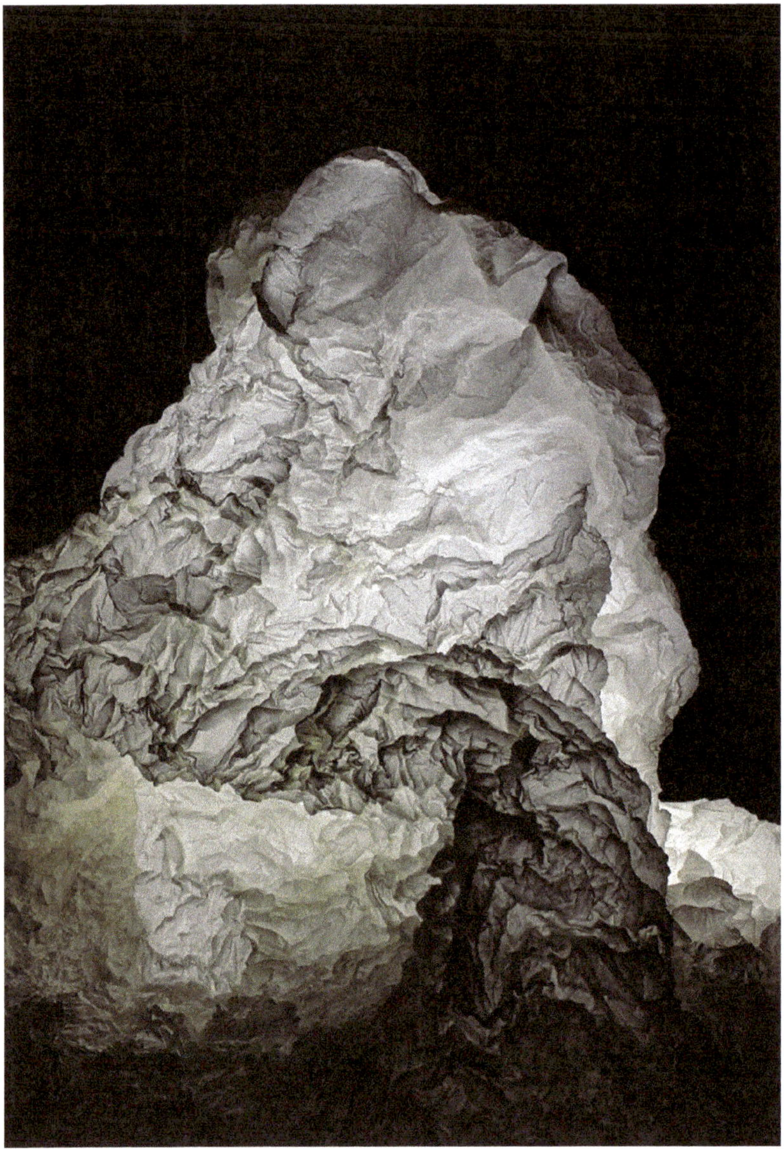

FIGURE 5.1: Chaotic crumpling; made from a large sheet of baking paper (45 g/m^2), this type of crumpling has been made by hand, through repetition of (almost) random folding. © CRIMP.

from plants to minerals, these structures unfold in myriad analogic shapes, through a process sometimes known as morphological convergence (Figure 5.2). Beyond the aesthetic, folding emerges as a consistent and intentional 'pattern' of adaptive strategies.

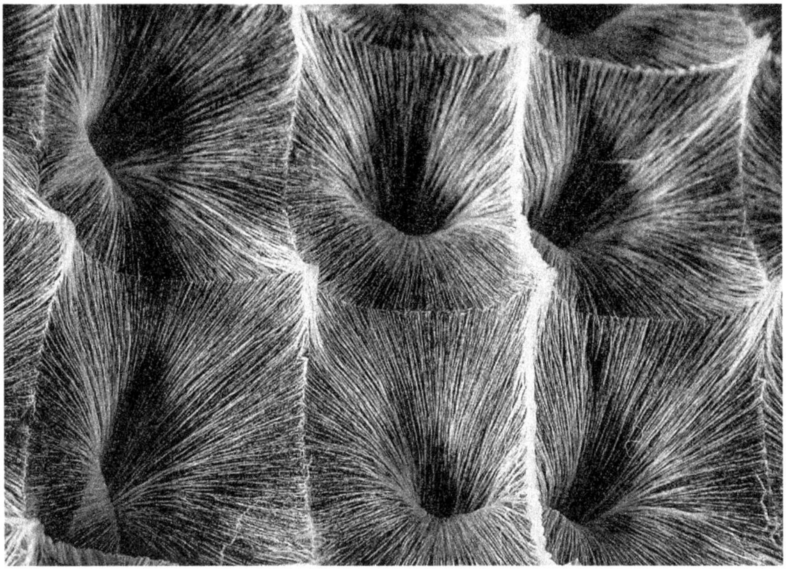

FIGURE 5.2: Made from a 200 cm × 200 cm sheet of thin, Japanese mulberry Tengu-jo paper (5 g/m^2). Original model by Vincent Floderer, the compressed form of which is a sea urchin with multiple spines (2000). Collection Jonathan Baxter, Southeast Origami Festival. Image, Romain Chevrier for CRIMP. © CRIMP.

6

Somewhere Between Weaving and Painting

Geoff Diego Litherland

In a field in Derbyshire, in 2017, weaver Angharad McLaren and I sowed and nurtured a crop of flax. We harvested and processed it into hand-spun linen yarns and finally wove it into a canvas for me to paint on. This collaboration set out to foster a more interconnected and ecologically aware relationship between artist, designer, materials and environment. Having agency over every stage of the process also allowed us to integrate woven patterns into our canvases (Litherland and McLaren 2017).

The patterns we developed have been designed using interlocking circular and geometric shapes, derived from hexagonal honeycomb structures, leaf forms and other repetitive patterns found in nature (Figure 6.1). These patterns, used as the core weaving structure for the canvas, act as both a metaphor for interconnectivity and, through their very structure, as a material embodiment. These initial

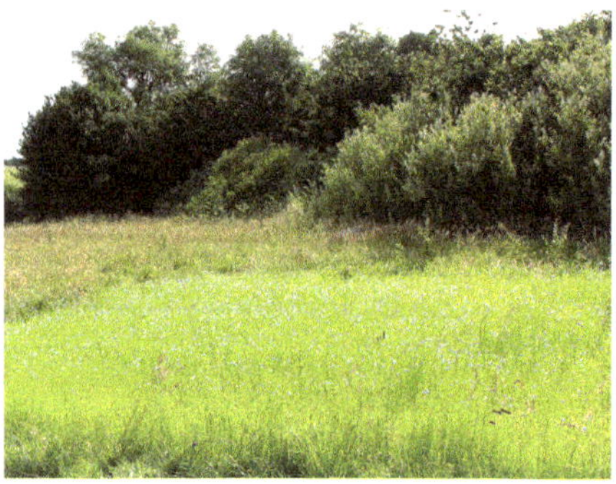

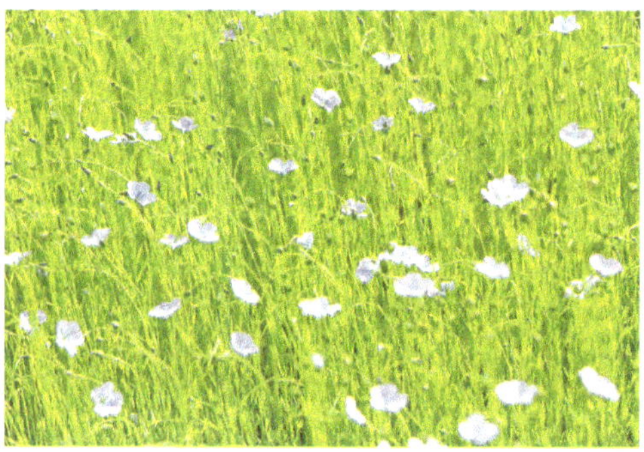

FIGURES 6.1 AND 6.2: A field in England, Derbyshire, summer 2017. All images © Geoff Diego Litherland.

configurations were derived from a shared regard for the geometric fabrics of Amerindian cultures whose designs and processes often embody deep cultural meanings. In reference to this, Beverly Gordon says that '[s]ometimes in their very layout or conception, textiles also served as actual embodiment of the way makers see the world – as what I think of as "cosmological maps"' (Gordon 2013: 223).

There has been a historical precedent for painting on patterned canvases that proved a useful counterpoint to the project, processes that the research of weaver Helena Loermans has been uncovering (Loermans 2021). She has been re-creating handwoven historic patterned canvases, known as *mantelillos* (Zuccari et al. 2005: 31–32), that were occasionally used by painters such as Velázquez, El Greco and Titian. Although the textural qualities of these *mantelillo* canvases aren't visible due to thick priming and paint coverage (Pacheco 1649), X-ray analysis shows the canvases to have large damask diamond type repeats: patterns at the time that would have been inspired by both eastern and Islamic traditions (Rodés Sarrablo 2012).

Although research hasn't proven why these patterned canvases were used, they, like many other fabrics, are an example of how cultural meaning can be embedded in textile patterns – a well-established notion in cultural and textile theory (Kraft 2004). Decoration in textiles can be seen as transformative: their visual and tactile qualities play an important role in sensory perception, social function and our understanding of the world (Igoe 2010).

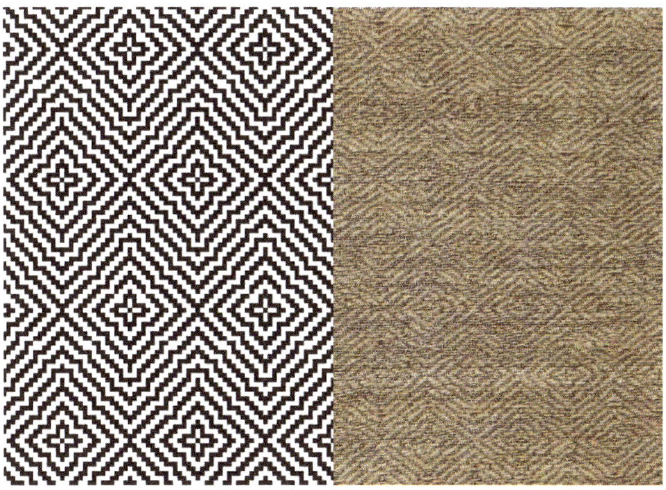

FIGURE 6.3: Geoff Diego Litherland and Angharad McLaren, *Hex-Acid Weave Pattern*, 2020. Digital weave pattern (left) and handwoven flax canvas (right).

Given the conceptual depth and meaning evident in these textile-led approaches to pattern and material construction, it could be seen as somewhat wasteful that in these historic paintings the canvas is typically completely hidden from sight due to being primed, painted and framed, with the intricacies of the woven process lost from view. As we have grown and handwoven our own patterned canvas, it would seem counterintuitive to not respond to these material qualities and nuances of the fabric and pattern when developing the final paintings.

In the process of making the work, weaving and painting have become joint investigative tools: every process and step involved influenced the final outcomes of the paintings. The two seemingly opposing practices of weaving and painting offered opportunities to explore the parameters of visual chaos and order. Weaving is a regimented and orderly process that involves considerate planning, preparation and time. Painting can also be meticulous in its preparation of surface and layering, but always has a capacity for speculative marks and for mistakes to become intrinsic.

I embarked upon an iterative trial and error process, responding intuitively to the materials and patterns of the canvas and to each layer of the painting process at a

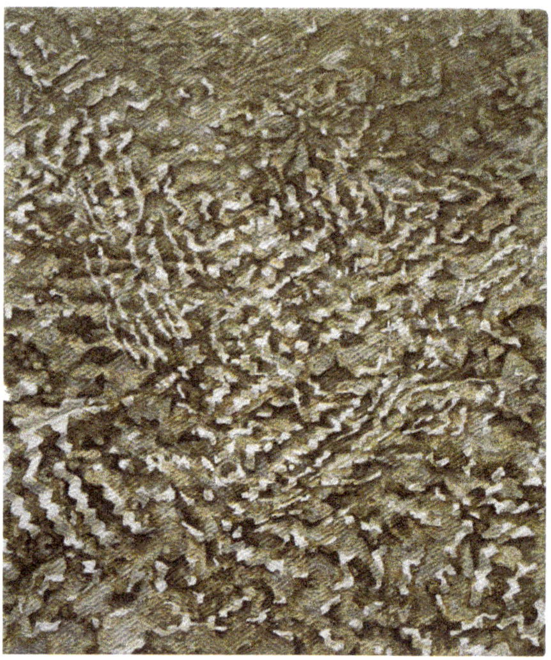

FIGURE 6.4: Geoff Diego Litherland, *Slow Ascents* (detail), 2021. Oil on handwoven, patterned, linen canvas, 32 cm × 38 cm.

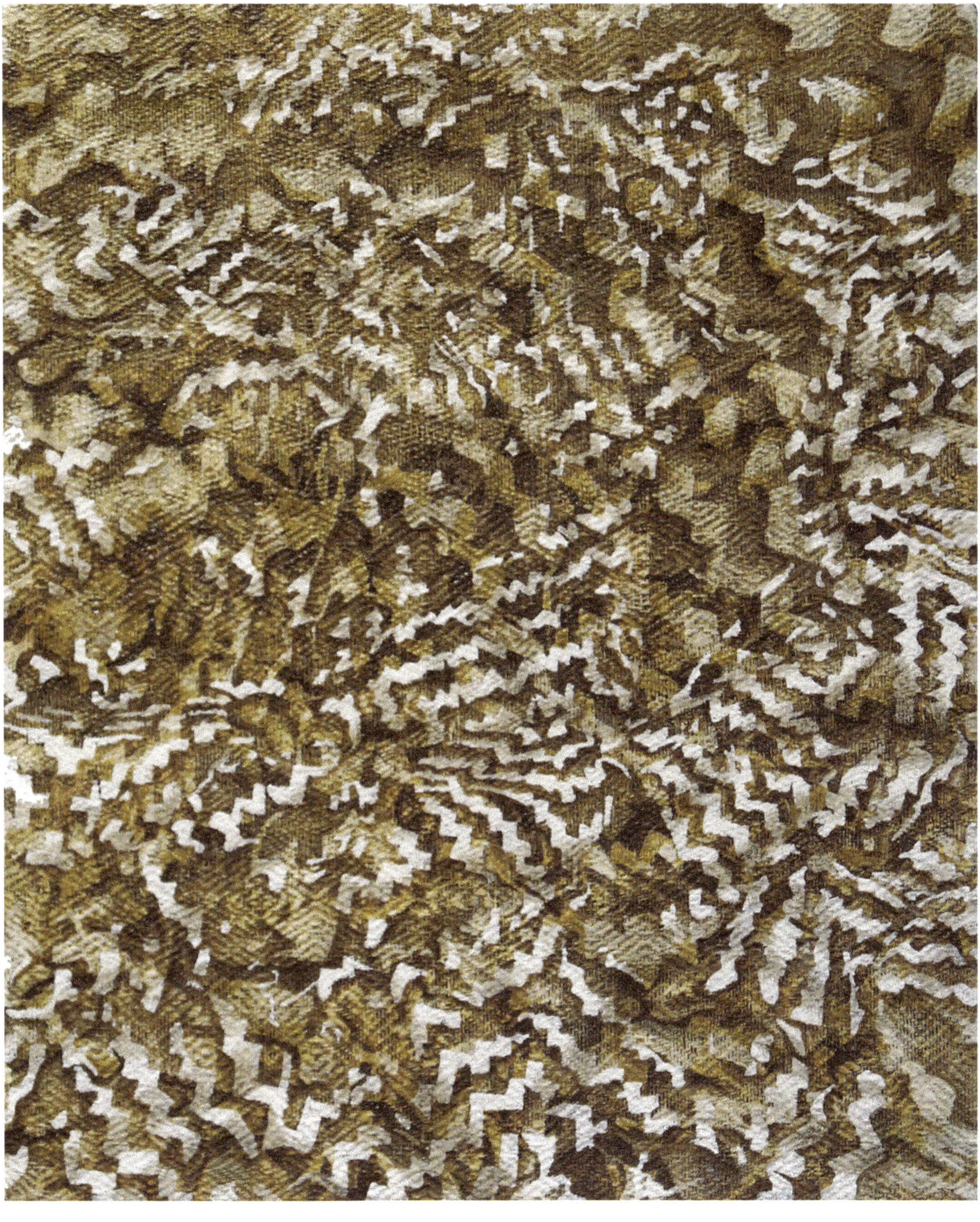

FIGURE 6.5: Geoff Diego Litherland, *Common Ground*, 2021. Oil on handwoven, patterned, linen canvas, 32 cm × 38 cm.

time. This is like anthropologist Tim Ingold's (Ingold 2010) ideas around the textility of making, where the forms of things arise within fields of force and the flows of material; practitioners make things by intervening in these force fields and following the lines of flow. Material and formal decisions were encountered and considered holistically to the works' conceptual ambitions.

These artworks explore the relationship between the patterned woven canvas and 'ground'. Here the term 'ground' has multiple and interconnected meanings: firstly, the historical ground (the surface) of a painting, which is the mid-tone wash of colour from which an image emerges; secondly, the ground (the earth) that cultivated the flax/canvas; then, the pigments that are taken from the ground (material) to make paint; and, finally the visual representation of the ground (landscapes) we have traversed, which are often used as starting points from which the paintings emerge.

Images of the ground (landscapes) painted in earth pigments are intermeshed with the canvas's pattern, creating a mutated hybrid that wavers between order and chaos. In each individual painting, various tensions are turned up or down to explore variations between digital and analogue, macro and micro.

The paintings have an unfolding perceptual journey for the viewer that seeks to exaggerate the wavering duality of painting, that of material and illusion. The works have no clear focal point: shapes and textures tessellate in and out of focus, drifting in the visual periphery, while ambiguous landscapes flitter between abstraction and figuration, flatness and space. Like the experience of foraging for wild food, the paintings entice the viewer to find recognizable tropes and order among the layered complexities of the environment.

The patterned canvas amplifies the physical material qualities of the surface, which are further pronounced by the unpainted deep edges of the canvas stretcher (see Figure 6.4). It is here that the viewer can see the unpainted canvas and clearly pick out the woven canvas's pattern. Ingold's concept of haptic vision feels pertinent in respect of how these paintings can be encountered, where perception is not only on the surface but within the very structures that permeate and give rise to them: in dwelling on the surface everything is seen and encountered. Ingold adds that, 'surfaces are important not in themselves but for what they potentially open up, and for what they disclose. But they are also important for what they hide, and for the deceit that they can practise on us' (2017: 102).

Meaning lies in and within the surface of these paintings, through their texture, visual complexity and combination of materials and processes that embody the slow labo-

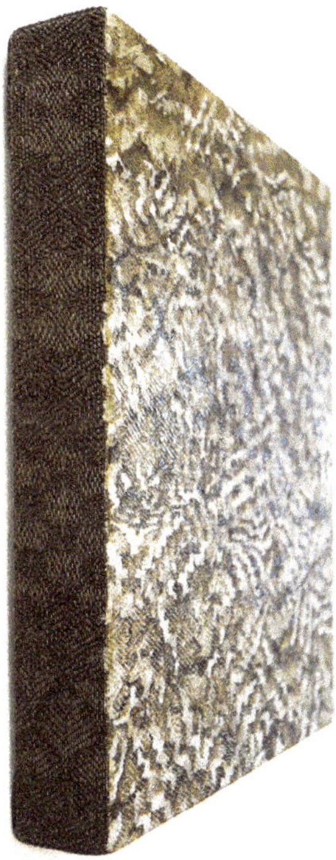

FIGURE 6.6: Geoff Diego Litherland, *Middle Ground* (side view), 2021. Oil on handwoven, patterned, linen canvas, 32 cm × 38 cm.

rious journey from the open field to the closed gallery. The measured time spent being in and with the environment and the focused manual labour involved have allowed undulating natural abstract forms to slowly emerge from within, born of an encounter where the material journey of the work has overtaken our subjectivity.

7

Knotting Across Species: Creating Order from Chaos

Eleanor Morgan

All straight threads want to get knotted. I begin each day untangling the cord of my earphones. Overnight, untouched in a drawer, it tangles itself again. One of my shoelaces is knotted so tight that I cannot unravel it and as I look down at my feet under my desk, I can see a knot of phone and laptop chargers that I have given up puzzling over. If these lines want to get knotted, I have decided to let them. To some people, this might look like messiness. But I no longer believe that time spent tidying up knots is time well spent or the sign of a tidy mind. Instead, let them wander. In doing this, in doing nothing, I am pleased to discover that I am in harmony with the Second Law of Thermodynamics (entropy), or as the mathematician Andrew Belmonte puts it, a 'law of nature [...] as applied to long floppy things: Long Things Get Tangled' (Belmonte 2007: 17243). In other words, knots tie themselves with or without our involvement.

What is a knot? If you asked me to create one, I would take a piece of string, loop one end, feed the other end through the loop and then tighten it. This is called an overhand or single knot. As with all knots, it involves a type of material (long and flexible), a learned action (the movement of my hands) and a specific pattern (the single knot). Alternatively, I could put the piece of string in my pocket for a day and it would likely become knotted, but I would have no control over what sort of knot it made. Creating knots is about trying to control the seemingly random knotting behaviour of long thready materials in small spaces. It is a method of attachment, a way of binding materials together which, unlike glueing or fusing, can potentially be undone without altering the original thread. As a subject for study, it begins simply, like the single knot I tie in my shoelaces, but soon becomes tangled.

My original intention was to create an elegant pattern of writing that linked together the ways that humans and other organisms use knots. I wanted to show the ties between species and how faced with specific materials and environments, different creatures have devised similar making processes. I would start with a runner bean plant as it winds its tendrils around a pole to reach the light. I would then follow a weaver ant as it uses its larvae like a silk shuttle to attach leaves together. Moving through sailors' knots and spider webs I would magically tie it all together with a brief study of knotted 'quipu' strings made by the Inca to record data. But once I got down to the real work, I discovered that very few species actually make knots. There are humans, a few birds and some individual higher primates. Other organisms bind, weave or stick together threads, but very few create true knots.

The reason for this scarcity of knotting might be because tying together simple materials, such as dry grasses or rope, requires complex behaviour and a certain dexterity (Hansell 2007). In contrast, animals such as spiders can create complex materials of tough and elastic, sticky or dry silk that only requires simple and repetitive behaviours to assemble into a web or a nest. To make a true knot, a spider would have to create a loop of silk and then crawl back through this loop and tighten the thread. Instead, a spider can join the threads of its web with a sticky glue.

But if knotting requires complex behaviour, then we share this ability with weaver birds. These animals collect strips of grasses and palms to create their hanging nests. The challenge is how to get dry strands to bind together. Using only their beak, weaver birds use a combination of weaving, knotting and binding to assemble their nest. The knots they use include the familiar overhand knot. Weaver birds must learn how to tie these knots; indeed, they improve with practice. They must also improvise and adapt to the specific building conditions, such as the qualities of the branch from which their nest will hang. They begin by attaching a circular strip of fibre to a branch, and then use the limited reach of their own body to form the shape of the nest, knotting new fibres until they have created the final tear-drop shape (Hansell 2007: 82).

Knot-making has also been observed in some great apes, but only captive animals who have been reared by humans. In my search for knotting in other species, these examples seemed to be anomalies that didn't quite fit with the 'natural' knotting behaviour of the weaver bird. Yet as the anthropologists Chris Herzfeld and Dominique Lestel argue, the tendency to treat the behaviour of nonhumans as only relevant when they are viewed in the wild ignores both the real effects of human/nonhuman interactions on the life of an individual animal and the potential creative adaptability of that animal in response to its environment. In other words, humans have focused too much on creating catch-all definitions of a species rather than looking at the different behaviours of individuals within that species (Herzfeld and Lestel 2005).

One example that they give is Wattana, a 25-year-old female Borneo orangutan. Born into captivity in Antwerp Zoo, Wattana lived in three different zoos before being moved to her current home in Apenheul primate park in the Netherlands. At the age of four, she began tying knots with materials in her cage. She was then moved to Paris, where Herzfeld and Lestel gave her a collection of thready materials, such as string, laces and hose. Using her hands, feet and mouth, Wattana created a variety of knots, tying materials together and even creating an installation of tied ribbons all over her cage. The researchers point out that as with humans, Wattana's knot-making is not planned in advance. Rather, it is a combination of initial observation and experimentation until, from repetition, it becomes a bodily memory. When I tie a knot, I follow a rhythm that my hands have danced for 40 years. The ease with which I can do this without thinking becomes clear to me when I attempt to draw a knot (Figure 7.1). I find it difficult to use my visual memory of a knot to draw from because a knot is both an object and a specific action. To create a representation of a knot, I hold my inky brush in my right hand and draw a single flowing line, looping from right to left, and left to right. It may look like a knot, but it follows a different pattern, a different dance.

When I tie a knot, my hands are the two points of contact with the thread. Wattana has developed a dance of five moving points: her hands, feet and mouth. She is entirely absorbed in the process, which she can focus on for more than an hour. Herzfeld and Lestel argue that Wattana's knot-tying is an example of new skills that can develop in individual animals from living in close contact with humans. Rather than being an example of 'unnatural' behaviour, it is evidence of the adaptability and creativity of other animals.

Looking at photos of Wattana and her knots, I think about the thready games I played when I was a child: cat's cradle and French skipping; knots to make pictures, to pass the time and to tell stories. But instead of these social games, Wattana reminds me of the hours of my life I spent as a receptionist on my own in a room playing solitaire on a computer. Her knot-making could be an activity that offers minimal, and mildly addictive, mental stimulation in an impoverished environment.

In my search for knot-makers, I was biased by what was visible to me. But within all organisms, knots are being made and undone in strands of DNA. Humans contain around two metres of DNA coiled up in each of our cells. As with my knotted earphones, the problem with putting long twisty things in small spaces is that it is highly likely to get knotted. And when DNA gets knotted, it can cause cells to die. The solution is an enzyme called topoisomerases. When the double helix of the DNA is over or under-twisted (think of the twisting threads in a rope), then this enzyme will

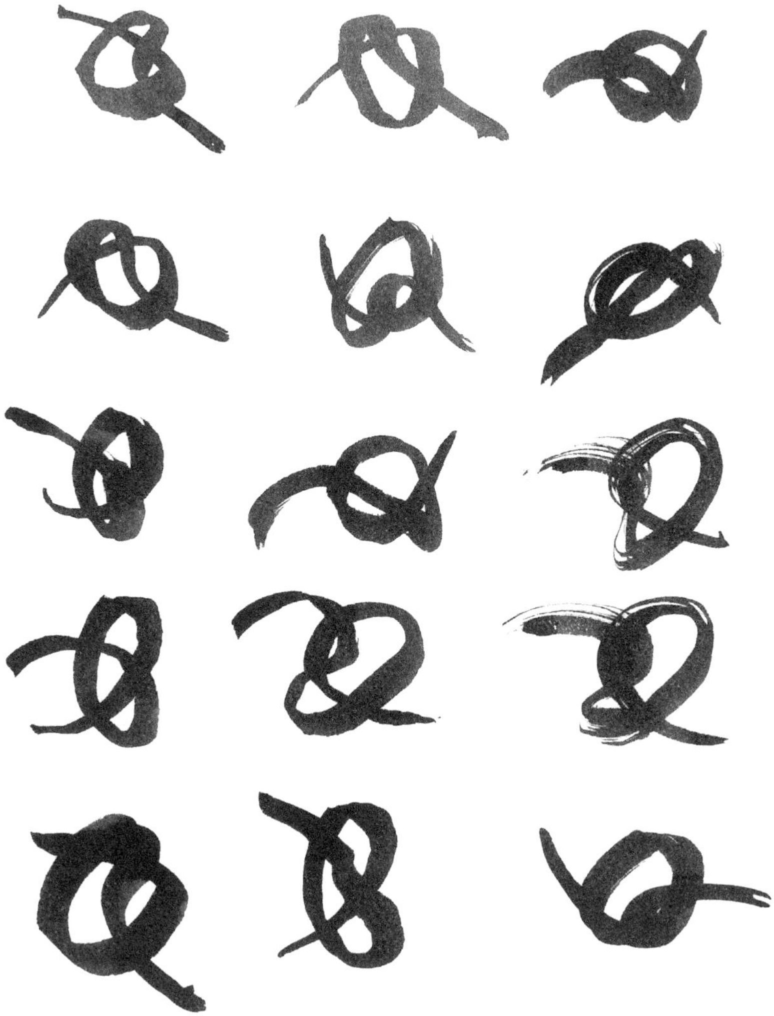

FIGURE 7.1: Author's attempts to draw a single knot, 2021. Ink on paper.

remove knots and remake the twist evenly. If knotting exists across species, this is the common strand that binds all living things: a process that controls the random, and potentially deadly, tendency of long threads to get knotted.

ACKNOWLEDGEMENTS
The author wishes to thank Dr Chris Holland and Dr Alice Pyne for their time and advice on spiders and DNA knotting.

8

Simplifying Complexity: The Visual Language of Neuroscience

Gill Brown

Many fields of science have their own specialized visual language, developed over decades or even centuries (Brown 2016), that requires trained judgement to be read and understood (Daston and Galison 2007). To anyone lacking such training, scientific conceptual figures can appear complex and incomprehensible, even though the visual elements that constitute the 'alphabet' of a visual language are themselves simplified representations of scientific theories or processes. To create these elements, scientists use specific shapes and patterns that are relatively straightforward to draw and easily identified by their peers as having particular scientific meaning. The same visual elements appear repeatedly in figures used within any one field of science, and the ability to recognize these elements, and what they represent, is the key to understanding the figures.

Graphic design practice is an important tool in the analysis of such figures, enabling them to be re-drawn and dismantled, and the scientific visual elements extracted (Brown 2019). The elements are then typologized based on their scientific meaning and collectively compared to identify those features that are indicative to a trained eye. This allows a nonscientist to begin to understand that language, and in the case of a graphic designer can open the door to true collaboration with scientists, to aid and improve their visual communication (*Neuroscience & Graphic Design*, 2022).

The field of neuroscience provides an accessible entry point into a scientific visual language and demonstrates how visual elements can be developed and employed. The neuroscientist Santiago Ramón y Cajal, working at the turn of the twentieth century, is famed for his beautiful drawings of nerve cells, or neurons (see Figure 8.1), made from life with the aid of a conventional light microscope (Newman 2017).

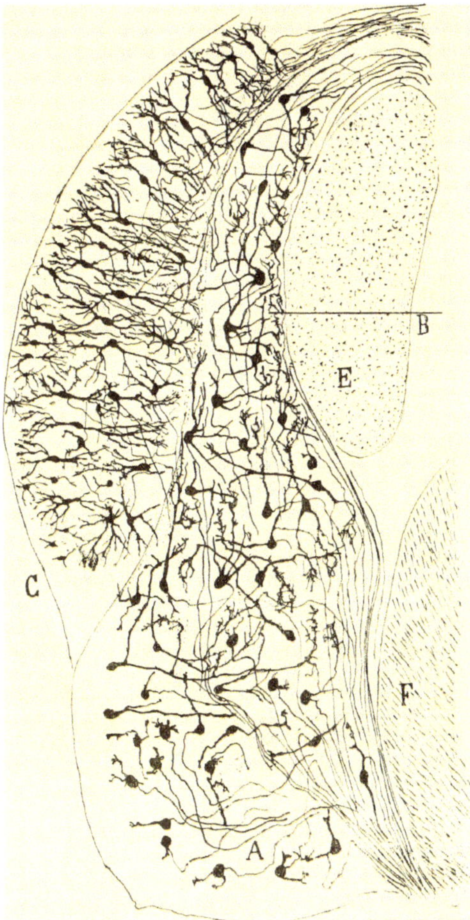

FIGURE 8.1: Santiago Ramón y Cajal, *Beitrag zum Studium der Medulla Oblongata, des Kleinhirns und des Ursprungs der Gehirnnerven* ('Terminal acoustic cores of a 4-day-old rabbit'), 1896. Book illustration, https://commons.wikimedia.org/wiki/File:Beitrag_zum_Studium_der_Medulla_Oblongata,_des_Kleinhirns_und_des_Ursprungs_der_Gehirnnerven_(1896)_(14578994027).jpg. Accessed 26 May 2021.

Few neuroscientists today have the time or ability to produce detailed drawings like Cajal's. Instead, they must create a simplified version of this complex reality if they wish to visually communicate their work. By extracting a single neuron from Cajal's drawing (see Figure 8.2), it is possible to identify some of its distinguishing features: a rounded soma (cell body) that contains the cell nucleus; branching dendrites emanating from the soma; and a long, thin axon that ends in branching axon terminals. It is these features that make a neuron instantly recognizable to

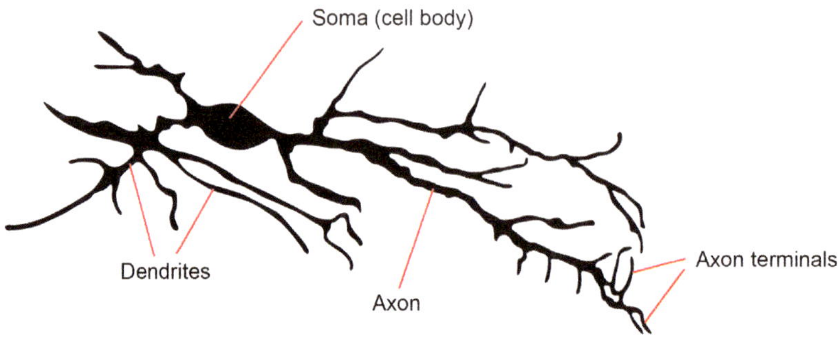

FIGURE 8.2: Gill Brown. A single neuron re-drawn from the image shown in Figure 8.1, 2021. Digital image. © Gill Brown.

a neuroscientist and, in their conceptual figures, a single neuron can stand in for the larger system.

The depiction of a neuron can vary greatly in its details and visual style, as a neuroscientist must work with the software and resources available to them. The examples of neurons shown in the left half of Figure 8.3 are from existing conceptual neuroscientific figures and were re-drawn to resemble their original style as closely as possible. What is immediately obvious is that neurons are not represented by any particular colour. By re-drawing the same neurons in monochrome (right half of Figure 8.3), it is easier to focus on the similarities between the shapes and pick out the distinguishing features; the rounded soma, almost always containing an ovoid nucleus (shown here in light grey); the branching dendrites; and the long, thin axon ending in the axon terminals. Small swellings, called synaptic boutons, are sometimes shown at the ends of the axon terminals.

Although not shown in Cajal's drawing, many depictions of a neuron also include a visual representation of the myelin sheath, a fatty insulating layer around the axon. This is shown as a series of short, sausage-shaped structures, lined up along the axon (see Figure 8.4). The colour of the myelin sheath is not indicative, and it is shown here in dark grey so that the similarities in shape are more clearly seen.

During Cajal's career, it was assumed that nerve signals were transmitted along neurons as electrical impulses, jumping across the synapse (an infinitesimally small gap) between nerve cells. In 1934, Henry Dale published a paper stating that chemical messengers, as well as electrical impulses, travel along neurons. These chemicals are carried within vesicles (bubble-like structures) and are released from the synaptic boutons, the small swellings at the end of the axon terminals (Dale 1934).

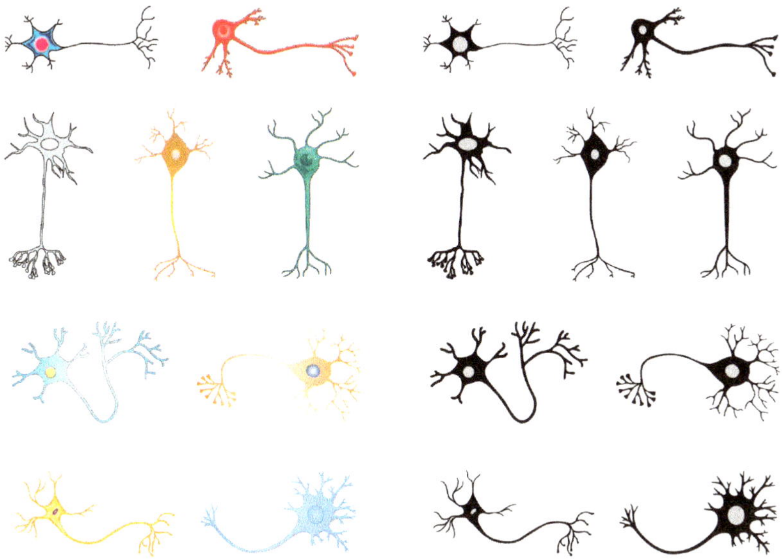

FIGURE 8.3: Gill Brown. Visual representations of neurons, re-drawn from existing conceptual scientific figures and shown both in full colour (left) and in monochrome (right), 2021. Digital image. © Gill Brown.

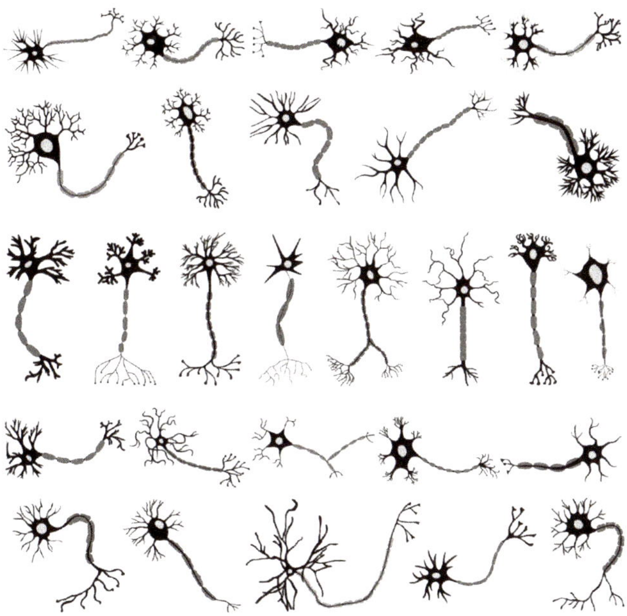

FIGURE 8.4: Gill Brown. Visual representations of neurons with a myelin sheath, re-drawn from existing conceptual scientific figures and shown in monochrome, 2021. Digital image. © Gill Brown.

A 'pre-synaptic' bouton releases the chemicals, and they pass across the synapse to be received by a 'post-synaptic' bouton of an adjoining neuron. This process is called chemical neurotransmission, but Dale used no figures in his paper, meaning there was no visual representation of the process. It was not until more sophisticated microscopic imaging arrived in the 1960s that a visual representation based to some degree on direct observation could be created (Guillery 2005).

A collection of these visual representations of chemical neurotransmission is shown in Figure 8.5. Again, colours are not indicative and have been replaced by black. Despite the variations in visual style, several distinguishing features are always present: a pre-synaptic bouton, containing spherical vesicles, is positioned above a post-synaptic bouton (without vesicles), with a gap (the synapse) between them. This arrangement of shapes, therefore, gives instant context to a figure, with a knowledgeable viewer recognizing the depiction of chemical neurotransmission.

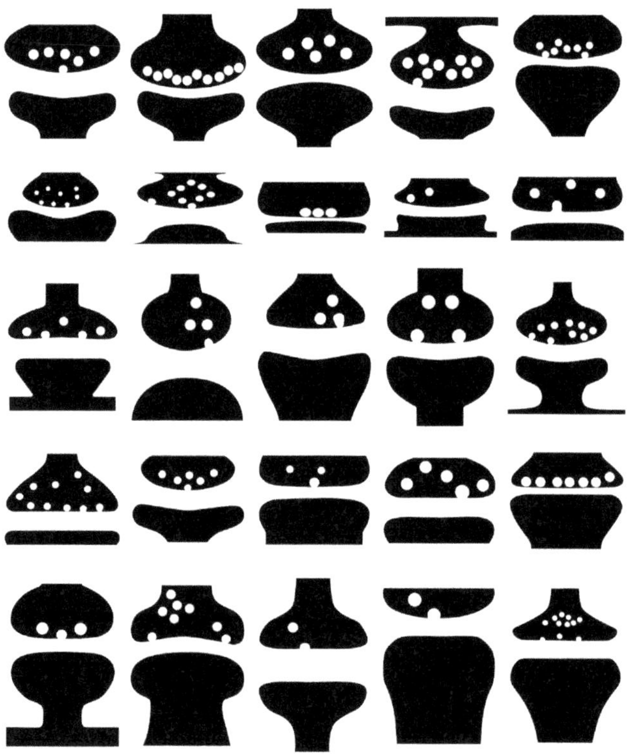

FIGURE 8.5: Gill Brown. Visual representations of chemical neurotransmission between two neurons, re-drawn from existing conceptual scientific figures and shown in monochrome, 2021. Digital image. © Gill Brown.

If a neuron is connected to a muscle cell, rather than to another neuron, the point of connection is described as a neuromuscular junction. In the visual representation of chemical neurotransmission at a neuromuscular junction, the post-synaptic bouton is replaced by a crenulated surface that represents the stimulated muscle cell. A collection of these visual representations is shown in Figure 8.6, with the muscle in grey. The crenulated surface is always present, even if highly stylized, and, in conjunction with the pre-synaptic bouton containing spherical vesicles, is indicative of chemical neurotransmission at a neuromuscular junction.

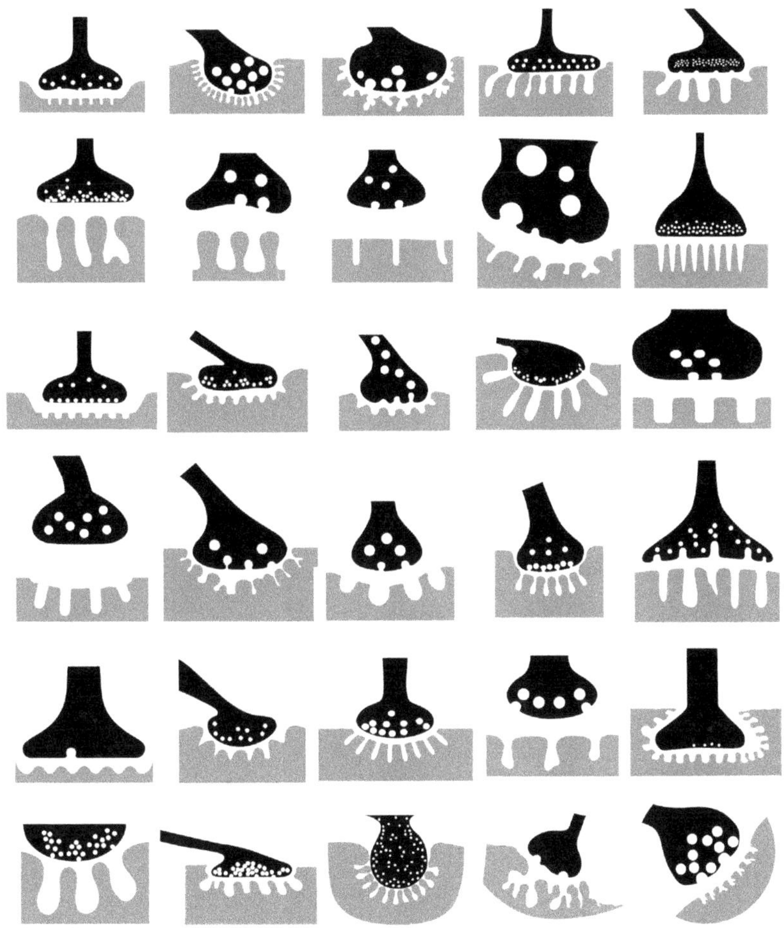

FIGURE 8.6: Gill Brown. Visual representations of chemical neurotransmission at a neuromuscular junction, re-drawn from existing conceptual scientific figures and shown in monochrome, 2021. Digital image. © Gill Brown.

The visual elements shown here form only a small part of the visual language of neuroscience, and each field of science has its own set of visual elements that enable the simple communication of complex ideas and processes. By identifying the shapes, patterns and colours that make up the visual elements, these specialized languages can be translated and understood, even by nonscientists.

PART 2

MORPH, FLUX, MUTATE

Introduction

Some of the contributions in Part 2, 'Morph, Flux, Mutate' could easily stem from the pattern dynamics that are so intrinsic to the essays in Part 1. The difference, however, is that the patterns developed here are subject to deliberate and continuing evolution – incremental changes that are gradually imposed upon a preselected set of variables to introduce new and surprising iterations effected over time. These contributions put aside the concepts of repetition and difference as binary and accept that both can occur at once or side-by-side as motifs are iteratively remade and reproduced, and so *x* leads to *y* leads to *z*. Similarly, a purposeful restlessness is produced when one seeks to repeat, rewind or return, or explore the diaphora of language, where a change of context provides a change of meaning in the repeated same but different.

When subject to incessant change the initial logic of pattern painted or drawn across a surface can result in continuous deviation as evident in the textile designs of Kate Farley, where the stripes, checks, ogees and diapers are continuously re-worked. Familiar motifs and passages are altered and transposed like musical rhythms and phrases. We see further hints of this in the butterfly forms that are defamiliarized in Rob Hillier's animations. These are patterns that morph towards the unpredictable and ambiguous, the life cycle of the butterfly hinting at our own mortality through metaphors of transformation and change. The passage of time is evoked again by Danica Maier as the reader turns the pages of her contribution to witness similar but evolving artworks that question her notion of the 'unrepeating-repeat'. Maier responds to a Jacobean bedspread depicting two intertwining versions of the tree of life and as the tree patterns are unveiled she reveals thoughts about the copy, difference and appropriation from her own and others' observations, offering a repository of gathered quotes exploring the reach of pattern–chaos from psychology, philosophy, linguistics and design. The overriding imperative in each of these contributions is the need to pay attention.

The sublime and the poetic are welcomed here as potential conduits for pattern and chaos in combination. This is seen in the creep of fire across the landscape in the intensely worked images of David Griffin where the repeated motif of the flame spreads across the paper. Similarly, it is manifest in the overwhelming and cloying potential of pattern as *horror vacuii*, the fear of empty spaces and the compulsion to fill in the gaps, explored in Lesley Halliwell's paintings. In these, there is a threat of eventual collapse and disintegration, a fall into chaos and loss of any sense of order.

In Katy Hammond's experimental and watery 'scapes' this creeping impulse creates beguiling images that slink to the edges of the page as if they might spill over into the entire book. They consume the whole surface upon which they are made, much as fire will eventually consume all available oxygen. Navigating the symbiotic existence of water and land is Glyn Brewerton's mapping of water and the shapes it creates in the Norfolk Broads. Using hand-drawn and machine-generated pattern, Brewerton measures and delineates the changing water levels of the Broads, corralling the seemingly chaotic movement into systematized order.

9

The Unrepeating-Repeat

Danica Maier

This contribution comprises three interwoven yet distinct parts: drawings (*Various Iterative un, re, peat(s)*), others' words (*Associated Thoughts on the Unrepeating-Repeat*), and my written voice (*Unrepeating-Repeat: reversal, once more, seek-seek afresh*). I invite the reader to experience each element separately through to its end and then return to engage with another. Each open page spread purposely holds these three elements together to allow the unfolding engagement with the unrepeating-repeat to spill out from one component and seep into the other(s).

FIGURES 9.1–6 (NEXT PAGES): Danica Maier, *Various Iterative un, re, peat(s)*. Six drawings. Pencil on mylar.

Associated Thoughts on the Unrepeating-Repeat

[A]n existential truth: that there's no such thing as exact repetition (Service 2016: n.pag.).

[E]vocative rather than descriptive, fragmentary rather than exhaustive, poetic rather than factual (Anon. 2007: 178).

If all you were able to draw from a first reading of *Difference and Repetition* was that it was a 'critique of representation' you would have already understood everything, provided that you know two things: what the word 'representation' meant and what the word 'critique' meant. This is the hard part (Hughes 2009: 1).

These famous lines from the beginning of T. S. Eliot's *Burnt Norton* succinctly convey the overlapping temporalities that are so central to this work: historical reckonings, revisions, re-visitations and reverberations, the eternal return that is never the same, always full of

---

Unrepeating-Repeat: Reversal, once more, seek-seek afresh

un- |ʌn| prefix. Denoting the reversal of an action or state; meaning not.
re- |ri:, ri, rɛ| prefix. Once more; afresh; anew.
peat |pi:t|. From Latin *petere* 'seek'.
-ing |ɪŋ| suffix. A form of a verb to show continuous or present continuous action.

(The above and all following cited definitions are taken from the *Oxford Dictionary of English*. Version 2.3.0.)

The unrepeating-repeat is a regular recurrence that contains variations and differences. It is found in configurations that seemingly repeat but include deviations or differences at some point within or throughout the recurrence. The unrepeating-repeat is a term I developed to describe what could also be named an iterative variant: 'the repetition of a process or utterance' including 'a form or version of something that differs in some respect from other forms of the same thing or from a standard'. It can be found conceptually and/or practically within objects, images and patterns; actions, sounds and music; and histories, cycles of life and beyond.

Un- as a prefix denotes the 'reversal of an action or state'. It folds the repeat back, reverses and brings a realization that it is 'not' the anticipated repeat. It can be experienced as the unexpected, a blip, a falter, a moment of surprise that the repeat is not as presumed. Not a smoooooooth pattern. *Un-* does not stop the repeat being, but reverses it. We go back, we seek again and experience it as not the same. *Re-* denotes 'anew' or more fitting 'once more'. *Re-* is the part of 'repeat' that entices one to look or experience the repeat 'afresh', while its meaning of 'once more' has the small prefix of *re-* holding the notion of repeat at its core. *-peat* derives from the Latin word 'petere' meaning 'seek'; introducing the 'attempt to find' which is core within the unrepeating-repeat. That we 'attempt

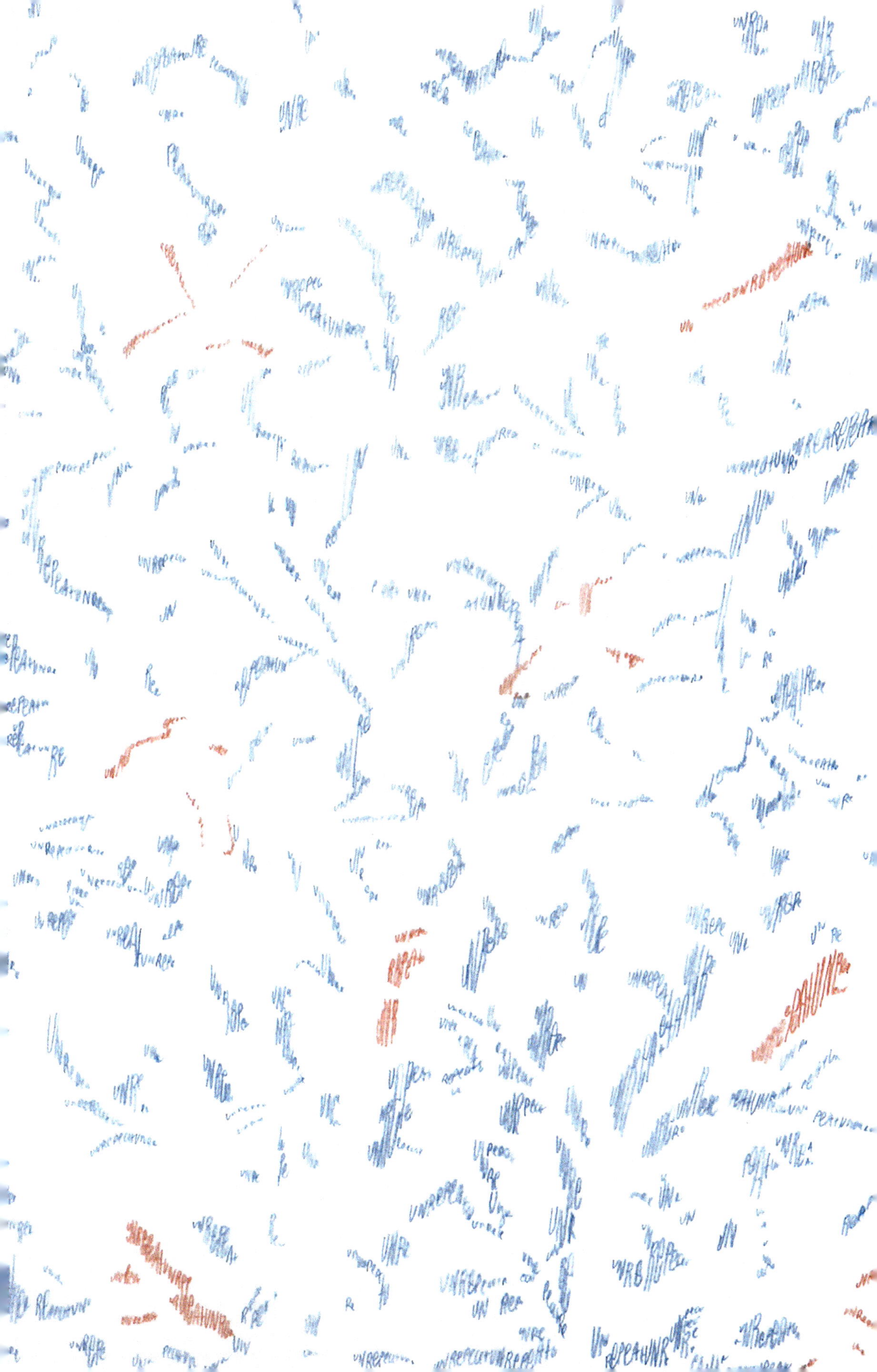

potential for change. The past is duration: the present is actualisation: the future is eternal return. But within all these, constitutive of them, is difference. Difference in kind constitutes duration. Actualised difference constitutes the present. The return of difference constitutes the future (T. May 2005 cited in Cook 2007: 37). We follow round this repetition of difference [in Marc Camille Chaimowicz's] *The Warsaw Suite* (1993–94) and *Here and There* (1978) (Cook 2007: 37).

Often people aren't aware that music is in motion in front of them all the time, that with each note it's either closer or further away, it's not in the same place that it was before. It can't be. If you play one note twice, it's going to have different implications, because you will have been altered by the first one. I don't really think exact repetition is possible, especially in performance: it's simply a different kind of movement [...] you've got a little older since it was first played, and you've probably started to think about something more, as we do when we get older. So the second time it comes, it has more urgency (Ades and Service 2012: 16).

---

to find' the moments of repetition; how they are similar but also different, a variant. When we combine *re*- with *-peat* we bring the notion of 'seek' into the frame with 'once more'. We 're-seek' or 'once more- seek' when we repeat. It is an action *and* an invitation.

The unrepeating-repeat is beautifully found within synonyms. Each word is similar, alike, comparable and akin to the other; they could potentially be substituted for each other. Sharing similar meanings, however, each is also different, specific. There is a space or nuance between the synonymous words. The gaps between speak and offer up something more, something different, something beyond the singular unit of one word or the other. I love a thesaurus, exploring the similarities and differences between words. Feeling their shapes and finding the parameters of meaning. Exploring how they fit together and are similar as well as how they are different, variant, unalike, non-identical, divergent, distinct, individual; how these words are the same, but also different. The mind compares between one and the other - exploring the gap between similar terms – finding the edges that describes the negative space between the words.

Iterative variants are found and celebrated within handmade or live works. Drawn, painted or made by hand, as well as actions or sounds performed live, cannot be repeated exactly; there will always be variations. These variants can be planned or unwanted – yet, they will always be part of the work. It is not possible to create an exact copy, replica, duplicate by hand or performed live – the unrepeated-repeat will always be found here. Further to the deviations that naturally occur in these instances, temporality aids in the unrepeating-repeat. Experience and memory of the first iteration change our understanding of the next, and the next after that. But can anything be repeated precisely – even when digital or mechanical means

Reading slowly and rereading, returning time and time again to read anew, we return, similarly, to the things in the world anew. Our slow and very local readings resist the all too familiar tone of those technological or instrumental readings that no longer share a relation with thoughts (Walker 2017: xv).

[R]epetition facilitates the interest response (Margulis 2014: 18).

In the copy, repetition and difference. Once a tactic of deconstruction or cool critique – of barbed pastiche, parodic emptying out – the practice of appropriation might also be performed with emancipatory intent, perhaps even with love. Reverent irreverence, subversive re-enchantment (Cocker 2015: 34).

It appears that a copy simultaneously presupposes and excludes both identity and difference, that it must be, at the same time, identical and different from the original (Mijuskovic 2009: 144).

---

are used? Walter Benjamin tells us that the most perfect reproduction is unique in existence in the place of experience. The moment of experience affects the repeat. Even when exactly the same, it is never experienced as such.

Memory of the iterative element changes through each repeated experience, aiding understanding, and can bring comfort from the recurrent familiar. Our recollections have been infiltrated by the reminiscence of the first iteration, or second, or third, or fourth, and so on. Before reproduction of music was widely possible, repetition was used by composers as an important element to hook the work into the minds of the listeners. So often we experience, view or hear most things before we engage with the original; our first real encounter being secondary to the copy, the simulation and the hyperreal. The original becomes the repeated experience.

The unrepeating-repeat can be noticeable or elusive – experienced as subtle variants within repetition and patterns that at first are not noticed. It has elements that correlate and pattern is found, yet each iteration has changes and variations while still remaining recognizable as similar to its predecessor. Attention is needed for the unrepeating-repeat as one must notice or seek it, to recognize the variant differences within the unrepeating-repeat pattern. One needs to pay attention, to seek and seek again. Re-seek (or re-peat) your attention to grasp that the assumed is not as expected; there is a difference, an anomaly, a shift between. You need to pay attention to catch the change and understand the deviations within the pattern. The unrepeating-repeat attracts our attention; it is the moment of disruption that pulls our attention back into focus. It acts as the metaphorical fly in the ointment or the skip of the record. At other times, attention is needed to experience and recognize that there is in fact a pattern, a rhythm, a duplication. That there is iteration. Through all these modes the unrepeating-repeat asks (and helps) us to hone our attention.

They further stressed that deep understanding occurs when the presence of new information prompts the emergence or enhancement of cognitive structures that enable a person to rethink his [sic] prior ideas. Likewise, according to Vygotsky (1978), through repeated experiences children learn covertly to plan their activities. Such repeated experience, he said, proceeds not in a circle but in a spiral, passing through the same point at each new revolution while advancing to a higher level (Lomibao et al. 2017: 2131–7).

Semantic satiation has sometimes been explained in terms of attention. Bored with repeated encounters with the same signified word, attention shifts to the only other available level – the constituent parts of the word – letters, phonemes, and so on (Margulis 2014: 18).

Repeating the units gradually changes their elements, and the contours also change until a new pattern defines the impression. Here, pattern formation is a medium of change (Mundry 2012: 91).

When considering the derivation of the verb appropriate, at its core is 'make one's own'. To 'make' something 'one's own' is creating a copy, but not an exact copy. It is developed and changed. The same but different from the source, the copy is an unrepeating-repeat of the original. The act of appropriation is a form of unrepeating-repeat – a copy, but not an exact copy. For me as an artist the appropriation process is one of spending time with the origin object, studying and unpicking it to know it in depth. Detailed examination occurring during the re-making process brings practical, phenomenological and tacit knowledge of the original object and beyond. Dismantling, undoing and unmaking to re-build, re-create anew, afresh; to bring new insight and knowledge to the source, while simultaneously seeking the origin object's core. I seek its essence, its soul. I do not know through being told but through experiencing, unpicking, dismantling, unfolding and exploring every aspect in new ways. This is how I discover; this is how I know; this is how I learn and this is how I host and honour.

Pattern is defined as a 'repeated decorative design', decoration, motif, marking, ornament, ornamentation; or 'a regular and intelligible form or sequence', series, arrangement, progression discernible in the way in which something happens or is done. Pattern, system, order, arrangement, method, sequence, structure, scheme, plan, form, format, framework, composition, constitution, shape, make-up, configuration, model, example, gauge, norm, guide, blueprint, prototype, original, template, sample, specimen, swatch. {*Stop, breathe*} The repeated aspect within the pattern is core to its *pattern-ness*. The fact there is repetition, reiteration, restatement, retelling, recapitulation, reciting, rehearsing, recurrence, repeating, redoing, retelling, replication, reruns, restate, reiterate, retell, replay {*Stop, breathe*} distinguishes it as a pattern. I love a pattern that is not exact.

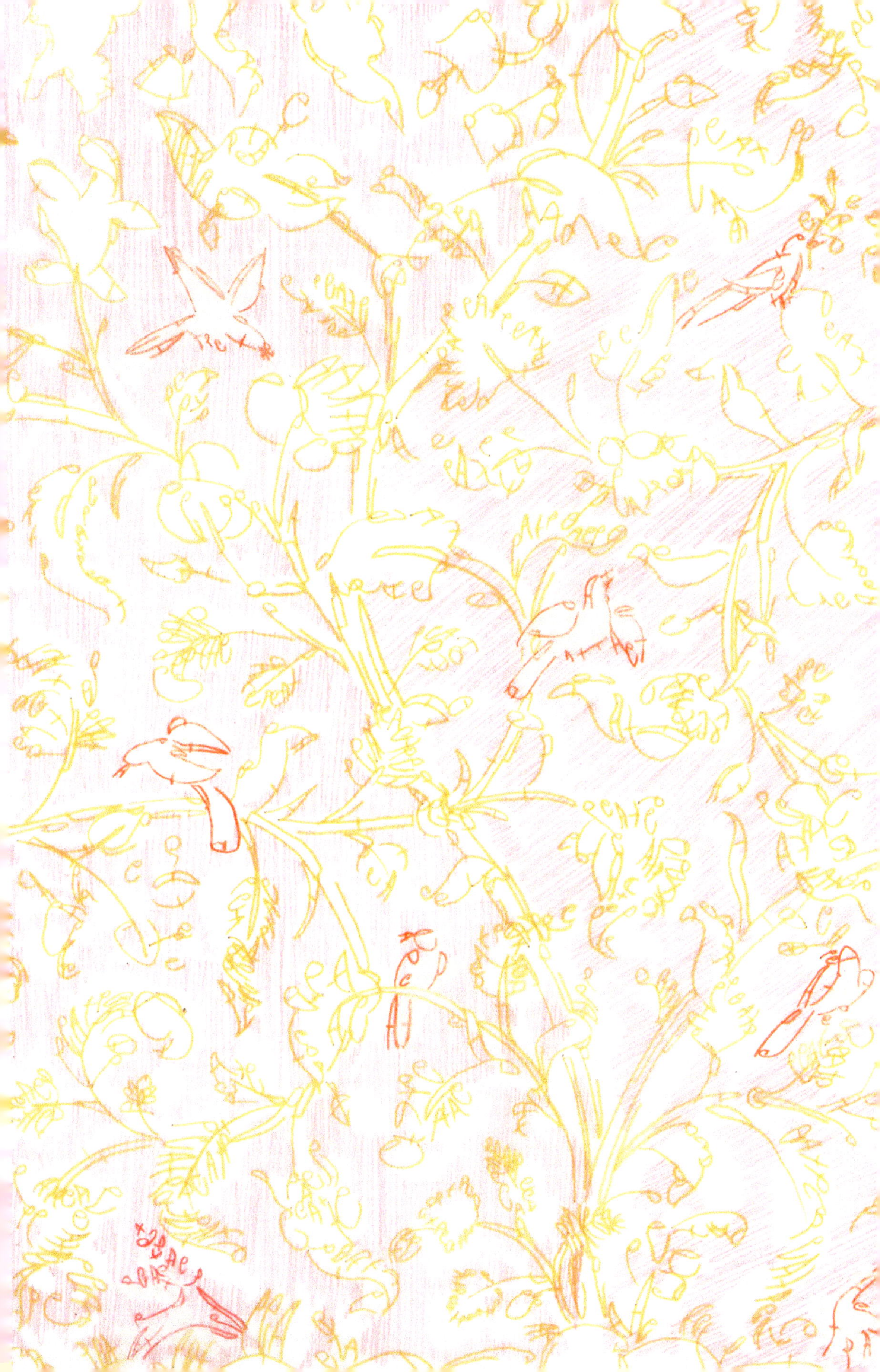

Several Criss-Cross artists, [Clark] Richert in particular, were also drawn to aperiodic patterns. For the preponderance of mathematical history, periodic tiling predominated in geometric thought. The concept of an aperiodic tiling scheme – a set of shapes that can tile a plane without repeating – is surprisingly young: until the 1960s, mathematicians believed that it was impossible (Skafsgaard Lowery 2019: 117).

Penrose tiles are remarkable to contemplate: they impart an immediate – but illusory – sense of symmetry. Small passages of symmetry are local, and evaporate when compared against neighbouring passages. Heterogeneity emerges: across infinite expanses, Penrose patterns are infinitely variable (Skafsgaard Lowery 2019: 117).

A fractal is a never-ending pattern. Fractals are infinitely complex patterns that are self-similar across different scales. They are created by repeating a simple process over and over in an ongoing feedback loop. Driven by recursion, fractals are images of dynamic systems – the pictures of Chaos. Geometrically, they exist in between our familiar dimensions. Fractal patterns are extremely familiar,

---

Repetition is the building block of a pattern. Whether decorative or in a discernible sequence, repeating creates pattern. It is the recurrence of an action or event repeating, echoing, parroting, quoting, copying, paraphrasing, plagiarizing, appropriating. Repetition can be a physical act, performance, engagement, spoken, printed, done and undone. It can go back and forth, back and forth, forth and back, back and forth. Habit. Habits, habits, habits, habits, habits, habits, habits, habits, habits are un/conscious acts of repetition with which we engage.

Pattern and Chaos. Chaos and pattern. The chaos of pattern. Patterned chaos. The overwhelming confusion of seeing, experiencing or finding patterns everywhere in everything. Even in variants. Chaos. Chaos, daos, eaos, faos, gaos, haos, iaos, jaos, kaos, laos, maos. The word sounds tumble around the mouth in a satisfying cartwheel. When is the moment that the unrepeating-repeat lapses into chaos? *When* – **is** the moment that the unrepeating-repeat, lapses into chaos. I seek and always find the pattern within the pattern within patterns. Looking and always finding patterns. Always looking. Always seeking patterns. Always seeing patterns in chaos.

Going in and coming out, going in and coming out, going and coming, in and out, in and out, in, out, in, out. When one can't help but see the patterns, when pattern *seeks* you, *plays* to you all the time – one can fall into chaos. Fractal's fracture into macro and micro. Micro and macro – asking for attention simultaneously – seeking in and seeking out. {*Stop, now read aloud*} Woods and trees, thickets and leaves, forests and plants, macro and micro, in and out, out and in, same and different, matching and not matching, back and forth, habits and rabbits. Rabbits who wear habits, habits who are rabbits. Habits, habits, habits, rabbits, habits, habits, rabbits, habits, habits. Rabbit. Habit. Rabbits and habits and abbots. Rabbits, habits, rabbits, habits, abbots, rab, hab, rab, hab, rab, ab, dab, rits, bits, tits.

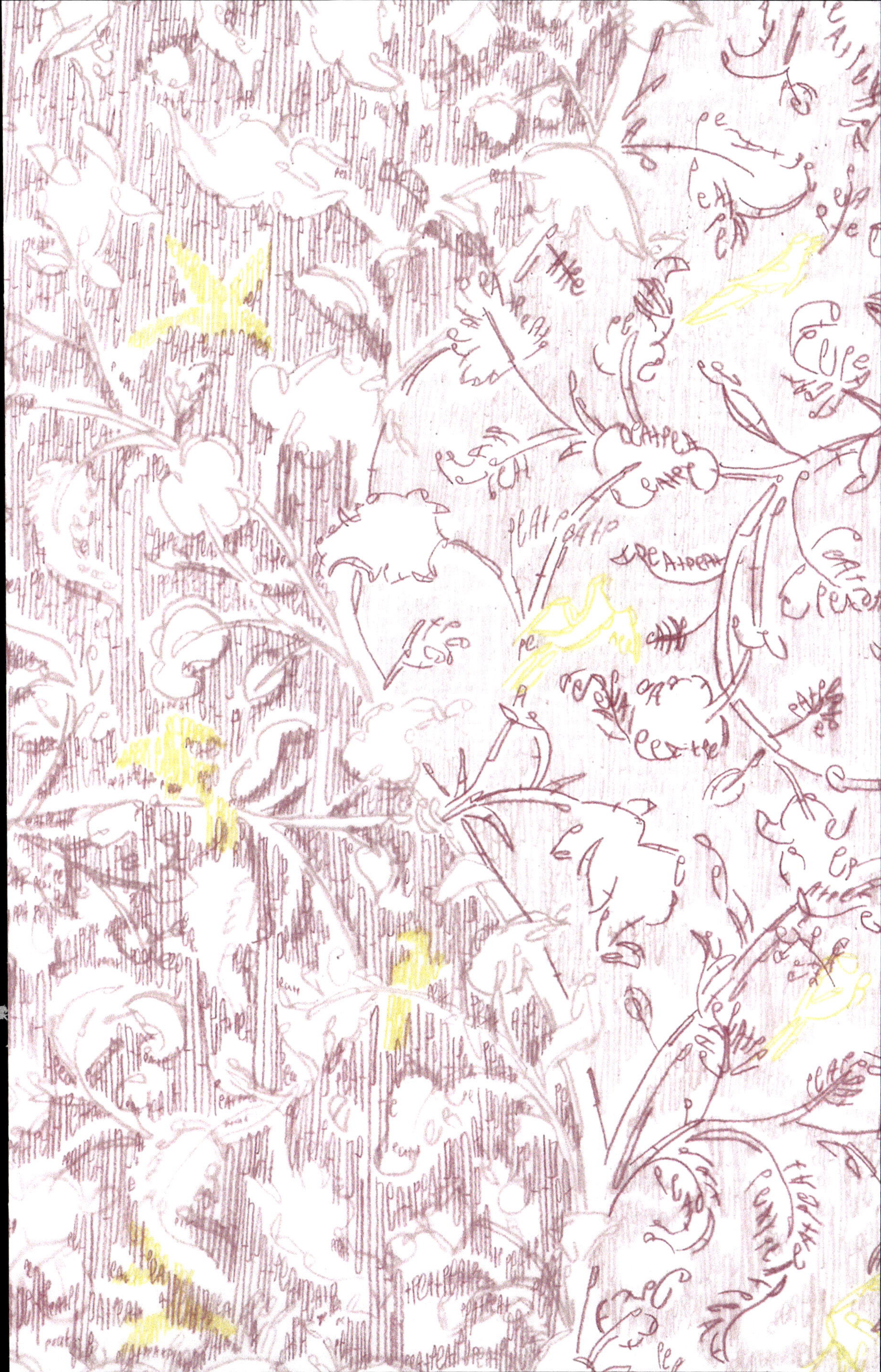

since nature is full of fractals. For instance: trees, rivers, coastlines, mountains, clouds, seashells, hurricanes, etc. (Fractal Foundation n.d.: n. pag).

A similar phenomenon [...] where repeated viewings, utterances, or hearings of the same word cause it to seem to degenerate into nonsense. Not just any nonsense, but a nonsense in which the semantics vanish and are replaced by a sort of super-salience of component parts – letters, phonemes, syllables (Margulis 2014: 17).

The voice, recorded and transmitted, the sound of the sigh of the listening woman, present and cyclical. Every beginning a sequel each beginning a sequel equel quell uel el l. Contingent cyclical logic in the space of listening, repeat this and repeat. Here I am to; begin, in a beginning cyclical (Cascella 2015: 22).

'Before unearthing this letter, I had questioned myself about the ways in which a book can be infinite. I could think of nothing other than a cyclical volume, a circular one. A book whose last page was identical with the first, a book which had the possibility of continuing indefinitely' (Borges 2000: 51).

{*STOP. Breathe.*}

A written illustration. The unrepeating-repeat is found within a crewel embroidered Jacobean bedspread comprising two trees of life. (This bedspread is part of the textile collection in The Collection and Usher Gallery, Lincoln.) At first glance, it seems a chaotic cacophony of stitched shapes without a clear pattern. Slowly, one can make out tree branches, leaves, and birds. At the bottom, we locate mounds that form two distinct grounds: this is our clue. Trees rise from the centre of each mound, a leaping deer on the left and a bird on the right of each trunk. As we follow the stems, we find both trees turn and sweep in similar directions. We look anew from one tree to the other and spot two birds in the same location on each tree, both eating a berry. Seeking further, the leaves are in similar places and shapes as they too are matched from one tree to the other. Looking back and forth, we see that each tree is the same, yet, on closer inspection each is distinctive. Individual birds and leaves have a similar shape and behaviour; however, each is uniquely stitched. Similar but variant, the two trees of life embrace the unrepeating-repeat. With further attention, the repeat fully reveals itself as it leaves the frame's edges and cycles around both left and right, top and bottom. The limbs of one tree end at the frame and are found folding round, coming across to the other side. Remove the ground and the trees will slot into each other from top to bottom, transforming into an ever-evolving and expanding rhizomatic ivy. Left to right, top to bottom; the image can be repeated out infinitely. Each time the same and each time different.

Now, go back once more, afresh, and look again.

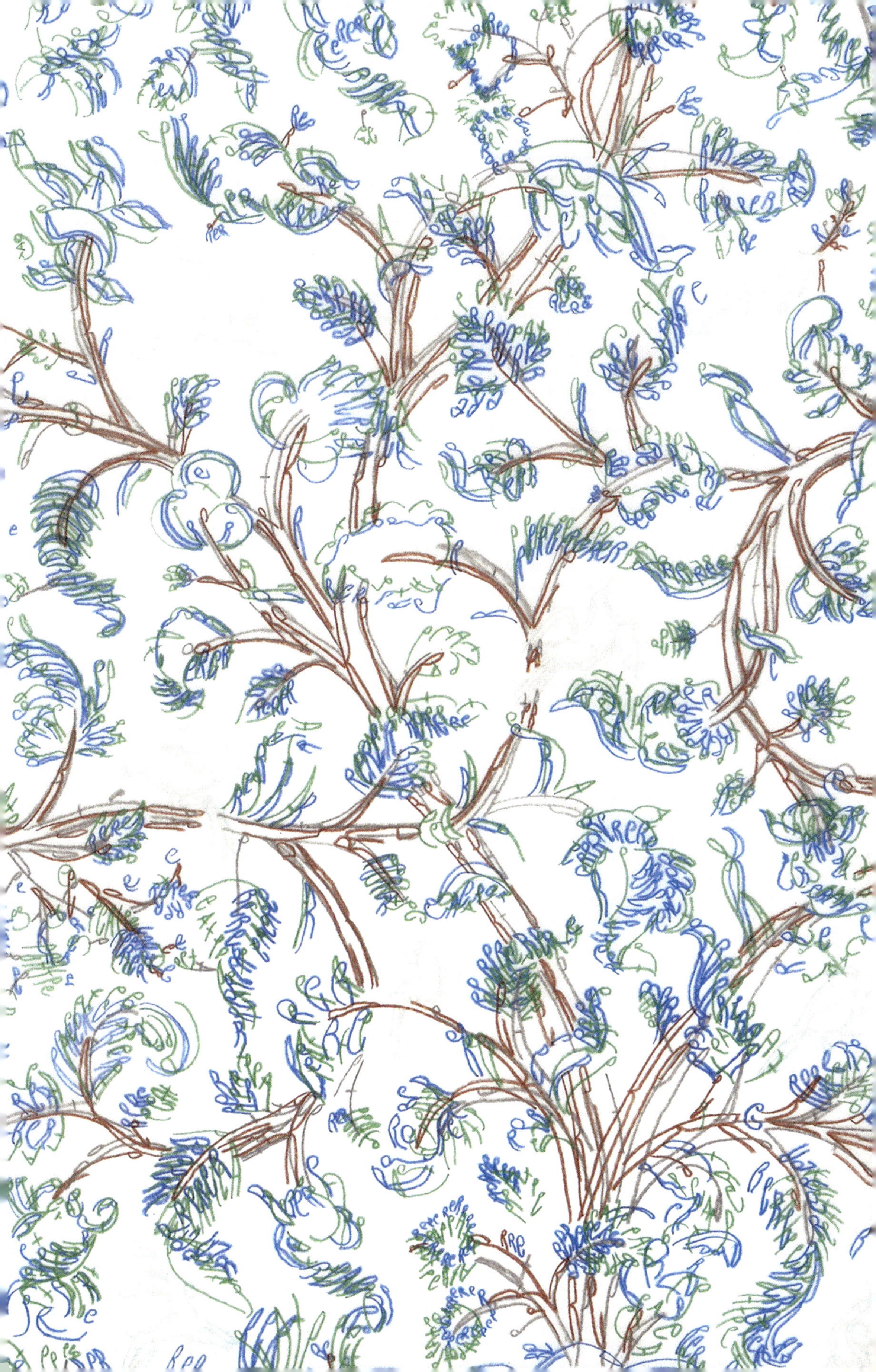

10

Pattern Evolution

Kate Farley

Motifs have been combined and arranged into pattern for centuries, applied to surfaces in relation to architecture, interiors and fashion. Formal pattern structures provide frameworks for the placement of motifs, defining the visual composition and rhythm of the overall pattern. Motifs can be placed by the designer within the scaffolding of pattern support, be it ogee, stripe or check, for example.

In this practical research project, the potential to combine and blend pattern structures within a single drawn composition was investigated; the structures evolving from one to another. Patterns were created and interpreted through pencil drawn studies with compositions initially reading from left to right, transforming from motif to motif, including stripes, checks, ogees and diapers (Figure 10.1).

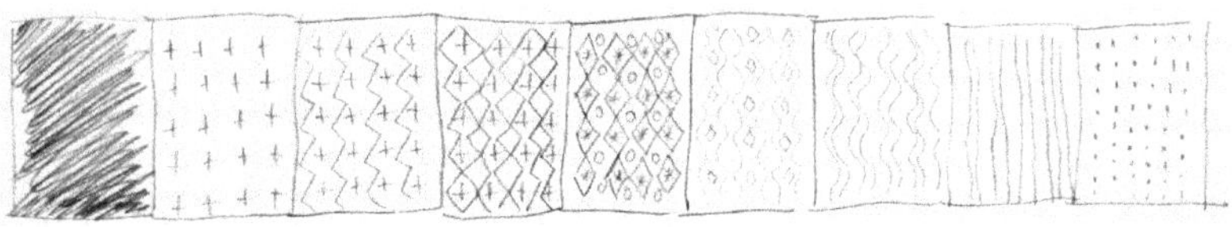

FIGURE 10.1: Kate Farley, initial sketch of pattern structures evolving, left to right. Graphite on paper. 297 mm × 80 mm. Image credit and © Kate Farley, 2021.

The formality of a motif within a structure was established before being broken down through a series of steps across the page, finding the simplest detail of each pattern element before evolving to the next through the evolution of marks. Small graphic motifs were plotted and spread across the page to form larger clusters, creating composite geometric rhythms. A template of a formal grid was used to locate and anchor the drawing as the narrative was built, retaining and guiding the evolution. Lines were lengthened and shortened, angles softened and shapes drawn on sheets of tracing paper, guiding the evolution and providing the design direction (Figures 10.2–10.4).

From top to bottom:

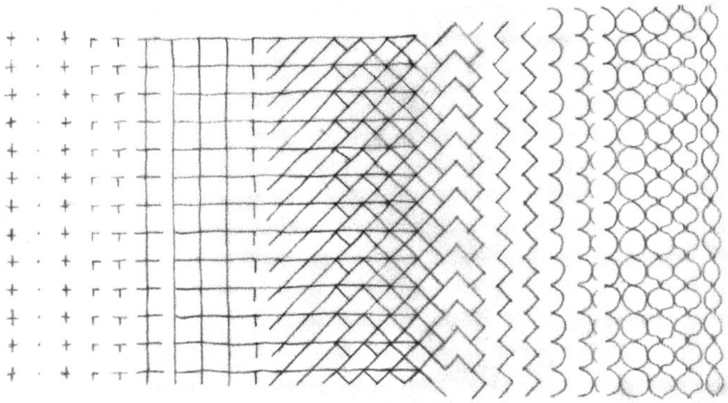

FIGURE 10.2: Kate Farley, testing the grid template to formally evolve the motifs. Graphite on tracing paper. 297 mm × 210 mm. Image credit and © Kate Farley, 2022.

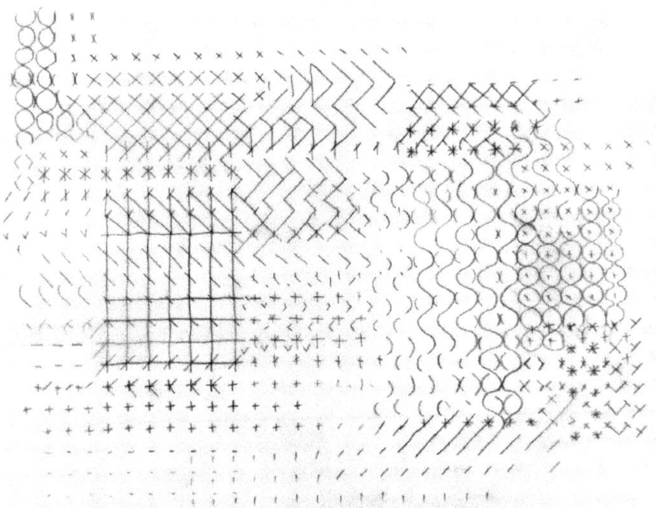

FIGURE 10.3: Kate Farley, increasing the pattern language in all directions. Graphite on tracing paper. 420 mm × 297 mm. Image credit and © Kate Farley, 2022.

The manipulation of motifs and rhythms can be considered in relation to composed musical themes and variations. Just as a motif played on a piano recurs, transforms and evolves before returning to a familiar musical phrase, drawn pattern motifs arrive, ebb and return. Identifiable but different, motifs are arranged, elements

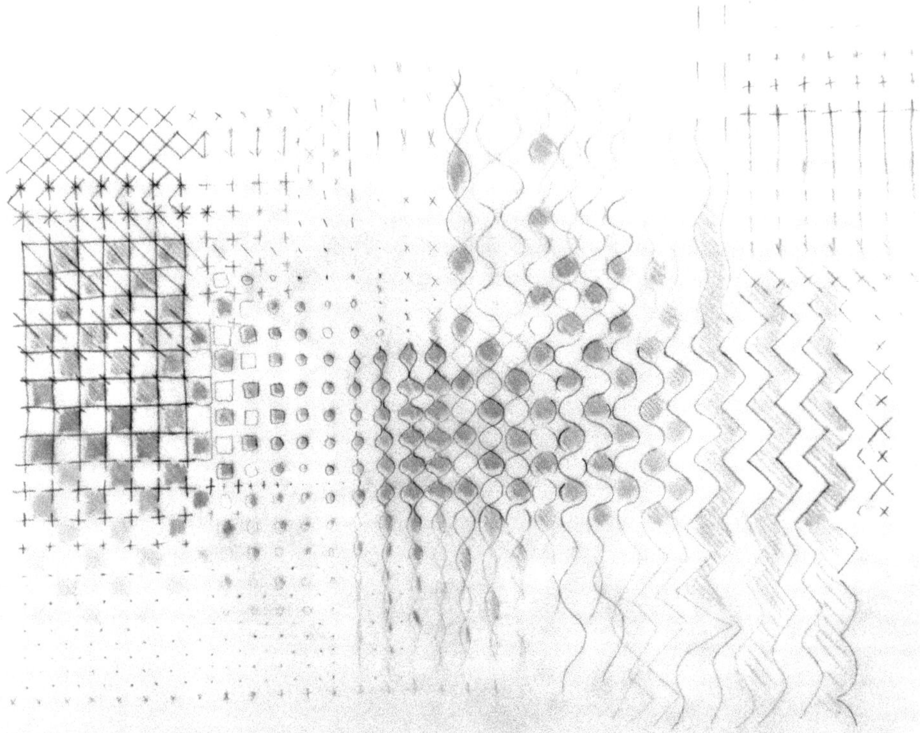

FIGURE 10.4: Kate Farley, leaving the grid structure behind. Graphite on tracing paper. 420 mm × 297 mm. Image credit and © Kate Farley, 2022.

deconstructed and the forms rebuilt to enable the rhythms to flow. The layered sheets of tracing paper are the sum; the upper layers present the melody while the shading beneath holds the gravitas of the bass line.

Each new drawing delivered possibilities and clarity of purpose, with awkward transitions and difficult mutations presenting themselves to be morphed to the next structure. Detaching the motifs from the grid structure liberated the composition to explore scale and rhythm in the lines and shapes flowing across the surface. Tone was introduced, the paper rotated, lines erased and reinstated. Recurring melodies appear in new surroundings, across the drawing, arrived at through alternative notation and contexts. Scale and density of marks in clusters or open space provide tonal variation, providing a depth to the experience, as quiet and louder refrains (Figure 10.5).

Greater dynamic arrived with greater directional transformation, morphing the drawing from a central point, to the left and right, and top to bottom, identifying a

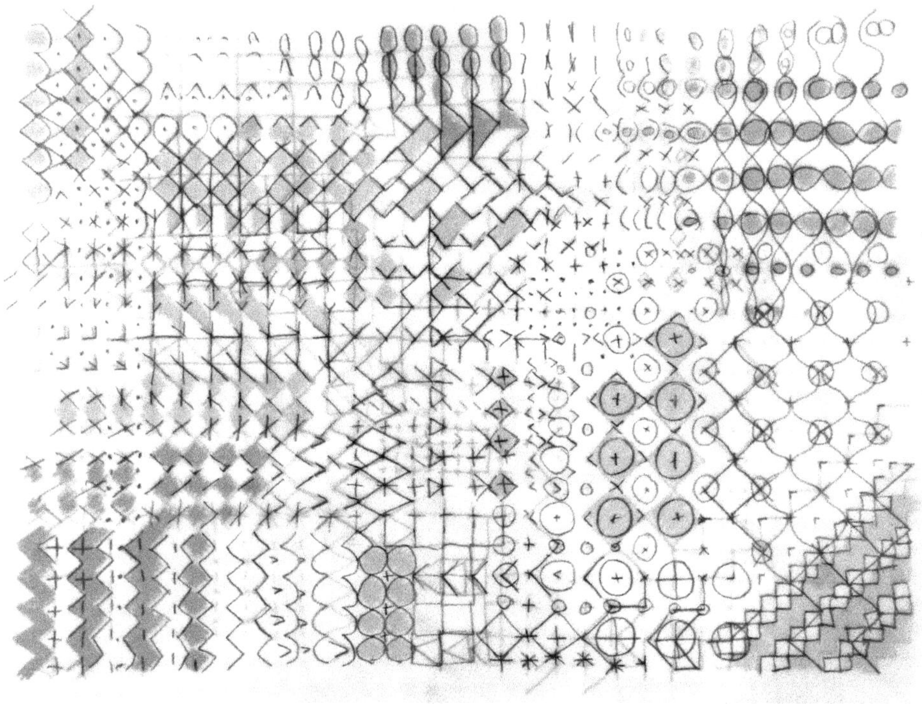

FIGURE 10.5: Kate Farley, leaving the grid structure behind. Graphite on tracing paper. 420 mm × 297 mm. Image credit and © Kate Farley, 2022.

new centre and continuing to create; both building and dismantling pattern formality. This enabled me to introduce more pattern structures from a single area, growing motifs as well as mutating existing ones. Single dashes as punctuation connected to phrases of more established rhythms resulted in a more complex variation of the theme.

The transformation of motifs in single drawings remains filled with potential for exploration. Repeat was not a consideration here but remains a further option. Colour and texture will add new ingredients to put into flux. The drawings continue.

11

Geomorphology: Mapping the Land, Above and Below Water

Glyn Brewerton

The riparian landscape in the Norfolk Broads is formed and shaped by measured human activity and wild natural phenomena. The drawings of Wheatfen Broad and a stretch of the River Yare near Polkey's Mill in Reedham Marsh in Norfolk reflect these contrasts, demonstrating the ways that drawing and pattern making can be used to make sense of a wild wetland landscape that is sometimes inaccessible or difficult to decipher.

Using observational drawing, aerial drone photography and data collected from an acoustic Doppler profiler radio-controlled boat, the resulting drawings of the Broadland landscape above and below water, reveal a succession of patterns formed from what initially appear as wild and chaotic landscapes into systematized and recognisable order.

Above

Wheatfen Broad is an ecosystem of fauna and flora corralled into pasture, meadow, fenland and carr; fragmented by the dyke and channel networks that were forged to combat river flooding from the River Yare, creating boundaries between pasture for grazing, and habitats for resident wildlife.

The drawn sections of reedbeds, sedge and carr are formed by lines and arced patterns that echo the ebb and flow of water. Juxtaposed as interlocking sculpted shapes, the drawings map a landscape divided, shaped and smoothed by tidal water (Figure 11.1).

FIGURE 11.1: Glyn Brewerton, *The Shapes of Wheatfen Broad, Norfolk*, 2019–20. Pen and ink. 80 cm × 135 cm. Courtesy of Glyn Brewerton.

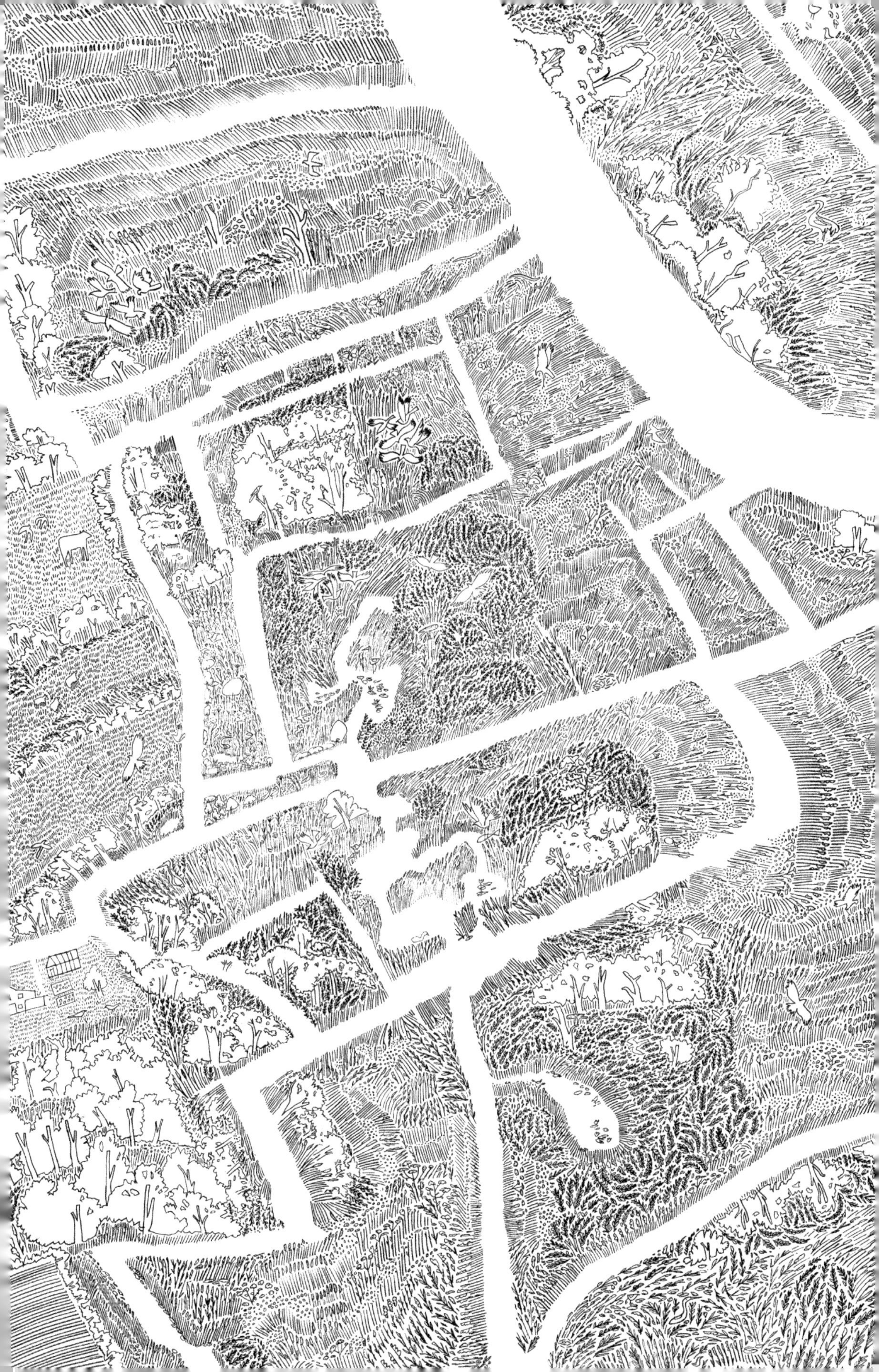

Within these shapes the drawn patterns convey a dense vegetal landscape that is hard to immediately understand when we look at it from human eye level (Figure 11.2). The drawings require an alternative flattened view from above, in order to edit the landscape into contrasting marks that illustrate the evolutionary succession of land formation in the Broad.

Below

In order to map the landscape beneath water, an acoustic Doppler profiler measures the water current with sound on an aquatic radio-controlled boat and is used to digitally draw a cross-section of the River Yare's riverbed; a landscape that the naked eye cannot see. The boat can be navigated in an undulating pattern across the river surface, scanning the riverbed using sonar to calculate the river velocity, the shape and the depth of the river channel cross-section.

From an illustrator's perspective, this manually operated boat acted as a new tool for drawing. Its linear path reflects the ripples of tidal water on the surface (Figure 11.3); its sonic pulses scour the depths of the sub-aquatic landscape, creating an unpredictable image of the underwater landscape, beyond a literal visual translation of fact.

The resulting graphic shows a river channel cross-section. The multi-coloured cells within the cross-section relate to various water velocities recorded within the channel at the time of measurement. The water velocity scale bar shows purples/blues as slower speeds and yellows/reds as faster speeds. The image shows a variation of water velocities present. It illustrates how shape, velocity and depth of the riverbed are digitally visualized, simplifying the image of hastening water across the river channel, into a kaleidoscopic pattern of multi-coloured squares (Figure 11.4).

Above and below

The process of responding to the Broads through digital mapping helped to see beyond the surface layers of vegetation and water that confronts a viewer looking into the landscape.

FIGURE 11.2: Glyn Brewerton, *Penguin Dyke Near Pool Marsh, Wheatfen Broad, Norfolk*, 2019–20. Pen and ink. 80 cm × 135 cm. Courtesy of Glyn Brewerton.

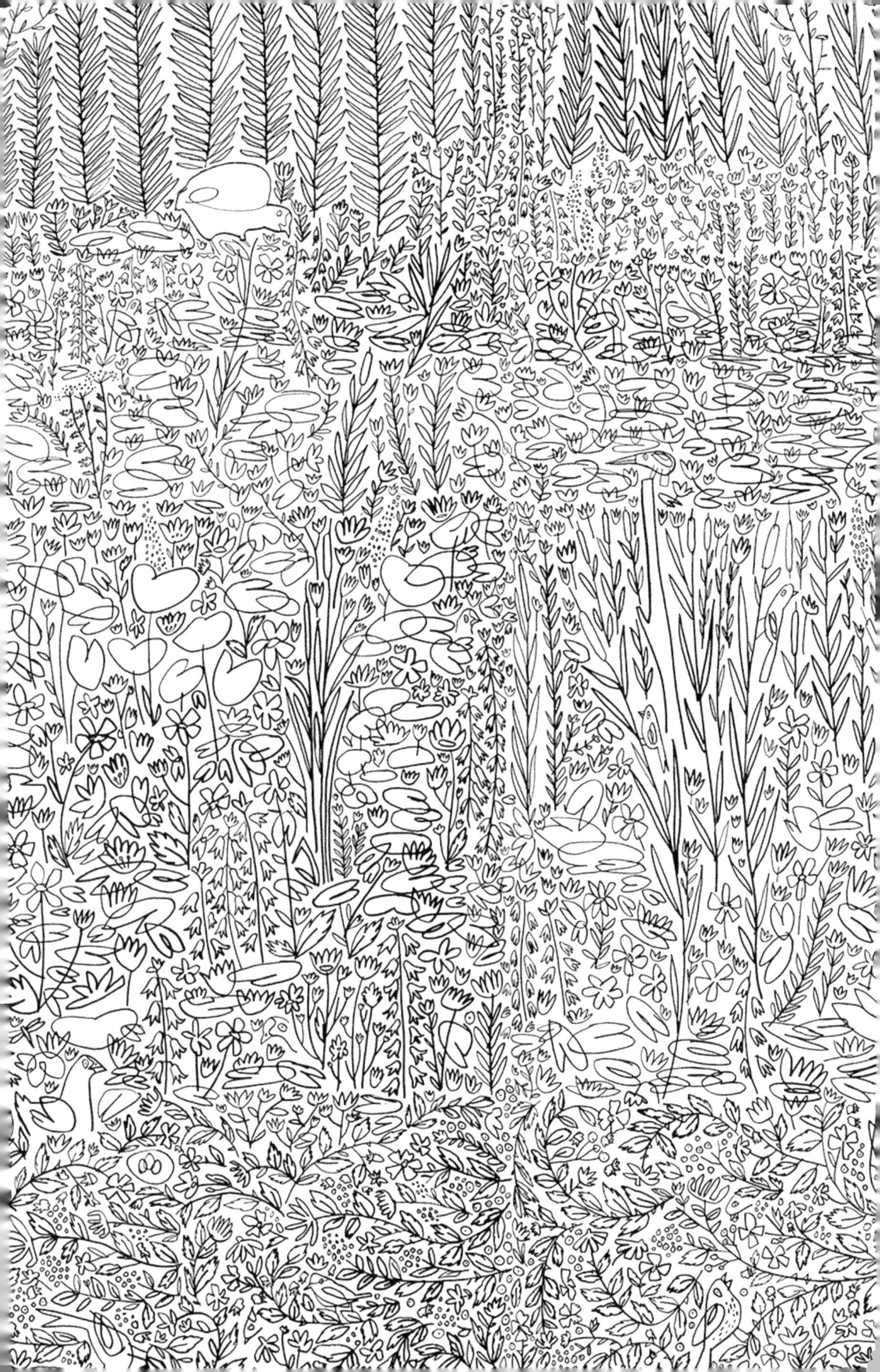

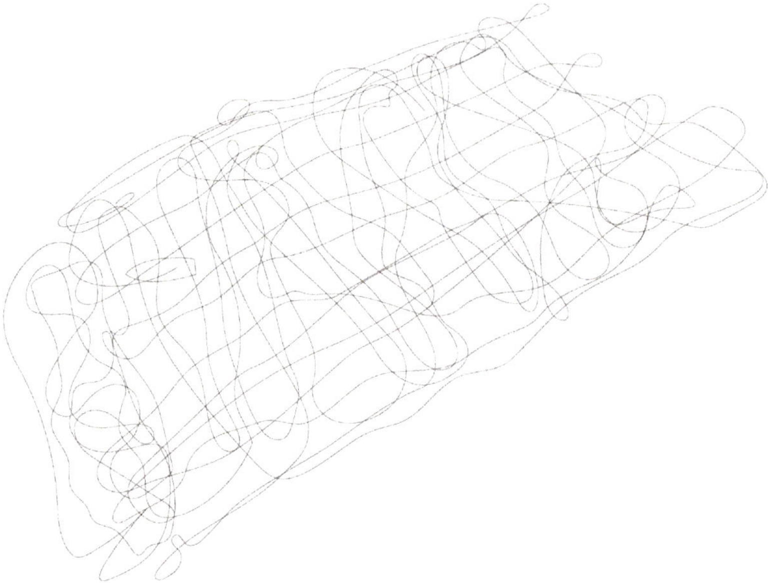

FIGURE 11.3: Glyn Brewerton, *Acoustic Radio-Controlled Boat Drawing Showing its Path Across the River Yare*, 2019. Courtesy of Glyn Brewerton.

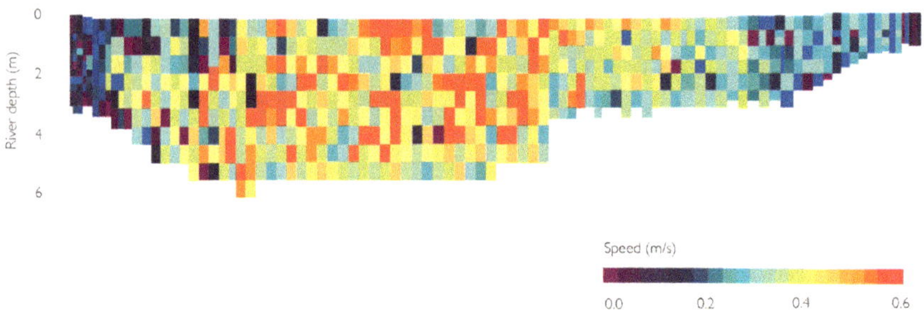

FIGURE 11.4: Glyn Brewerton, *Acoustic Radio-Controlled Boat Bathymetric Survey Scan Showing Measurements of River Velocity, Depth, and Shape of a Cross Section of the River Yare, 2019.* Re-drawn in Adobe Illustrator. Courtesy of Glyn Brewerton.

The use of digital wayfinding and data collection became an alternative drawing tool for the illustrator to map the unseen landscape; influencing thereafter the way the hand drawings were created, using pattern as a means to appreciate and understand the fabric of the wetland landscape above and below water.

12

Flux

Katy Hammond

Flux is an ongoing investigation of Edmund Burke's aesthetic theory of the sublime (1757) through fictional 'scapes' and fabricated narratives. Utilizing alternative photographic processes and materials, the work attempts to explore the fluidity and slippery nature of images. These photograms are arguably chaotic, the physical process forcing a relinquishment of control and in turn giving way to randomness and uncertainty. There is no preconceived order or meaning. Yet, with closer inspection, pattern can be discerned and picked out of the layers of texture, line and form, perhaps born from the innate human desire to seek order in the chaotic. Their ambiguous quality attempts to encourage shifting perspectives that can flip between pattern and chaos, having the potential to be, at the same time, both and neither and ultimately leaving the work to sit most comfortably in a liminal space where the commonly perceived binary pair meet and overlap.

FIGURES 12.1–3: Katy Hammond, *Flux*, 2019. Photograms. Courtesy of Katy Hammond.

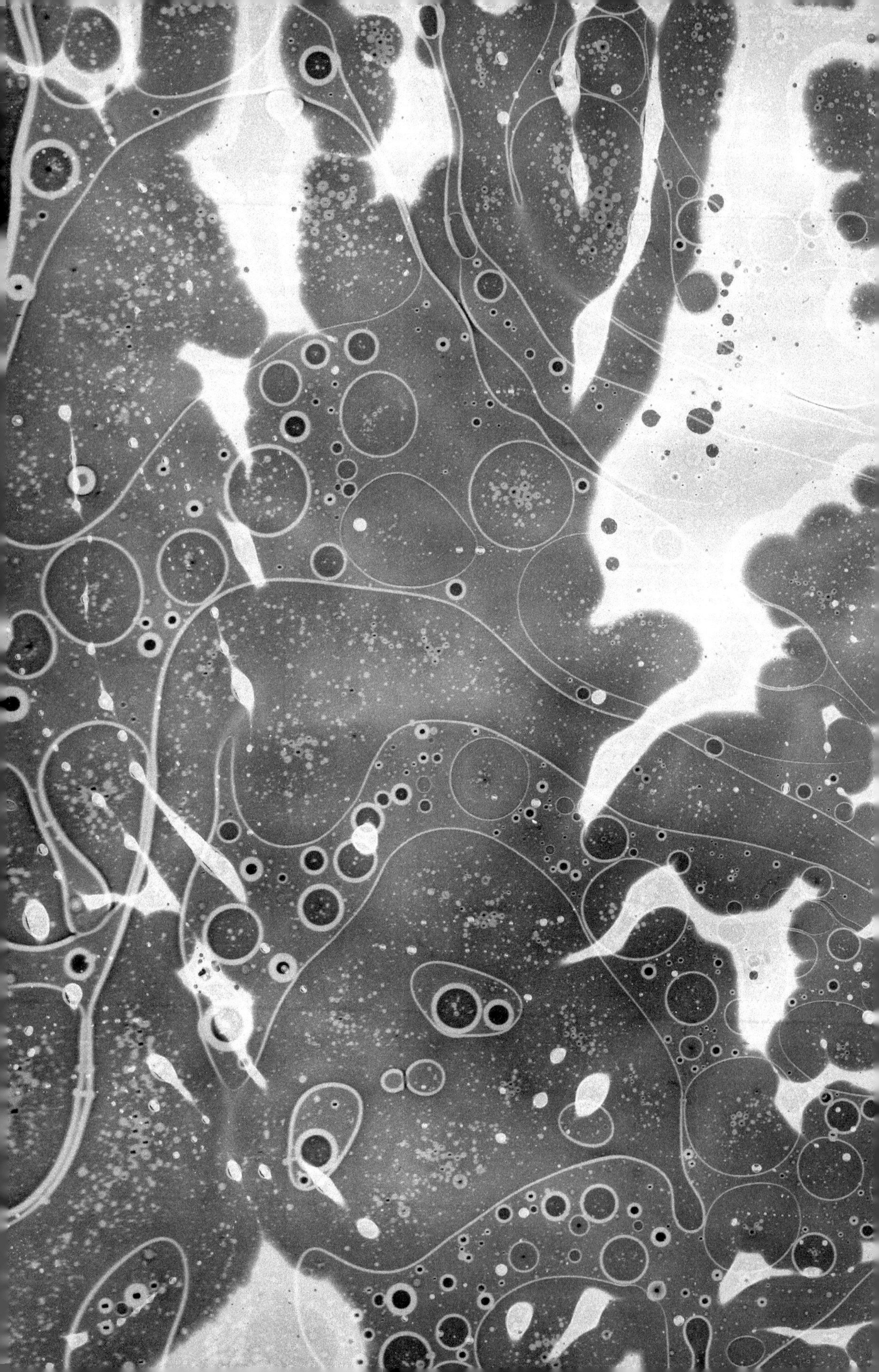

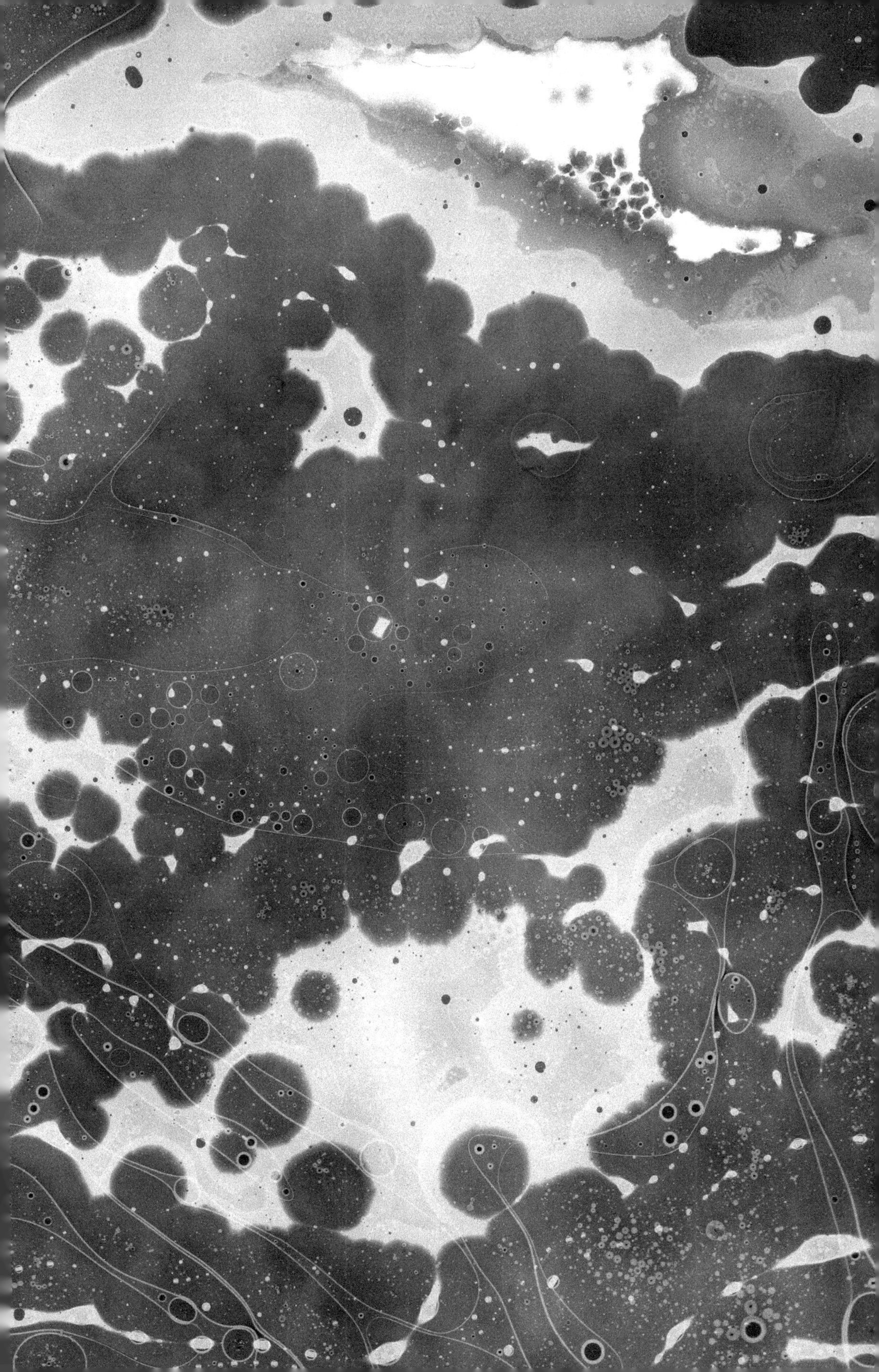

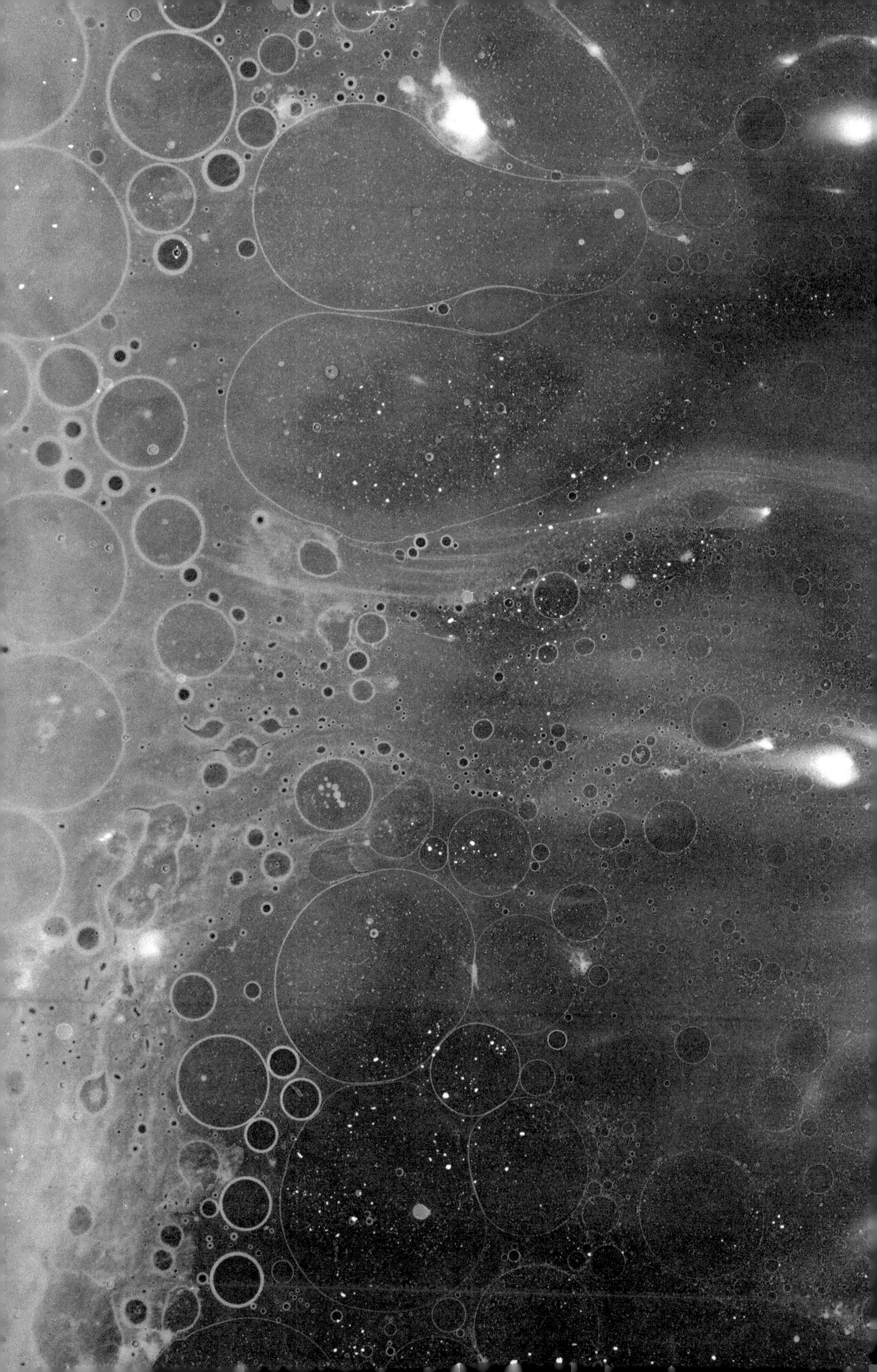

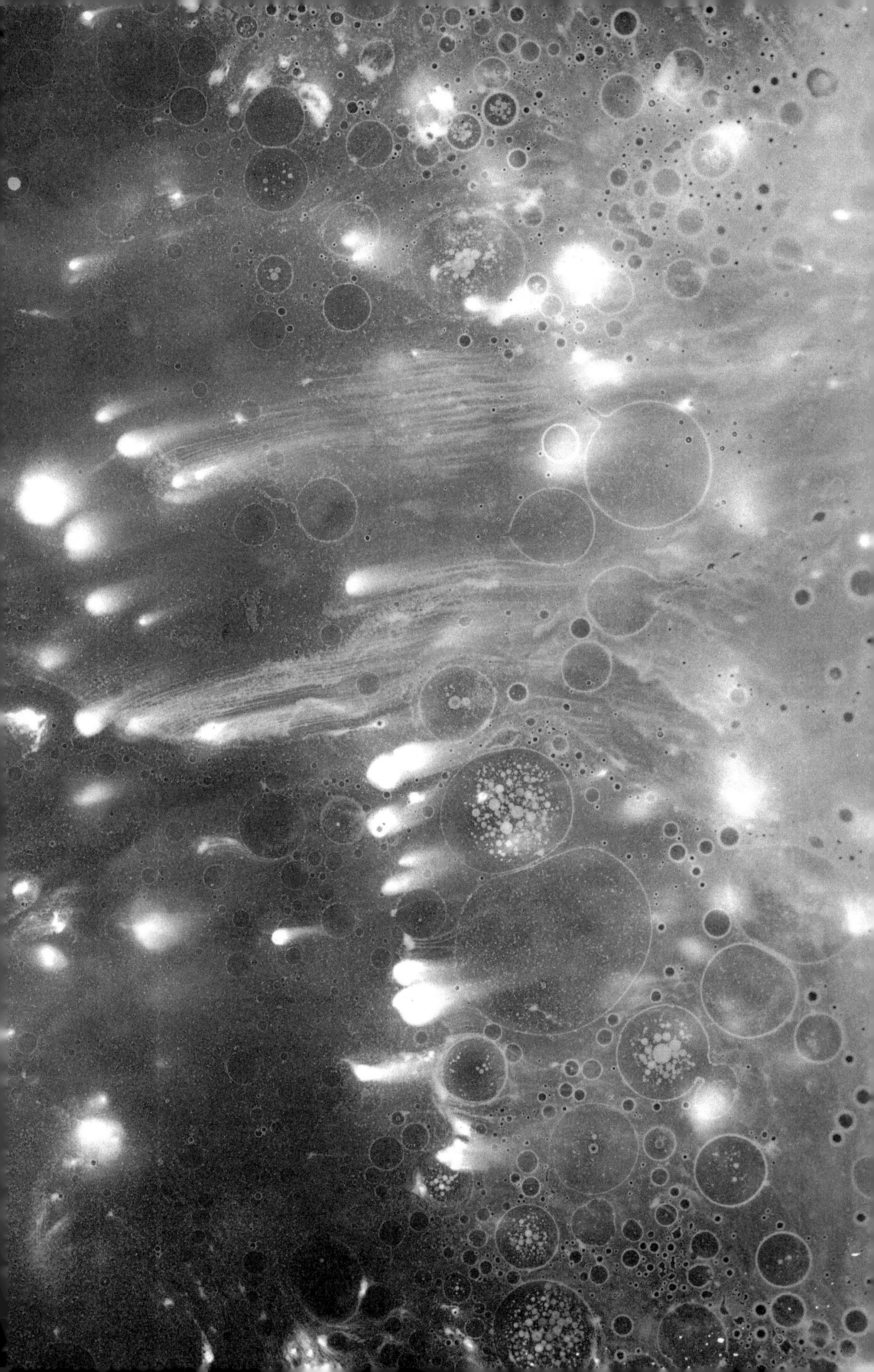

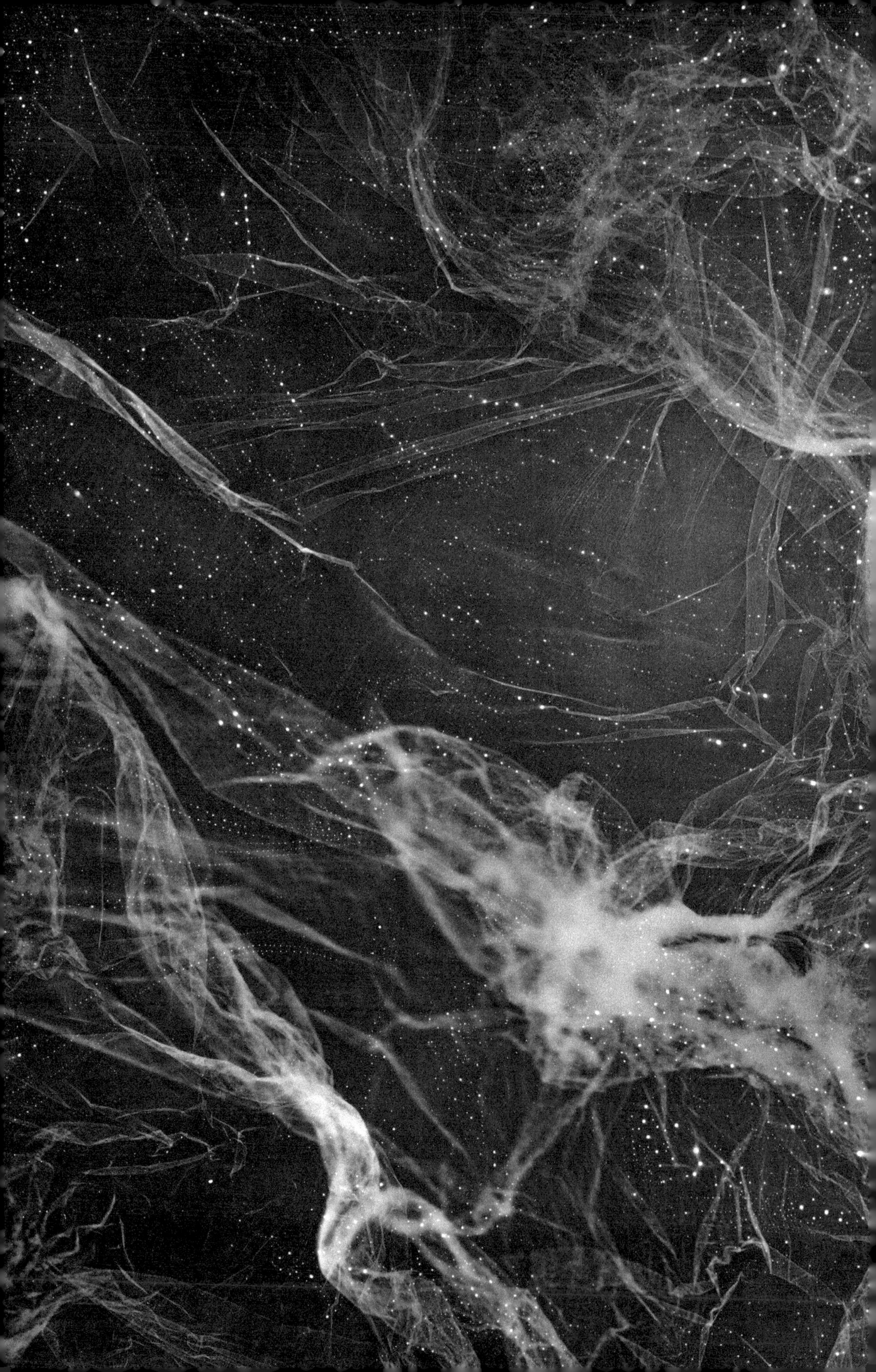

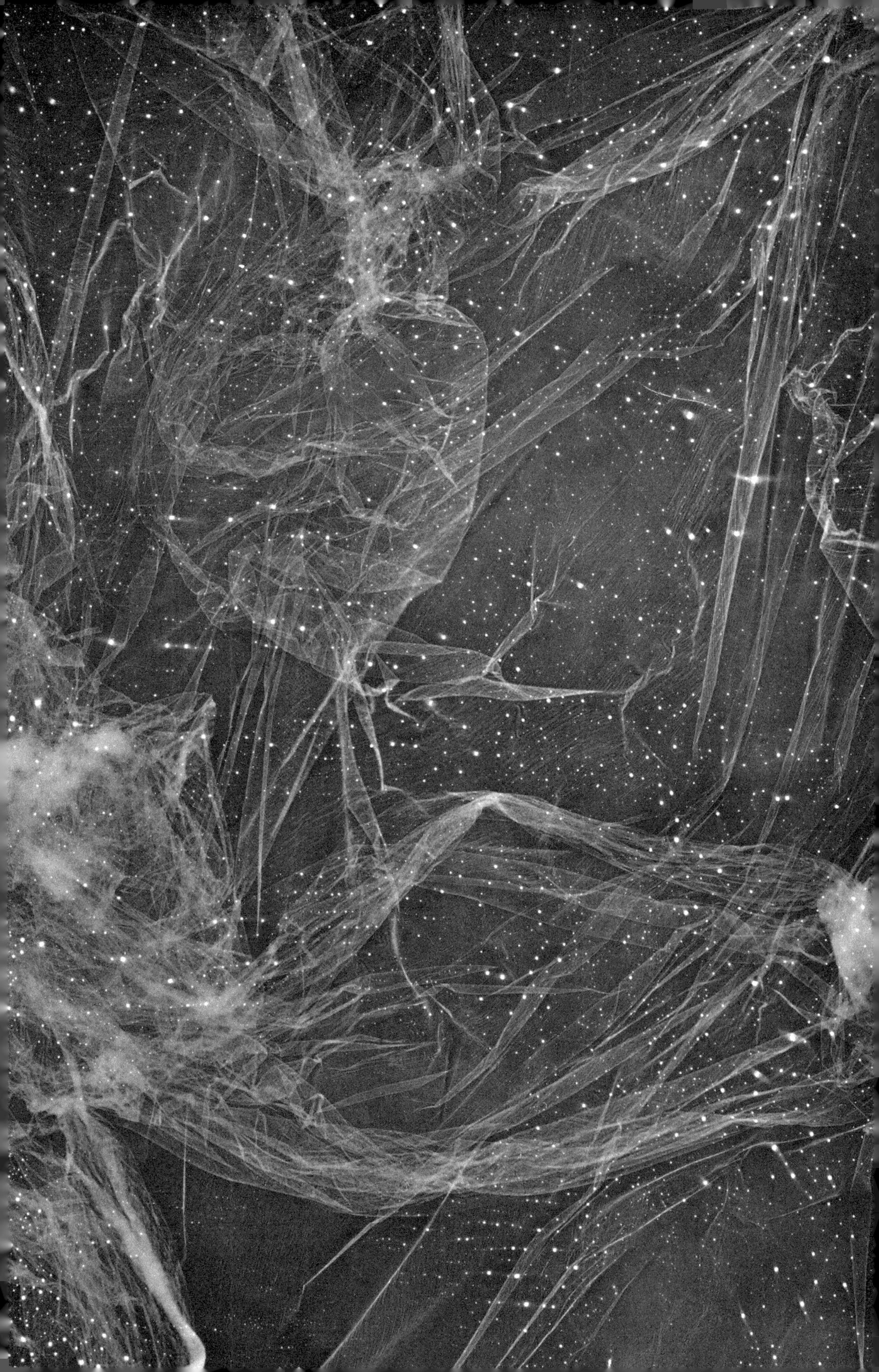

13

Drawing Fire

David Griffin

To watch things burn is to see them unravelling in a convection circuit that breathes heat into carbon, feeding back carbon to heat, until all other meanings are obliterated. This predatory looping in the dark light of the firepit undoes the fractal geometries of branching wood, and the joinery of entangled objects moulded by our pattern-seeking inclinations (and hands and fingers). We chitchat and circle the pit, watching the collapse of old chairs, turned lumber, paper documents marked over with texts, or bits and pieces of metal (which may survive a smaller fire's transformations, but which will afterwards lie still like bone). The fire turns between us. We stir and encourage it because we want it just there. And as we connect around the compacted release of light and heat, is this flame not also a demonstration of underlying harmonies in our concepts of order and disorder?

In spite of our perceptions of the flux in its spaces and shapes, the smear of its colours, and its mysterious oscillations, the fire that consumes stuff is not, so to speak, an object. We may speak or draw pictures of fire in embodied terms, as if it were just some thing, like those things it works upon in the firepit, but really, fire is a convergence. It is passage, unfolding like music. Like a song, it takes up space without giving us any real surface to reflect upon. Also like music, it transforms what it touches. But it lacks intent: it is fire.

This fire that we sit near as we are cold and in love is flux incarnate. It is all untraceable surfacing, quickening gaps, crackles and creases, and scaling self-similarity. Among the classical elements of earth, air, fire and water, it is fire that represents endings, woven into the others but not interacting with them in quite the same generative way. Its consummate purpose is to empty the bodies it fills. In fact, if all the things of our world and our colossal cities emerge from the relations between those four essential substances, then fire is the depletion of our manufactured complexities, it is reduction and return, the ultimate chemical

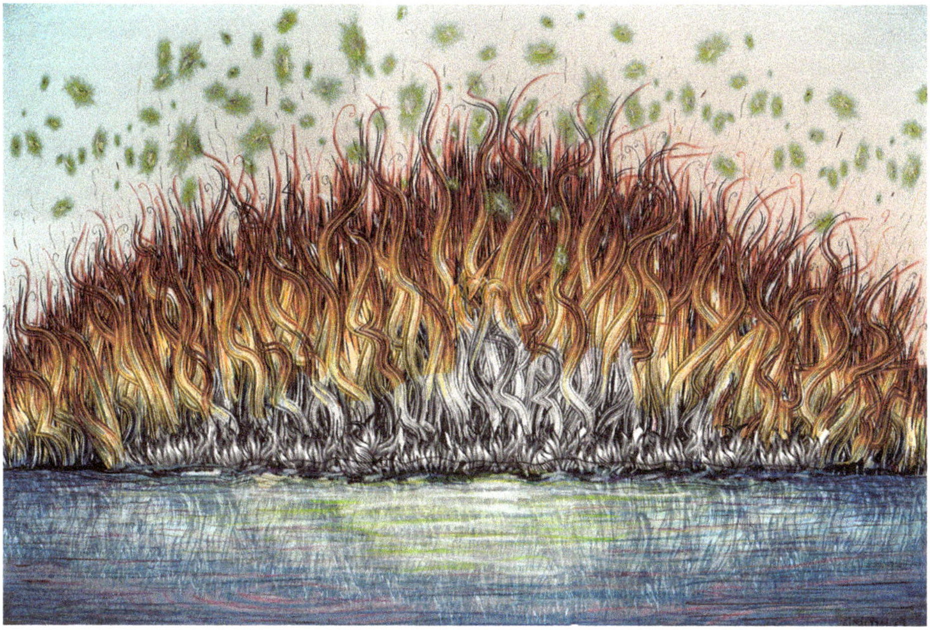

FIGURE 13.1: David Griffin, *Brightly Burning Breath*, 2019. Crayon, ink, pencil, acrylic. Digital image on paper. 60 cm × 40 cm. Courtesy of David Griffin.

death. The other elements supply all the choreography for our flame's disordering performance.

Stepping back just to look, the smoldering wreckage at the fire's edge shows that we are ruthless in our pattern-seeking urges, such that we will squeeze or cast it from whatever is at hand. Do we do this to calm our rippling anxieties? The poet Wallace Stevens suggests as much in his 'Connoisseur of Chaos' (1971: 216): in spite of our worldly discernments of order against disorder, he proposes that the two ideas are really one, 'chalked on the sidewalk so that the pensive man may see'. For the sake of symmetry, we can use the carbon remains from the pit as our mark-making tool, instead of chalk.

In fact, a pinch of that crumbling black is a fitting choice for a marking tool for a few reasons. In his volume of elemental biographies, Primo Levi – chemist and survivor of Auschwitz – describes carbon as the first and final state of all organic matter, a life-giving chemical byte, wandering freely and productively across all the layers of the world (1985: 227). Yet, he notes, the carbon at the heart of our burnt wood and our trembling bodies is also 'congealed in an eternal present' (1985: 229). And that presence – timelessness – is a state that seems to turn Levi's pensiveness to dread. Carbon suffuses everything at all scales. It dirties our fingers as we handle it; it colours our thoughts and language as we mark with it. And the contemplation of it leads Levi to this final, gloriously graphical point:

> It is that which at this instant, issuing out of a labyrinthine tangle of yeses and nos, makes my hand run along a certain path on the paper, mark it with these volutes that are signs: a double snap, up and down, between two levels of energy, guides this hand of mine to impress on the paper this dot, here, this one (1985: 235–36).

Making fire

In a universe still uncoiling from astonishing, inscrutable beginnings, disorder may just be order misunderstood. There is a relevant physical model of time (that dimension of our experience so poorly captured in memory and foresight) that imagines it as an algorithmic movement from lumpy to smooth states (Carroll 2008: 48–52). The arrow of time bears all things from knotted fields of galaxies, stars, people, cats and viruses to a cold terminus of dispersal and disorder. If this model of temporal experience makes any sense in our handheld world of coffee cups around campfires, then a couple dancing a paso doble is as illuminating a feature of the universe as the moon's multiscale, looping pathways, or for that matter, the fires we build on geometries of growth and communion.

And just as time in that technical view is an entropic process, or just as earth, air, water and fire are vitally, productively entangled substances, or just as silence is the actual space of musical performance, what we seek is humbly and incompletely represented in symbols of creamed coffee, a paso doble or the life of a fire: ignition, consumption and diminishment. We light a campfire around which to smoke, speak or sing and all those gestures join into the turbulent atmosphere generated by the flame.

Circling the pit, we are encouraged by each other to look into the dark and see, to pull meaning from the noise. In that exercise, we may realize a bounty of images and pass them round the circle; a willful arranging of the chaotic eddies we glimpse, or in which we ourselves spin, provide sense and structure to the flow of the evening, the drinking and the company.

Can the jubilant flickering of fire be measured? Does it have volume? Does it possess some internal, indecipherable geometry? Is there symmetry in its licking? Can the eccentric eddies and volatile flow of its dance be modelled? Can fire be drawn?

It leaps and dances, it flickers, spreads, shifts, transforms, creeps, smolders, consumes, muddles, dissipates. Fire aggregates, penetrates and integrates. Moreover, it behaves. Fire as metaphor explores, assesses, weaves, tangles, favors, forgets, recalls, criticizes and renders. It chatters and pops, it spits vulgar glory.

Indulge for a moment an alternate binomial for our species: *Homo Ignis* ('fire' in Latin [DeVaan 2008: 297]). This is a label that suggests that the emergence of our primate lineage can be distinguished by the central role of fire in what we are: that fire is somehow ours, in spite of its wildness. No other animals at any scale have managed a method to control the chaos and destructive potential of flame. So, this is a narrowing of the notion of *Homo Faber*: the 'maker-human', an old label for our family used in roughly the same way more by authors such as Hannah Arendt (1958: 153) and music theorist Pierre Schaeffer (1985: 15). We make fire and are made by fire.

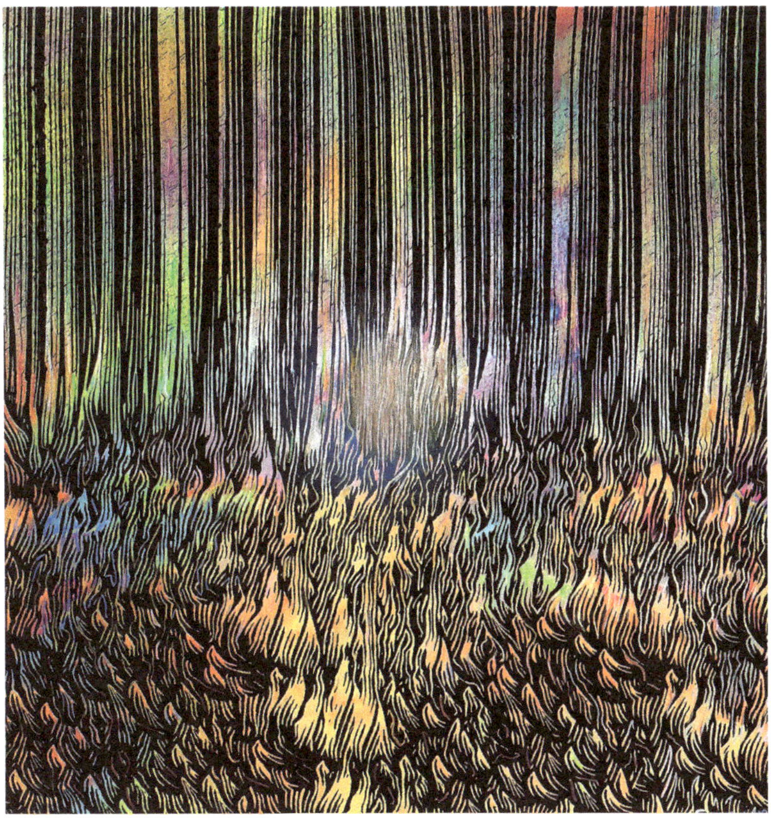

FIGURE 13.2: David Griffin, *Twilight Colours by the Fire*, 2019. Crayon, ink, pencil, acrylic. Digital image on paper. 90 cm × 90 cm. Courtesy of David Griffin.

Fire is the toolmaker's tool (a subtle recursion worth contemplating): we exploit its untidy chemistry to integrate the base elements found around us – alloying iron and carbon, for example, to create something weirder than either substance on its own, we make steel. And then we may brighten that already useful spark cast in our forges by pouring, shaping and intensifying it in pursuit of more feats of modulating the turbulence. Fire shapes the tools we use to make tools.

Speaking of the end, we have also employed the finality of fire as an intimidating, mythically fearsome and cataclysmic weapon. The house is burned down, flame is put to the ramparts, missiles are fired and we torch the heretic and the witch. Far too late, we may recognize our victims are just people, members of our own family with their own songs around the firepit, and their own longings and misgivings. Hopefully then, feeling the inescapable cold in the heat, our social codes themselves are softened and can be digested and transformed, but sadly slower than those we burned in our violence.

Fire is a beacon, an alarm that denotes a call to action or flight, lighting our dark communities. The primal experience of heat, smoke and furious sparking, and the imminent threat signalled by crackling flame ensures that no matter how we may practice its semiotic primacy, fire also stands for itself in a pure signification. It is what it is but more. We use fire for all the things itemized in this text, and we echo the behaviour of the flame itself in analogy: spark, fuel, growth, spread, diminishment, death.

Drawing fire

Michel Serres suggests that '[i]n the beginning is the noise; the noise never stops. [...] It is our apperception of chaos, our apprehension of disorder'. Living in our noisy world of fractional messages and broken rhythms, drawing is an activity that uniquely engages us – mind and body manifest at the tip of a pencil. In our endless search for salience, the things we say and make – that is, our utterances – are winnowing performances on the noise that is/in us. As a species of utterance, drawing reflects our pattern-seeking habits of mind in a great looping surmise. It is also a linguistic act (Tversky 2002), encouraging us not only to visualize but also to compose with the ceaseless drift of light and heat, then to amend, annotate and share our impressions with our comrades. But how do I trace the hungry pace and heat of flames in the dark? How do I develop a visual meditation on flux incarnate?

Gaston Bachelard offers us a philosopher's view of fire as a Rorschach inkblot, interpretable along a continuum from 'pre-scientific' to post-scientific thought and language (1964: 49–50). In his project, fire is a kind of principle by which to conduct a 'psychoanalysis of objective knowledge' (1964: 10–12). Yet peering into the flame, into its flickering expressions of chaos and order, I just want to draw it – to fix its dangerous intensity to the page; to understand it 'by hand'.

Fire is a chaotic system, but unlike similar things (a murmuration of birds, for one picturesque example), it is also, crucially an existential encounter. Fire is immolation and incandescence; everything it touches is absolutely changed, emptied of other meanings, roughly translated into ash by the licking flame. How do I draw all this? What if my sidewalk-chalking does nothing to quiet the murmur and hiss?

To visualize a roiling chaotic physics that we only know for certain burns everything down, we track across the realms of our limbic systems, those threshold spaces in our primate brains that link basic neurology to species-specific behaviours and

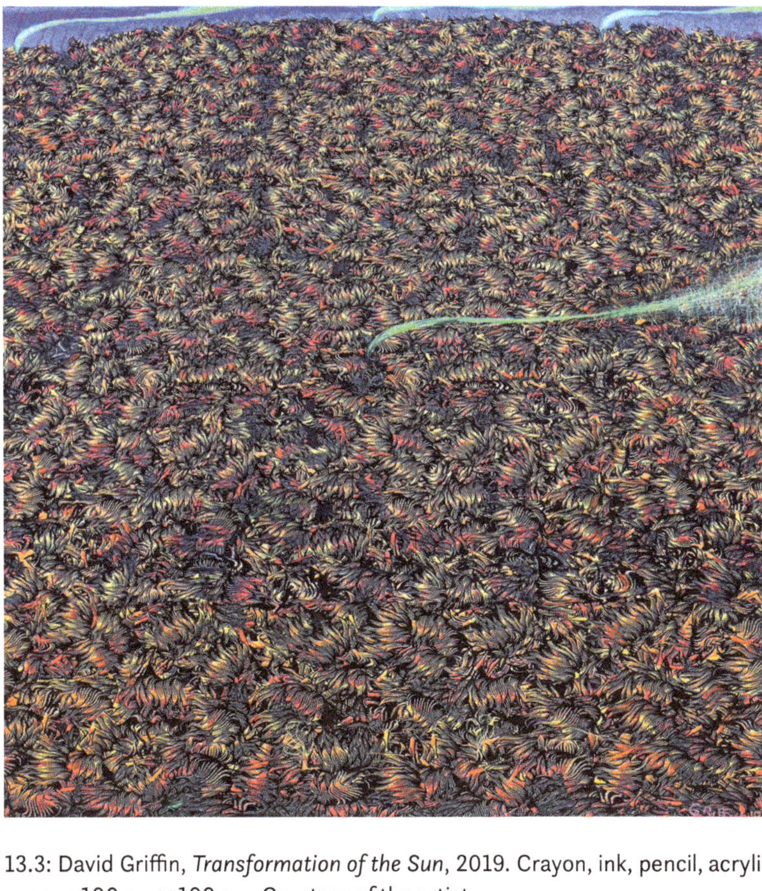

FIGURE 13.3: David Griffin, *Transformation of the Sun*, 2019. Crayon, ink, pencil, acrylic. Digital image on paper. 100 cm × 100 cm. Courtesy of the artist.

emotion (see Mega et al. 1997, for a brief on limbic functions). In a muddled, bespoke mechanics, our incipient drawing may then circuit through the fusiform gyrus, where aspects of language are integrated with shape recognition and colour. Meaning is breathed into the primal smoke and light, and the heat and pain. We recognize fire and we recognize that it changes everything.

But of course we must finally take up the carbon and mark-up the page. There is enough recent scholarship on drawing as a psycho-physical exercise that we should not doubt its values in translating, coding and inventing (e.g. see Fish and Scrivener 1990; Tversky 2011; Griffin 2016), with the proviso that representing any particular thing is an improbable task. In the matter of communicating certain primordial experiences, for instance, how would you represent the colour red? We see red, and it is a sign to run, or maybe stop, or maybe put in your mouth to taste. As artists and designers, we are at play with shapes and colours, enabled thereby to compose our thoughts with an exciting range of implications and desires. And yet,

> Q: What is red like?
> A: Red
> Q: What colour is red?
> A: Red

Fire is the same: it is alive in the present. It is a fleeting, flickering glimpse of chaos – all eddies, flux and recursion. How do we represent something that significantly precedes songs or pictures? How do we represent that indispensable image, the primary technology, the toolmaker's tool? How do we draw something so deeply rooted in us that it is a kind of conduit metaphor (Reddy 1993), a communication-key, rather than merely some trope or parable?

Fire is a peerless symbolic talisman: light and heat, spark and ash, consummation and cleansing. Fire starts, it grows, it breathes, it sweeps, it travels, it licks, it dies. We are dust. A fire is in my head. She is incandescent; love smolders; I am in heat. Here is another metaphor in an endless procession: fire is a time-machine. On the ludicrously tiny scale of campfires and conversation, the flames we fan in the pit accelerate those objects in their grip towards the cold disorder faced by everything else in the universe. Fire is the cosmic ultima Thule – an existential threshold, the end of what we know. Fire is chemistry and poetry all at once. How do we draw such a thing? We must invert the analogy: the best, most complete drawing of fire is the one that burns itself to ash.

14

Imago Images

Robert Hillier

Introduction

This chapter draws on an outcome of my recent design practice, *Imago Images*, which was inspired by the butterflies I frequently observed in my Norwich garden. The outcome comprises two distinct visual elements, a limited edition set of artist books and a self-generated animation produced by creative coding. The animation provides a companion counter-piece to the books. Viewed together, the books are fixed and static whereas the animation is constantly evolving and unlimited. In describing the creative process and the visual methodologies associated with *Imago Images*, ideas of unpredictability, ambiguity, visual change and altered states are explored. The contrast between the books and the animation can be likened to a butterfly's transformation from the immobile chrysalis, to the active and moveable adult (or imago) stage. It shifts from representational images of particular single species to the construction of heterogeneous forms through the combination of observation and conjecture. In doing so, imaginary butterflies populate an imaginary world. *Imago Images* reflects contemporary interests in the interface between writing, drawing, film, typography and creative coding in the making of new and alternative connections through hybridized visual design practice.

Context

Imago Images is a creative design-led research outcome which depicts visual representations of a number of butterfly species. The work was informed by a previous project named *List Landscapes* (Norwich Arts Centre 2018) which comprised the making of lists as a daily life organizing principle. An important aspect of these lists includes a way of marking time through a long-standing interest in the observation and recording of natural history species and associated phenomena.

One such phenomenon which is key to *Imago Images* is a butterfly's transformation from pupation to the adult or imago stage.

Imago Images is not a work of natural history. However, it alludes to the natural history illustrations of H. D. Swain and Richard Ward. Swain was responsible for the illustrative content of *British Butterflies* (Riley 1938), a pocket-sized book that formed part of the Observer Book series, whilst Ward both wrote and illustrated the similarly titled *British Butterflies* (Ward 1963). This more recent work takes the form of an album designed to house butterfly picture cards that were given away in packets of tea. As a child, as well as diligently collecting the 50 cards in the series and carefully glueing them into the album, I also found the pictorial depictions and textual descriptions of each butterfly species to be both familiar and strangely unfamiliar. This combination is disconcerting. *Imago Images* also echoes similar notions of the familiar and the unfamiliar and provides an oblique commentary on my own life cycle. However, rather than that of a child with a blossoming interest in natural history, the perspective taken here is that of an adult graphic designer coming to terms with new technologies associated with hybridized visual practice. The familiar in *Imago Images* is represented by ten limited edition box sets of self-authored, pocket-sized artist books (Figure 14.1), whilst the unfamiliar takes the form of a separate self-generating animation, co-produced with Scott Grandison through the

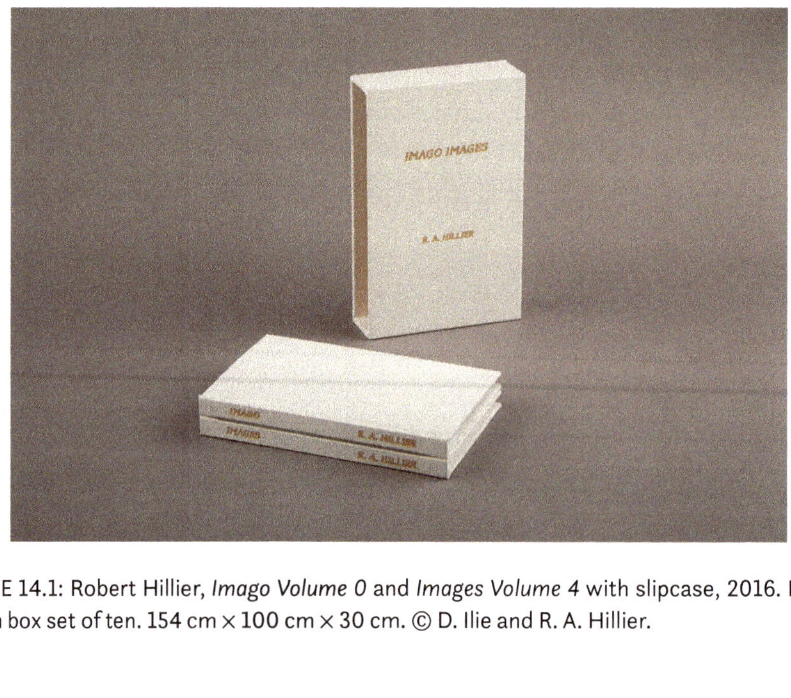

FIGURE 14.1: Robert Hillier, *Imago Volume 0* and *Images Volume 4* with slipcase, 2016. Limited edition box set of ten. 154 cm × 100 cm × 30 cm. © D. Ilie and R. A. Hillier.

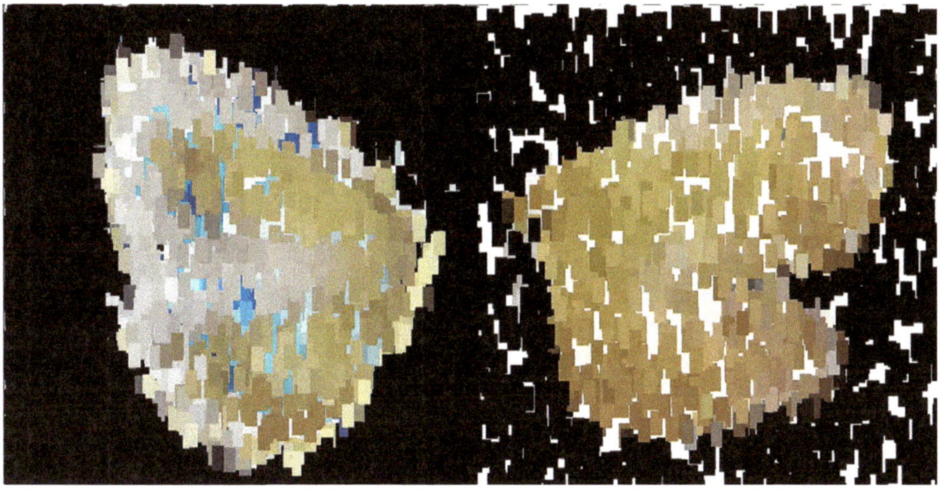

FIGURE 14.2: Robert Hillier and Scott Grandison, *Imago Images Animation*, 2018. Computer screenshot. © R. A. Hillier.

application of creative coding (Figure 14.2). Viewed together, the books are fixed and static, whereas the animation is unlimited, constantly evolving and can never be the same twice. In connecting traditional design practice with contemporary visual media, *Imago Images* creates alternative, transformative and hybridized worlds of its own. They are imaginary butterflies from an imagined world.

Initial visual and conceptual development

Imago Images depicts manipulated drawings of butterflies seen in my Norwich garden over a ten-year period from 2006 to 2016. A conceptual catalyst for the transformation of the original drawings was the varied and ambiguous meanings associated with the Latin noun 'imago'. These include the last or perfect state of insect life and an image or optical counterpart of a thing (*Oxford English Dictionary* 1964: 604). In progressing this concept, *Imago Images* examines transformative and oppositional ideas through a series of diverse perspectives. These include the juxtaposition of contrasting tropes such as Latin text with dyslexic typography (Hillier 2012), positive space with inverted imagery, the kaleidoscopic with the monochromatic, the hand painted with digital representation and the finite with the infinite.

The title *Imago Images* is important to the visualization process and forms the foundation of the work. It is comprised of two words that both start with an uppercase

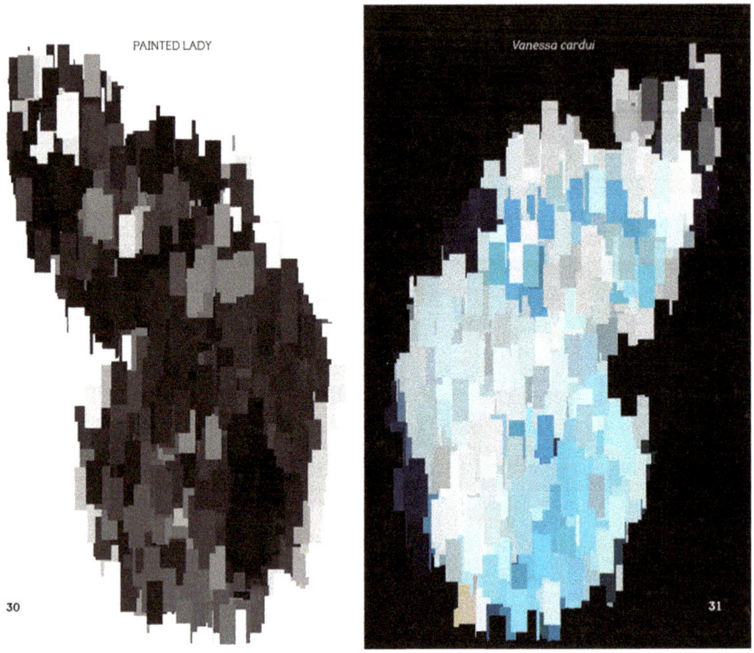

FIGURE 14.3: Robert Hillier, double-page spread from *Images Volume 4*, 2016. Limited edition book from a set of ten. 142 cm × 176 cm. © R. A. Hillier.

letter 'I'. In its sans-serif form, the uppercase letter 'I' can also be viewed as a simple vertically orientated rectangle. Through the application of creative coding, my original drawings of single butterfly wings were converted into rectangular blocks of colour and tone – similar in appearance to a multitude of letter I's. The random generation of these blocks formed the basis of the *Imago Images* animation, as new butterfly wing images were created from an infinite number of uppercase letter I's (Figure 14.2). Screenshots of some of the images captured from the animation were then subsequently used to illustrate the *Images* books (Figure 14.3). The images in these books can be regarded as being finite, depicting a single moment in time. The animated digital outcome offers temporary and flexible ideas concerning the potential of infinite form.

Representation and physicality

Why should each volume that makes up a limited edition set of book-based artworks be visually identical to each other? With this question in mind, the two volumes within each of the ten *Imago Images* box sets markedly differ from each other. The visual representation of the butterflies in the *Imago* books is identical throughout all ten volumes (these volumes are collectively numbered 0). Each *Imago* book therefore contains identical monochromatic images of a selected species and can be viewed as being consistent and homogenous with each other (Figure 14.4). Conversely, the second book, *Images* (volumes 1–10), contains multi-coloured digitized images of the butterflies that differ in each volume (Figure 14.3). Consequently, the ten *Images* books can be viewed as being heterogeneous because each visual representation is unique and non-identical from each other.

The format of the *Imago Images* books references the pocket-size format of *British Butterflies* (Riley 1938). In doing so, a historical link is formed between the subject of butterflies and the physicality of the *Imago* books. A further physical connection can be made in that, as the pages within the *Imago Images* books are turned, the flapping action of butterfly wings is evoked. As such, the pages become analogous to butterfly wings. Another conceptual connection that might be made between a butterfly and the printed book concerns anatomy. For example, within the language

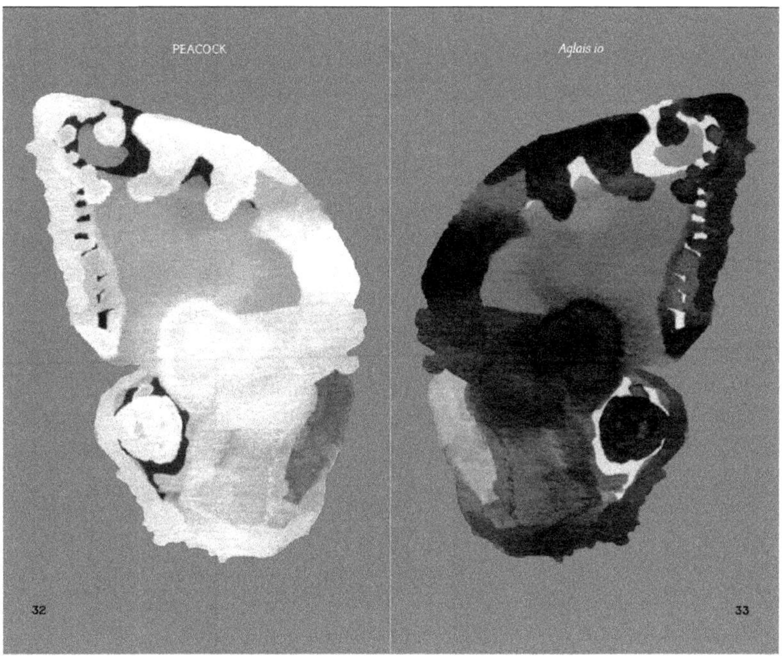

FIGURE 14.4: Robert Hillier, double page spread from *Imago Volume 0*, 2016. Limited edition book from a set of ten. 142 cm × 176 cm. © R. A. Hillier.

of printing, the main text of a book is often referred to as the 'body' (Peacock 1995: 406). With the exception of the preface in the *Imago* book and the foreword in the *Images* book, both books contain no other body (or text). Similarly, the body of the butterfly is also missing from the visual representations within the *Imago Images* books as the insects are only pictorially depicted by their wings. Nevertheless, the semblance of a whole butterfly can be recognized – something that is perhaps greater than the sum of its parts (Figures 14.3 and 14.4).

In developing the anatomical links between insect and book, it is apposite to note that a butterfly wing, like a printed page, has both an inner and outer margin. The anatomy of a butterfly wing and the anatomy of typography also share many common terms. A butterfly wing contains, as does an uppercase letter 'A', a component called the apex. In some butterfly species, such as the Peacock butterfly, the wing also has a section called a tail. The descender on the lowercase letter 'g' is similarly called a tail (*Typography Deconstructed* 2010). It is interesting to observe that *Imago Images* includes visual representations of the Peacock butterfly, otherwise known by its scientific name of *Aglais io* (Figure 14.4). This Latin name for the species also contains both an uppercase 'A' and lowercase 'g'. There is an intrinsic ambiguity and duality at play here, two names for the same insect and two references to the apex and the tail.

An additional conceptual connection that can be made between butterflies and books is through the notion of signatures. In printing parlance, a signature is 'a synonym for the folded paper sections of a book' (Peacock 1995: 466). Book signatures can be seen as a stage in pre-production, usually taking the form of pages in multiples of four. There are four signatures within the *Imago Images* books, just as there are four stages within the life cycle of a butterfly. In further progressing the notion of signature, yet another meaning of the word can be given to a person's name written in a distinctive manner in order to authorize a document. For James McNeill Whistler, a butterfly motif was often used as a signature on his paintings (Sutton 1963). The signature of my name appears on the book cover (Figure 14.1) and animation credits of *Imago Images* in uppercase Sylexiad Serif Medium Italic. It is a typeface I designed in 2006 for dyslexic readers and, as such, identifies the work as being part of a wider historical canon of my design practice. Sylexiad is an anagram of the word dyslexia. For some people with dyslexia, phonological difficulties can be indicated through the phenomenon of mirror writing and the reversal of letters or words when reading (Hillier 2012). *Imago Images* offers similar dyslexic perspectives in that the butterfly wing images are inverted, mirrored and reversed versions of earlier iterations of my drawings.

Transformation, life history and symmetry

My original butterfly wing drawings were initially constructed from simple blots of coloured ink and then digitally scanned. As the visualization process progressed, some of the scanned drawings were inverted and then converted to monochrome versions (Figure 14.4). The transformation of these images from the hand-drawn to digital forms suggests a palimpsestic idea of stratification where the new replaces the old (Hillier 2016). This process of development is also evidenced throughout *Imago Images* as the single wing images are replaced by paired images of rectangular blocks of colour and tone (Figures 14.2 and 14.3). Indeed, the word 'image' can be substituted by the word '*image*' (and vice versa) because the Latin translation of *imago* is *image* (*Oxford Latin Dictionary* 2005: 50). This process of replacement echoes the metamorphosis experienced by a butterfly during its life cycle from egg to pupae to chrysalis and subsequently to imago. Life history is therefore represented in *Imago Images* through the metaphors of transformation and change.

When I reflect on my own life history, the wing has always been a significant aspect of my artistic visual practice. As a child, the first images I remember drawing were of spitfire airplanes, scarlet macaws and imagined butterflies. All of these subjects have wings in one form or the other and, as such, they all have the ability to fly. In primary school, I remember creating imagined butterfly images by using two distinct methods. Firstly, by painting a single blob of colour on the right side of a sheet of paper and then folding it to produce a double blot. In doing so, a whole butterfly could be represented. The second method was to create images through potato printing. The potato was halved, and simple rectangle forms were then cut out and dipped into poster paint. An image of a butterfly was then generated through repeated application of blocks of colour. Both visualization processes bear resemblances to those used in *Imago Images*. The visualization process utilized in the books is reminiscent of my childhood method of image production through the creation of blobs (Figure 14.4). The animation, however, echoes some aspects of the second method I have described, in that a comparable technique of repeating and replicating simple rectangular forms has been applied. However, rather than using potato prints, creative coding was employed to create the imagery (Figure 14.2). A symmetry connects my childhood methods of image making with that of my current practice as an adult man.

Aurelian

A lepidopterist is a collector of butterflies. Another, more archaic description of a lepidopterist is an *aurelian.* The term is derived from the word chrysalis which relates to the Latin word for gold – *aurum* (*Oxford Latin Dictionary* 2005: 21), referencing the golden colour a butterfly gains before emerging as an imago. I now aspire to become a contemporary aurelian insofar as *Imago Images* is a collection of hybridized ideas and design practices that are connected with butterflies. Like all collections, the content of *Imago Images* is selective and often incomplete, with many gaps to be filled. *Imago Images* also acts as a metaphor for my own developing visual practices progressing from one design discipline to other more experimental and hybridized forms. These have offered the opportunity of personal transformation as a designer and, perhaps, the spectacle of a chrysalis turning golden.

15

The Chaos of Delight: Spatial and Temporal Interruptions

Lesley Halliwell

My interest in the intersection between pattern and chaos comes from a long-standing pre-occupation with the decorative surface, repetition and geometric form. It is founded on a large body of drawings (2000–15) where I used the spirograph motif as a constant element through which pattern and process were recurrently explored. Here, the drawing process was systematized with decisions about motif, scale and colour determined before each drawing began. I then decided on my approach; for example, whether I would start from the centre of the paper and work out in a series of concentric rings or begin in the top left-hand corner and work in gridded rows, like knitting, from left to right, right to left. As each drawing was executed, the repetitive process unfolded. Meanwhile, pens ran out of ink, paper fibres wore away, concentration wavered, and my hand slipped. The tension that arose between control and accident were assimilated into the practice. Evidence of the making process took on greater significance with process-led variations remaining visible and acting as a counter to the formal structure and systematic process. The moments when ordered precision broke down became a desirable outcome, redolent with meaning.

While pattern is generally based on the orderly repetition of a single motif, designed so that recurring components fit together in sequence, I recognized in my own work an alternative approach where pattern is formed and reformed by the nuance emerging from the human hand, process-derived inconsistencies and deliberate intermissions. In this essay, I explore the significance of the interruption and its contribution to the viewer's experience of pattern. I propose that, rather than seeing pattern and chaos as binary positions, we should view them as fluid concepts that are intimately connected, with unstable boundaries that intermingle and overlap. Focusing on different forms of spatial and temporal interruption in specific artworks, what follows is an exploration of the tension between pattern and chaos from a maker's perspective.

Pattern coalesces surface through strategies such as the grid, repetition, tessellation and rotation, and provides structure on and across a surface. However, art critic Amy Goldin proposed that the defining characteristic of pattern is not one of repeat motifs but rather the 'constancy of interval' (1975: 50). In Goldin's scenario, it is the negative spaces that articulate the pattern, which only comes into phenomenal existence when there are enough space-interval repetitions to establish it clearly as a unit. In the painting *Horror Vacui* (2019) (Figure 15.1), I play with the irregularity of the interval so that the movement between the pattern and the potential for chaos is more ambiguous.

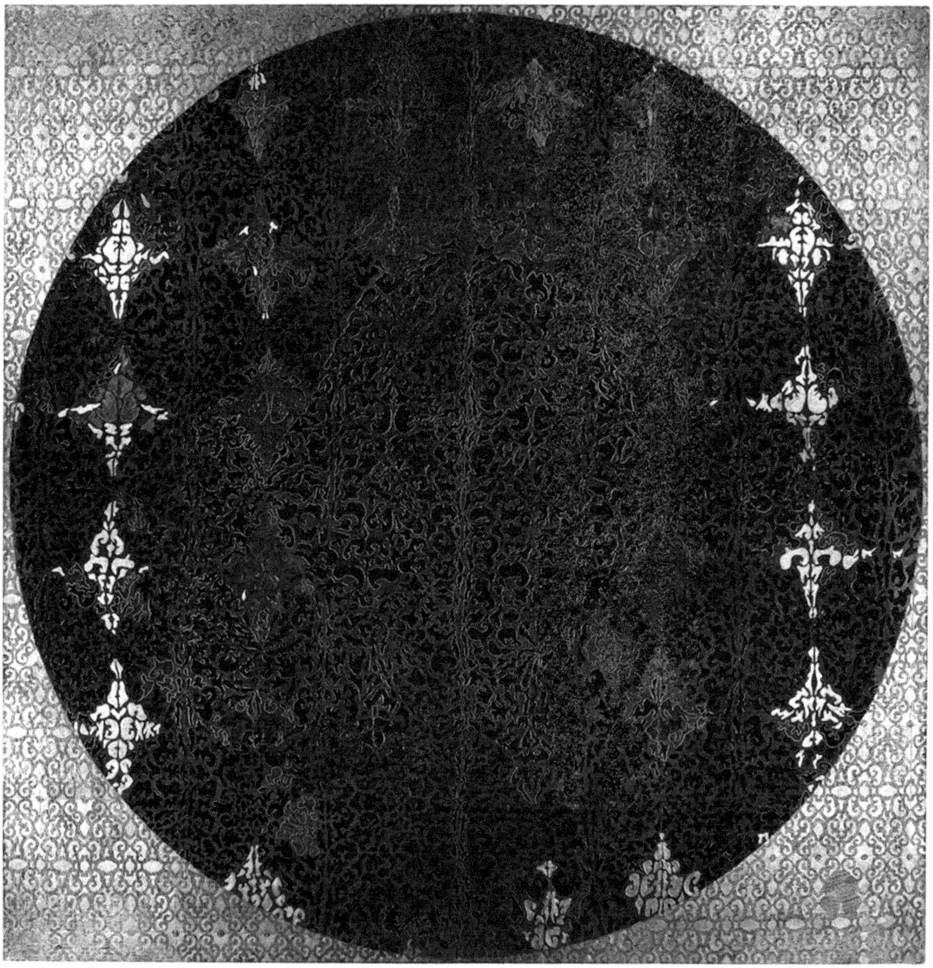

FIGURE 15.1: Lesley Halliwell, *Horror Vacui*, 2019. Casein paint, pencil and lacquer on board. 120 cm × 120 cm. Copyright Lesley Halliwell. Photo credit Tony Richards.

Horror Vacui (from the Latin 'fear of empty space') consists of a large circle positioned snugly within the frame of the painting panel. While the structured metallic pattern around the edge contrasts with the looser 'disintegrating' pattern within its centre, a central line splits the image in half, a bit like a walnut, and provides a sense of symmetry. Reminiscent of my childhood colouring-in books I chose the title to reference my constant striving for completed-ness by filling the entire surface of an artwork with detail. Here, almost every piece of available space was worked into and over; a filling-in that reduces depth and creates an initial sense of impenetrability. However, the irregular spacing between the little motifs also created points of disjuncture, punctuating the otherwise pervasive pattern and offering the viewer a point of entry where above-ness and below-ness of surface are experienced.

The painting's cloying all-over-ness resonates with the anthropologist Alfred Gell's notion of 'unfinished business'; a particular reading of pattern in which it 'remains "unfinished" inasmuch as the technical, formal and specifically ornamental complexity of the art object exceeds the viewer's ability to organise the visual field' (Rampley 2004: 14), and thus remaining incomplete in the imagination of the viewer. In this scenario, the patterned surface reveals itself over time, resulting in a 'delay, or lag, between transactions that [...] never result in perfect reciprocation, but always in some renewed, residual, imbalance' (Gell 1998: 81). Gell argues that this 'imbalance', where the complexity unsettles the viewer's comprehension of the pattern not only slows down the process of looking (which in turn promotes an attachment between image and viewer) but also adds to the viewer's enjoyment of the pattern. Gell describes this as a pleasurable frustration (1998: 80) and is what I refer to as the chaos of delight.

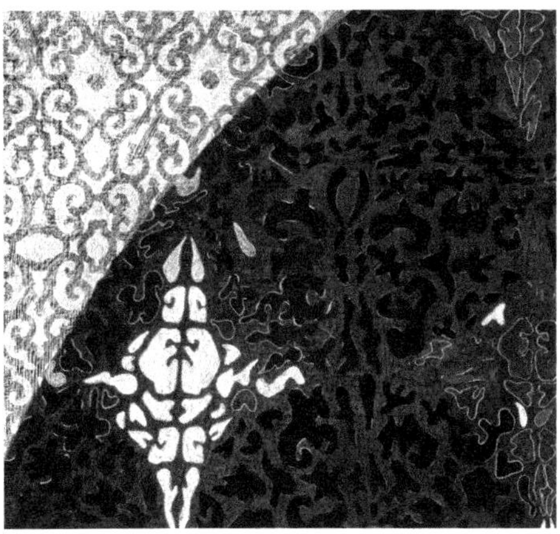

FIGURE 15.2: Lesley Halliwell, *Horror Vacui*, 2019 (detail). Casein paint, pencil and lacquer on board. 120 cm × 120 cm. Copyright Lesley Halliwell. Photo credit Tony Richards.

Gell's unfinished business stands in contrast to Roger Fry's 'purposeful order'. In 1909 Fry wrote, '[t]he perception of purposeful order and variety in an object gives us the feeling which we express by saying that it is beautiful' (Harrison and Wood 1992: 83). Unlike Gell's unfinished business, the compositional balance of Fry's ordering principle regulates the surface and prevents the viewer from becoming overwhelmed or disorientated. In his discussion on the deregulation of pattern, artist and academic Jason Hoelscher (2013) suggests that the contemporary viewer's experience of Fry's purposeful order is different from what it was in the 1900s. With a fast-paced digital era having transformed our global understanding of time and space, Hoelscher argues that in this 'culture of slick, binary-coded and focus-grouped order, a degree of painterly *non-order*' now serves a similar role (2013: 24, original emphasis). Thus, for the contemporary viewer, an element of uncertainty is a pre-requisite for the assimilation of pattern.

Hoelscher's painterly non-order prevents the viewer from entering what film-maker Hito Steyerl refers to as 'free-fall'. Steyerl uses the term to describe the 'groundlessness' of modern life '[a]nd [...] the departure of a stable paradigm of orientation, which has situated concepts of subject and object, of time and space, throughout modernity' (Steyerl 2011: n.pag.). As defined by linear perspective, space is described by Steyerl as 'calculable, navigable, and predictable'. Think, for example, of the role of the horizon line in visual navigation and how its certainty fixes the viewer's position. In contrast, when there is no vanishing point, we perceive the artwork as offering only a continuous present. Rather than providing the viewer with a secure visual field the flattened patterned surface prevents 'any sense of above and below, of before and after, of yourself and your boundaries' (Steyerl 2011: n.pag.). My research has shown that the viewer looks for ways to off-set the falling, floating and weightlessness of such all-over patterning. Moments of interruption are essential because they offer the viewer visual stability in order to (re)ground themselves within what would otherwise be a depthless plane.

By employing a degree of irregularity, *Horror Vacui* hovers on the edge of pattern, alluding to orderly interval but without fully manifesting it – a form of unfinished business that invites the viewer to decipher the grammatical structure embedded in the work and prolong the viewing experience. Hoelscher describes a pattern like this, which references and then relaxes its sense of strict orderliness, threatening to break at any given moment, as 'experiential deregulation' (Hoelscher 2013: 23). *Horror Vacui* contains elements of Goldin's constancy of interval and a move towards Fry's 'purposeful order' but then veers away, presenting as orderly in some sections and contradicting that order in others. As both maker and viewer of *Horror Vacui*, its maze-like structure continues to unsettle me. I scan the surface of the image trying to make sense of its symmetry, which, once within my grasp, appears only to

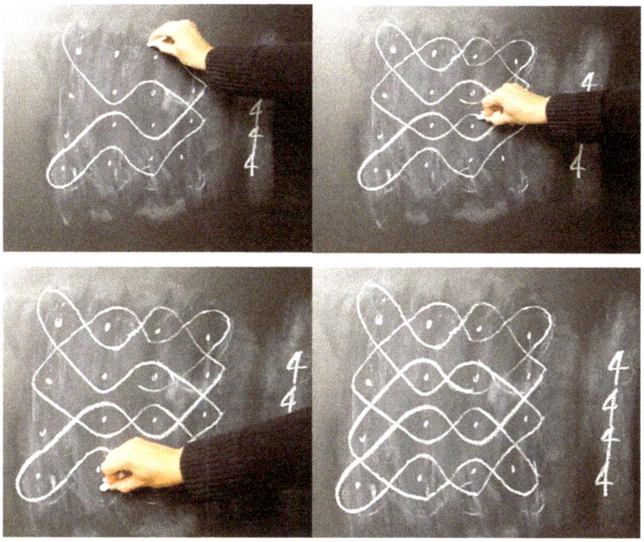

FIGURE 15.3: Lesley Halliwell, *Additive Trace*, 2016. Video stills (7 min 37 sec; 7 min 48 sec; 7 min 50 sec; 7 min 56 sec). Chalk on blackboard panel. Copyright and photo credit Lesley Halliwell.

slip away again. Perhaps, like Alice of Wonderland, I too have to learn to accept the image's 'nonsense' for it to make perfect sense (Deleuze 1997: 21).

Temporal interruptions are evident in my video *Additive Trace* (2015) (Figure 15.3), a piece of work that recorded and re-played my hand movements while I attempted to master the art of Southern Indian *kōlam* patterns. Exposing my drawing errors as the chalk squeaked, scraped and snapped clumsily over the board, I hesitated, thinking about where to position the chalk, which direction to take the line and how to navigate between the positioning dots. When I withdrew the chalk's pressure, it hovered above, tracing an unrecorded path in the expanded zone of surface. Gliding across the paper, the retraction of the hand resulted in pause, realignment and return. A zone of energy rather than a mark, the movement is sensed in the completed work as a negative space.

Following a similar trajectory, writing about 'taking a line for a walk', anthropologist Tim Ingold explains that when a line is 'jiggled' about on the spot, it creates a dot where all the energy and movement is focused; an 'isolated and compact moment, broken off from those preceding and following it' (cited in Mall 2007: 72). Ingold continues that when he lifts his pencil away from the paper, before returning it to the next spot, he creates a 'transverse movement' that, while playing no part in the process of inscription, offers up an empty space. When the interruption is experienced as temporal, it becomes a pause. The lifting of the pencil or chalk is like the moment between inhaling and exhaling, a liminal space replete with potential.

In *Additive Trace*, these moments of uncertainty were recorded as a series of temporal interruptions in the flow of the drawn line. Cultural theorist Paul Virilio's 'picnolepsy' is useful here to understand the implication of these repeated 'tiny deaths' as momentary sensory deprivations (2008: 48). Virilio's idea is that the viewer reconnects these interruptions so that vision, despite being a 'montage of temporalities', is perceived by the viewer as complete. However, I realized that these 'cinematic stutters' cannot be too long or too big, otherwise the pattern is in danger of falling into ruin. The 'Principles of Approximation', in Walter Shepherd's *Glossary of Graphic Signs and Symbols* (1971), is a fascinating account of the degree of variant permitted before 'ideal forms' become (un)identifiable. Through an intricate measurement system Shepherd identifies when, for example, an incomplete circle is recognized as a circle (1971: 33), with the viewer being able to 'fill-in' the missing information. If the interruption is too large, then the shape/pattern is in danger of being lost. The dynamic between pattern recognition (order) and degree of variant (disorder) is complex – the two states are delicately balanced and intimately connected, with thresholds that merge and overlap.

As is evident in my experience of making, spatial and temporal interruptions are essential, both materially and conceptually. They are not only transgressions that prompt an active and productive tension, compelling the viewer to see anew, but are devices that enable the mind to understand surface topology in a new way. My artworks demonstrate that it is possible to experience pattern and chaos simultaneously within the same artwork. Richard Wollheim's notion of 'twofoldness' is useful here to further understand this phenomenon (Gaiger 2008: 53). Wollheim argues that surface and content are 'two aspects of a single experience [...] distinguishable but also inseparable' (Wollheim 1987: 46). In a similar manner, I illuminate ways in which the viewer can experience variations of order and disorder within the same image. In other words, the viewer can be visually aware of the pattern despite (or because of) the uncertainty brought about by spatial and/or temporal interruptions. In this scenario, the viewer, as an active participant, seamlessly negotiates the threshold between the two.

While I gravitate towards systematic processes and ordered compositions, I find ways to unsettle the patterned surface. This unsettling may be deliberate or accidental and achieved through process-derived inconsistencies, compositional design or material choices. Rather than the binary positions so commonly associated with pattern and chaos, the tension between order and disorder results in a zone of interaction, with thresholds that intermingle and overlap. These spatial and temporal interruptions are not passive but are experienced both in the maker's engagement with pattern and in the viewer's response.

PART 3

DECOMPOSE–RECOMPOSE

Introduction

If pattern invokes suggestions of order, what happens to this sense of regularity when subjected to interruption and intervention? In these contributions, chaos and disruption are used as purposeful creative strategies. Chaos is imposed upon known and controllable variables, and the fragmented or damaged results are recomposed such that traditionally held patterns of thought and form give way to new insights about subjects as diverse as technology, gender and sites of government. Conversely, in looking for latent pattern, or imposing structure on the fragmented or 'formless', new constructs are built through a process of re-composition. Fragility and instability are embraced as strategies for questioning the status quo as pattern shakes hands with chaos.

The exposure to decomposition is at its most extreme in the processes employed by Catherine Yass. By burying a photographic transparency in the river depicted within its image, for example, Yass creates subsequent representations of place that are scarred by physical damage, adding to their metaphorical and psychological resonance. The creative potential of decomposition is equally embraced by Chris Brown as he accepts the inadequacies of his digital TV in the production of quirky and unconventional photographs provoked by a repeated glitch.

In the work of artist Mark Graver, 44 sounds are gathered randomly then layered together, accompanied by the chance selection and layering of 44 found words or phrases forming a video and prints. That the work might be 'performed indeterminately using motion sensitive speakers triggered at random as people move around a space' seems especially fitting, as if echoing the impermanent flux from which pattern transpires. How language, letterforms and syntax might affect and be affected by pattern–chaos thinking is examined by Pauline Clancy as she seeks to undo the primacy of language. Clancy rearranges familiar letter forms, graphically and materially 'interrupting' their communicative role to create, quite literally, 'a type of chaos'.

Through a process of more controlled and less chance-driven separation and fracture Charlotte Hodes's delicate revisioning of a nineteenth-century plaster frieze from the Elgin marbles brings to life a female protagonist that 'promenades' through an intricate network of hand-painted fragments. This, and the shattering of Elizabethan form and imagery in the ceramic patchwork of Zoë Hillyard, re-orientates our understanding of received knowledge in the museum collection, along with a welcoming of reparation and renewal.

The re-imagination of history within the work of Hodes and Hillyard is a nod to temporality, also a recurring theme for Judith Stewart, James Quinn and Katarina

Andjelkovic. Stewart's composite images derived from video stills highlight personal and social observations of journeys made, whilst alluding to the complex dance between thinking and making and the thin line that separates the orderly and disorderly processes within art production. Quinn manipulates video stills from previously unused material where he collapses the temporal dimensions of video sequences so that they appear all at once, or out of order, 'collapsing the distance between video still and video sequence'. Finally, Andjelkovic examines the way Idris Khan's photographs can both animate form and question the durability of major monuments to stability such as the United Kingdom's Houses of Parliament or Tower Bridge. These contributions consider what is 'alive' in the lens-based image and demonstrate that philosophical debates about fallibility repeat and recur across time and across mediums.

16

Foment

Catherine Yass

FIGURE 16.1: Catherine Yass, *Damage (canal)*. Photograph.

FIGURE 16.2: Catherine Yass, *Cemetery 2020*, 2020. Photograph.

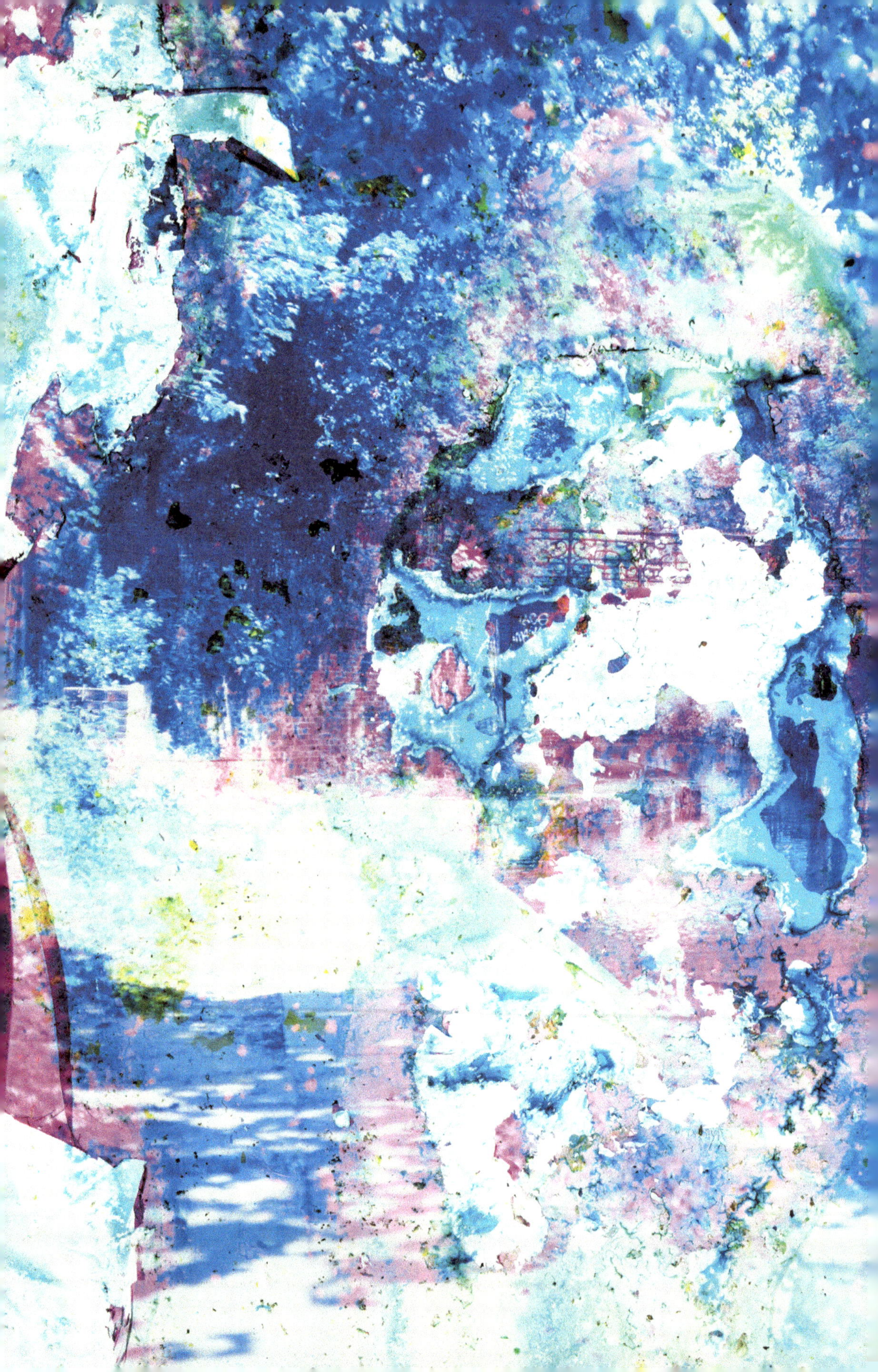

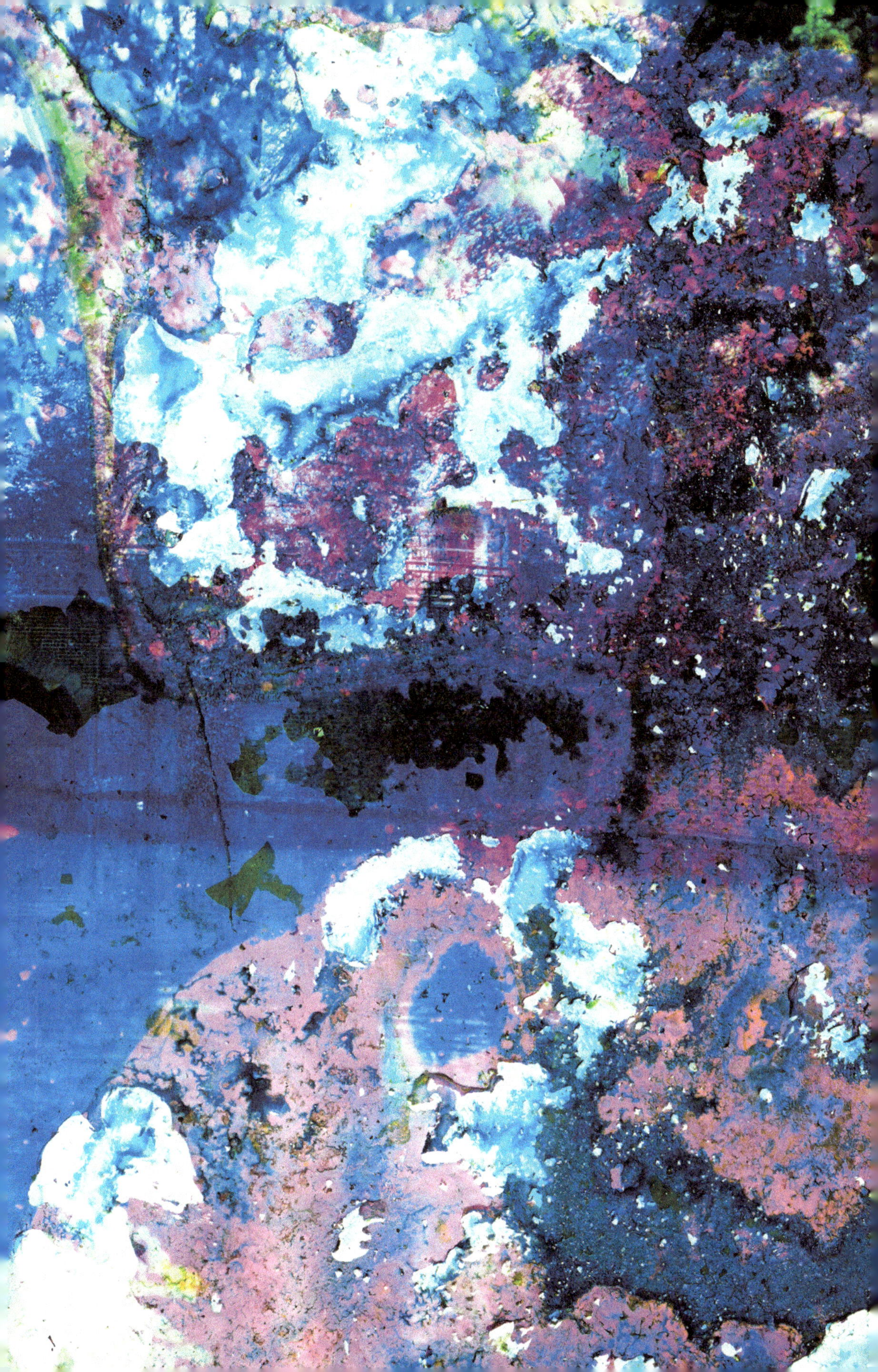

Cemetery 2020

In the cemetery, in the bright green of COVID-19 May, the pattern of leaves between trees shifts with every breath to re-form into a new configuration. The photograph would kill it dead and its pattern would never change, stiller than the bodies decomposing underground. But an underlying deep blue negative suggests other moments, contingent and latent images that escape being stilled. Abney Park Cemetery in London was founded in the 1830s as a nondenominational cemetery by nonconformist Isaac Watts. It was open to all regardless of religion. Abney Park was also an arboretum with 2500 varieties of plants. Since the cemetery fell into disrepair in the 1970s, it has been maintained as an overgrown ruin.

Damage (Canal) *(previous page)*

This is from a series of photographs taken of empty places around where I live. The transparencies are left on their own in the place where they were taken and then retrieved some time later, damaged by the material of their own image. In the canal the oil pollution broke down the scenery and split and swirled it into abstract pattern. The image decomposed leaving a residue of chemical marks spiralling through it. The marks are arbitrary but they appear to be composed, perhaps because we want them to fit into a pattern of order in the face of the image's entropic collapse.

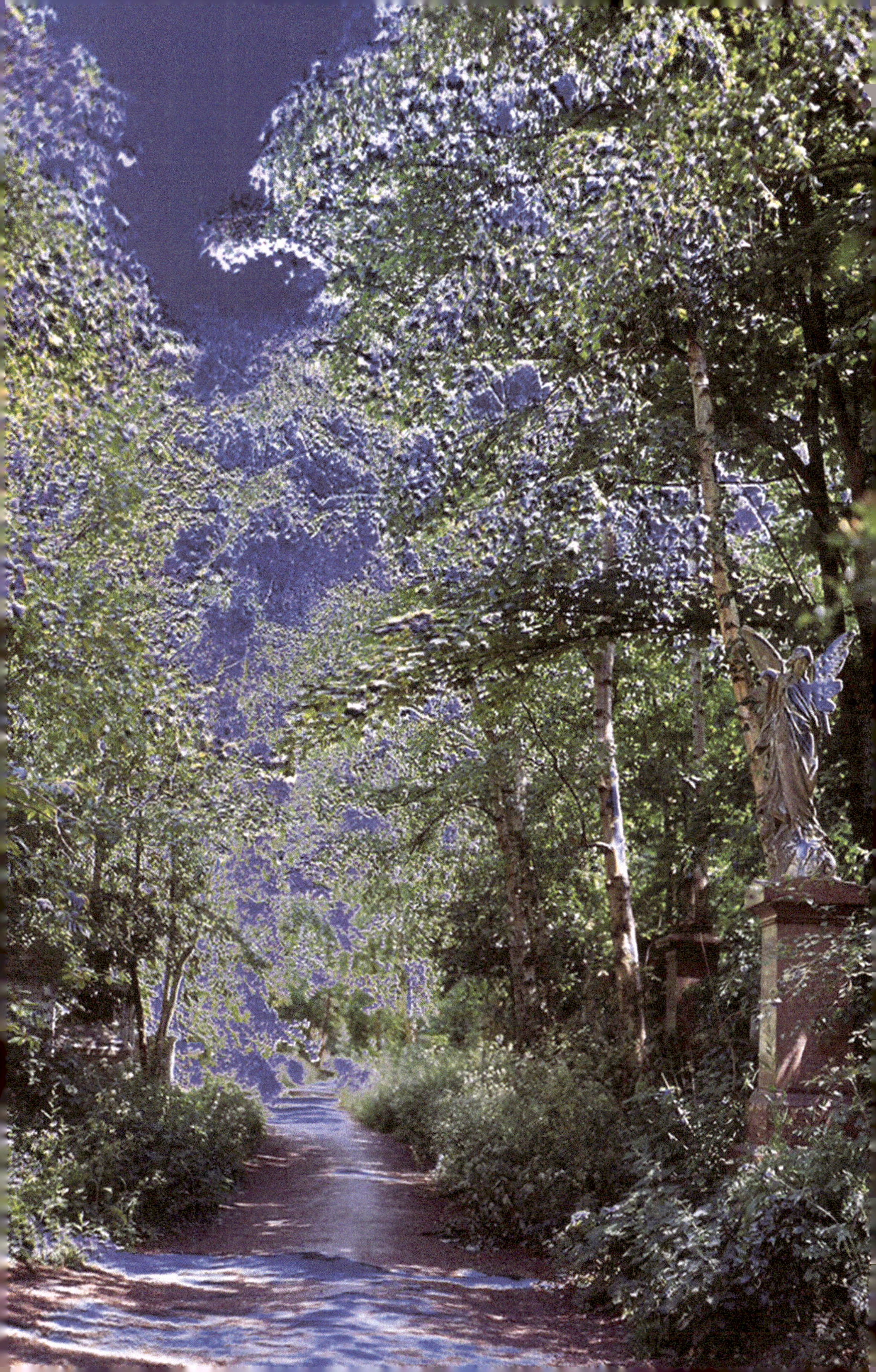

17

Meniscus

James Quinn

In his text *The Unspeakable Art of Bill Viola: A Visual Theology* (2014), Ronald R. Bernier states that by extending a moving image's duration, Bill Viola presents a mode of temporality that resembles Henri Bergson's notion of a 'psychological time', or a zone of perception that presents time 'not, in the mathematical, scientific notion of time, as a sequence of successive, atomistic, and discrete moments – the time of clocks – but as a multiplicity continually unfolding in "duration"' (Bernier 2014: 43). For Bernier, Viola's works 'extend the moment of viewing into a time of attentiveness and receptivity, wherein what is captured is, in a sense, the *invisible*' (Bernier 2014: 43). The same applies to my experiments – the depiction of an otherwise invisible or imperceptible phenomenon, rendering visible the string of continuity that proliferates a moving image.

How best to succinctly disseminate the running length of a moving image work without somehow misrepresenting the work to the reader, or perhaps more crucially, without encountering some form of loss with regards to subtleties present in the artwork? The frequent approach for a moving image practitioner is to adapt a moving image work into a series of video stills for inclusion in printed texts, operating as an indicative sequence that attempts to portray how a moving image behaves, but this raises challenging issues. This problematic relationship, between the moving image sequence (film or video) and its representation as a series of stills for documentation in the textual format, is thus the primary site of investigation here.

The works presented here as still images (Figures 17.1–17.4) aim for an alternative, ancillary format, whereby the temporal dimensions of video sequences appear all at once, or out of order. This practical outcome might offer comment on – or perhaps even collapse – the distance between the video still and the video sequence and its associated shortcomings. Beyond this, further points are considered – can the works present a paradigm shift – can the video still become a sum of the moving image parts? If so, does the video still begin to occupy its own space as an individual artwork, emancipated from its role as documentation for source footage?

Before discussing this experimental outcome, I will outline the technical steps taken to generate these works. Each individual frame from the full duration of the source footage is edited to overlap with one another, creating an assemblage of frames consolidated into a single image. The obvious comparison to make here is to the digital composite image – the layering of two or more separate images to form a new image. Unlike the atypical photographic composite image, however, these outcomes possess characteristics that pertain to their origins as moving image works, most predominantly, a distillation of the temporal dimension (and filmic characteristics such as camera or subject movement) of the source footage.

In the case of *Sannox* (Figure 17.1), the work's source footage presents a panning camera movement, depicting the two halves of a cliff face and the walking path taken from left to right over the course of ten seconds. The edited outcome instead presents this motion simultaneously, with aspects intersecting with one another. In *Peacock* (Figure 17.2), the source footage presents a camera zoom towards a stationary peacock butterfly, filmed over ten seconds. When studied frame-by-frame, the

FIGURE 17.1: James Quinn, *Sannox*, 2021. Digital composite image. © James Quinn.

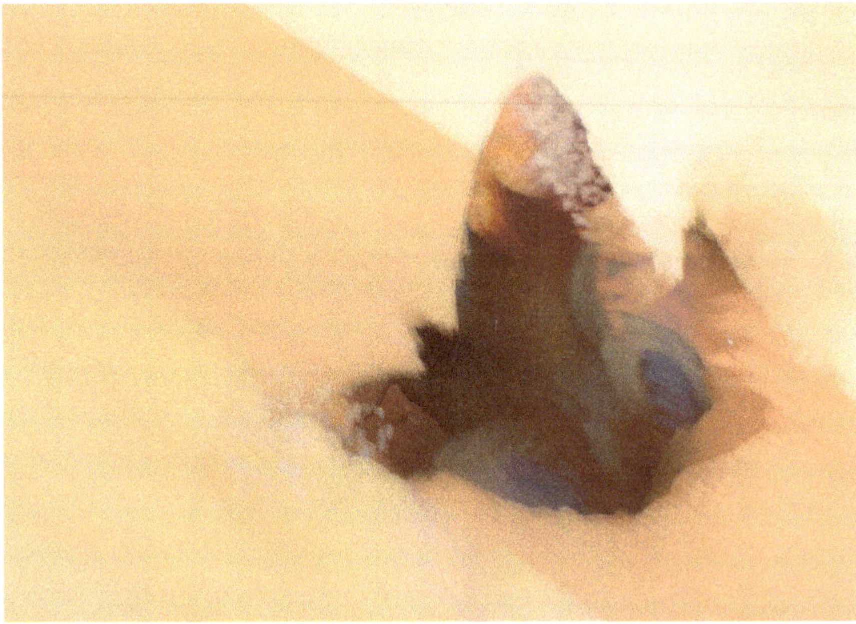

FIGURE 17.2: James Quinn, *Peacock*, 2021. Digital composite image. © James Quinn.

zoom presents the subject in a range of scales, starting from a distance and gradually increasing in size over time. Presenting these differing scales simultaneously results in an impression of the subject embedding itself within itself. *Swans I* and *Swans II* (Figures 17.3 and 17.4) were both captured with a stationary camera over ten and twenty seconds, respectively. When edited the work presents a simultaneous depiction of the trajectory of the source footage's subjects, resulting in an abstract representation of this movement.

Observations

The process of generating these works, as with any practice-led experimentation, allowed for unanticipated outcomes and concepts to present themselves. These works permit access, not to a sequence of successive frames, but instead, a collapsing of these frames into a layered whole. Layering each isolated frame of the source footage permitted the outcomes to tesselate from their point of origin, forming an expanded concertina of sorts – joined at both ends, the work concurrently signals the sequence of steps taken and steps to come, with any idiosyncrasies captured in the footage displayed in equal measure.

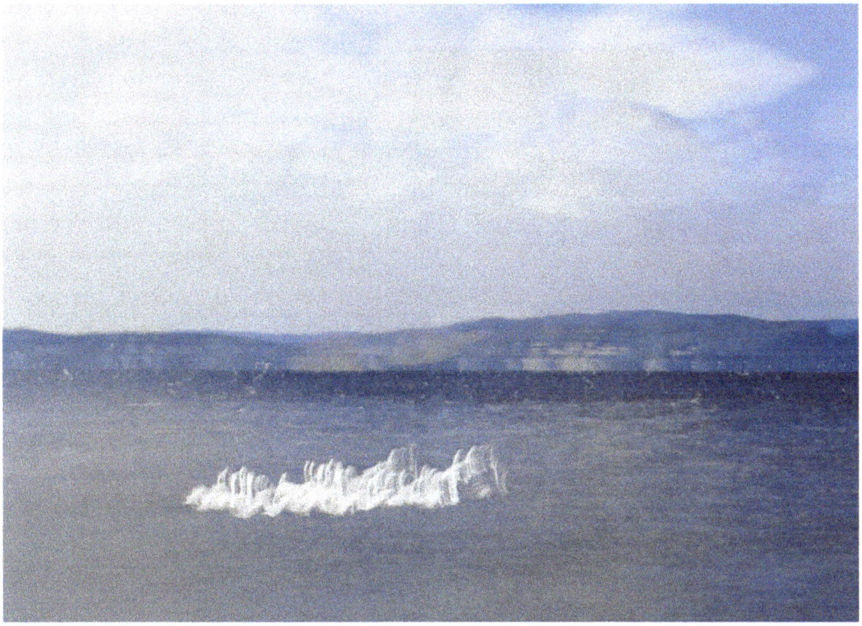

FIGURE 17.3: James Quinn, *Swans I*, 2021. Digital composite image. © James Quinn.

FIGURE 17.4: James Quinn, *Swans II*, 2021. Digital composite image. © James Quinn.

By signalling the site of these works' origins and the assemblage of the video frames that conjoin to create their running length, the outcomes share some symmetry with specific arts practices and theory. Various artists have repopularized the use of the celluloid film and its associated apparatus, such as in the unmistakable audio and visual presence of the film projectors in Tacita Dean's 16-mm film *Michael Hamburger* (2007) and *Antigone* (2018) or the 35-mm film projector in Luke Willis Thompson's *Autoportrait* (2017). Moreover, David Batchelor's 2015 Whitechapel show *Monochrome Archive* project (1997–2015) exhibited a supplementary catalogue of source images alongside large-scale projection of these works. Such approaches might draw some comparison to the key practitioners and concepts of Structuralist film – Peter Gidal notably states in *Structural Film Anthology* (1976), that 'in Structural/Materialist film, the in/film (not in/frame) and film/viewer material relations, and the relations of the film's structure, are primary to any representational content' (Gidal 1976: 1).

These references triangulate around the same core ideal – acknowledging the illusionary quality of the moving image medium. Referring to the works generated in the present context, exposing the singular frames that comprise the source footage is a decidedly anti-illusionary approach – something Laura Mulvey explores in her text *Death 24x a Second* (2006). Mulvey states that 'the process of delaying a film inevitably highlights its aesthetics and the illusion of movement, and the hidden presence of the filmstrip on which the illusion depends' (Mulvey 2006: 185). Mulvey suggests that delays or alterations to the temporal aspects of film offer a chance for viewers to shift their 'consciousness between temporalities' (Mulvey 2006: 184), namely, that diegetic temporality which is dictated within the narrative of the film, and the temporality concurrent 'beyond the constraints of the film frame to the society from which it is derived' (Mulvey 2006: 184). This anti-illusionary ability to stretch temporal aspects of the work is doubled down upon as Mulvey cites Raymond Bellour, claiming that disrupted temporality generates a 'pensive spectator' (Mulvey 2006: 186), a viewership that steers towards engagements with the nature of cinema itself. The works therefore might represent an anti-illusionary portrayal of the sequence of frames that substantiate their source material.

Sharing my outcomes (Figures 17.1–17.4) with a frequent collaborator resulted in a remark regarding their painterly quality, or the extent to which the work shared certain aesthetic monikers of painting practice. Initial impressions of a work such as *Sannox* might invoke the abstracted qualities in Gerhard Richter's photographic paintings. A further comparison might be drawn from Jane Dixon's body of work employing layered frottage techniques to generate images drawn from imagined or real architectural spaces. To some extent, these experiments share commonalities with the concept of the palimpsest – the continued reworking of the same visual plane with subsequent iterations of the written word by erasing or imposing each new addition upon the previous layer.

Meniscus

The visual outcomes in this article are but multiple lines of enquiry in a new experimental methodology, each yielding subtly different results. As previously stated, they depict a visual coalescing of the individual frames that substantiate a moving image. It is the conceptual dimension of the work that presents further avenues for research, however. Daniel Crooks and his work *Phantom Ride* (2016) assist in defining an essential quality in my experimental outcomes. *Phantom Ride* presents a two-way projection screen, with each side presenting a different image simultaneously – one side depicting a forward trajectory, the other the sequence of events that have already passed. To Crooks, this staging allows the screen to become a threshold between the past and the future; the screen becomes a 'meniscus' (Crooks 2016) in this sense or the site through which time passes. Perhaps this otherwise imperceptible meniscus is what is laid bare in my experiments – the temporal threading that strings a succession of singular frames together, each of which shapes a cumulative image that draws upon the sum of its parts not sequentially, but simultaneously.

18

Digital Dadaism

Chris Brown

This photographic project came about through the experience of frustration with my digital television. Like all things digital I was seduced by the world of advertising into believing that digital is better, that my life would be enhanced by the purchase of this latest technology. However, disappointment crept in as I found the signal seemed increasingly susceptible to atmospheric conditions and some nights, I could barely receive any channels at all. A new, malignant force, it seemed had entered my life, the *glitch*. As Hunt puts it: 'The digital revolution has [...] made the glitch, or more specifically the impact of the glitch, a more commonplace occurrence ...' (2018: n.pag.).

At first, I railed against the modern world of digital broadcasting and these daily glitch-induced frustrations. Eventually, however, I came to actually enjoy the serendipity of my evening's viewing, even to welcome the uncertainty of reception in this supposed multichannel idyll of on-demand televisual choice.

I had slowly but surely come to accept then love this visual bomb, for I had noticed, often in my frustration of surfing through the channels in search of an uninterrupted picture, that the glitches in the signal would cause the picture to stick and as the TV's processor struggled to catch up, create sometimes beautiful, sometimes ugly, but often intriguing distortions (Figure 18.1).

The fleeting, temporary nature of these televisual digital collages led to the idea of using what seemed a suitably antiquated digital single lens reflex (DSLR) camera to document these unfolding sequences of imperfect, randomly generated imagery and of using the digital camera's advantage of metadata to record and verify the narrative of unique viewing events seen only by the photographer (Figure 18.2).

The collages of fragmented images that resulted from switching channels seemed reminiscent of the works of the Dada artists of the 1920s, who utilized the

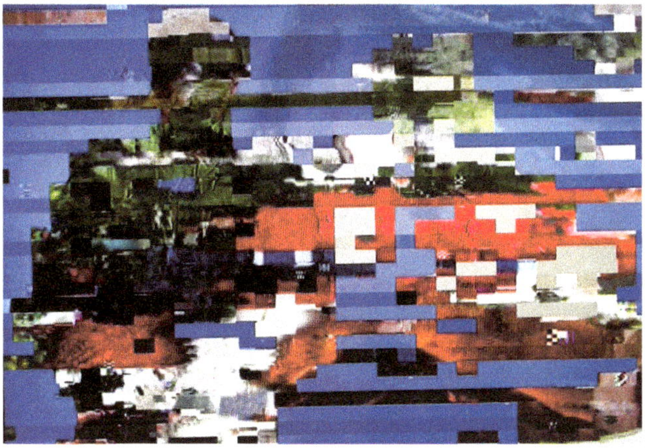

FIGURE 18.1: Chris Brown, *TV glitch* 08:05 on 24 May 2017. Photograph. © Chris Brown.

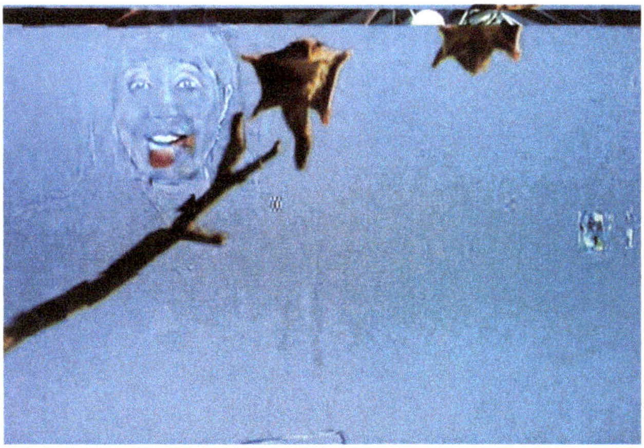

FIGURE 18.2: Chris Brown, *TV glitch* 22:52 on 13 June 2017. Photograph. © Chris Brown.

'methods of juxtaposition and surprise – the shock derived from seeing an unlikely or unexpected association of realistic elements' (Barnicoat 1972: 161). Indeed, the Dada artists are the group widely credited with first using photomontage as an artistic technique:

> The pasting together or otherwise combining of separate and disparate pictures to form a new visual entity was one of the most striking contributions of the artists of the 1920s [...] the men and women of the 1920s assembled pictures that widely varied in subject, perspective, scale, and tonal value (Newhall 1982: 210).

The Dadaists' photomontage technique was one of 'manipulating found photographic images to create jarring juxtapositions and chance associations' (Meggs 1992: 246). In a similar way, I found these digital televisual montages could be controlled to an extent by my own interventions, by switching between channels or rotating the aerial lead or socket and thus creating a 'loop' of imagery in the TV processor's memory. By selecting certain channels, ironic, sometimes amusing juxtapositions could be formed that seemed to comment on politics and consumerist culture (Figure 18.3).

In the 1920s, the work of Grosz, Höch and Hausmann became 'a biting political weapon' (Ades 1983: 121) and Heartfield's photomontages were 'a crushing indictment of Hitler and capitalist militarism' (Ades 1983: 121). The technique of détournement, used by the Situationist International and punk movements in the 1950s, 1960s and 1970s, respectively, also ironically juxtaposed imagery to comment on society, politics and culture.

Similarly, the glitched 'portraits' I captured (Figure 18.4), with their random disassembly of the all-too-perfectly manicured faces seen on television these days, recalled the savage caricatures of Georg Grosz, whose 'biting pen attacked a corrupt society with satire and caricature [...] [and] [...] advocated a classless society' (Meggs 1992: 249).

However, despite these occasional horrific mutilations, the images I recorded lacked such political intent and tended more 'to a poetic disorganization of reality'

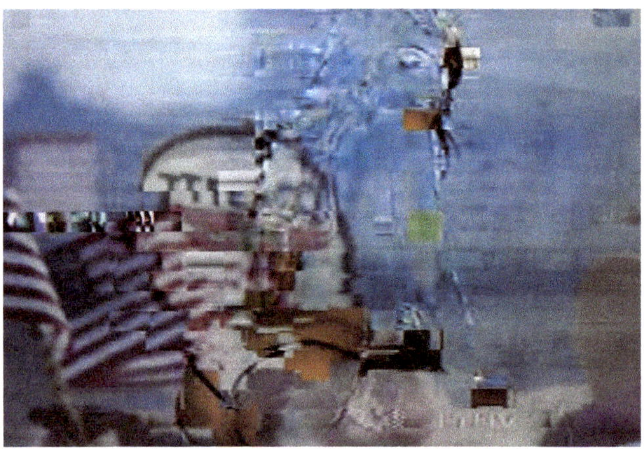

FIGURE 18.3: Chris Brown, *TV glitch* 23:15 on 21 September 2016. Photograph. © Chris Brown.

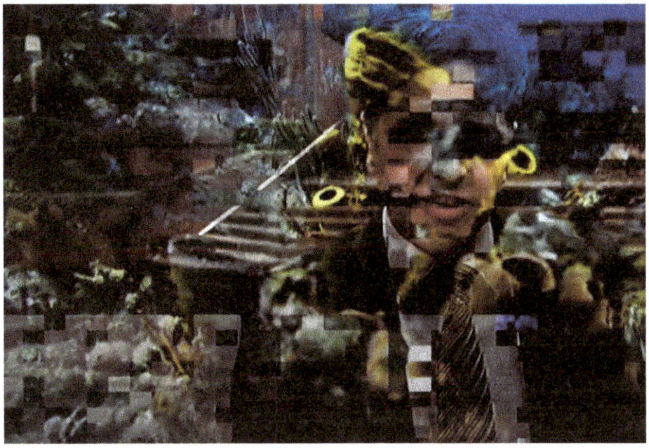

FIGURE 18.4: Chris Brown, *TV glitch* 23.00 on 13 June 2017. Photograph. © Chris Brown.

(Ades 1983: 121) akin to the collages of Max Ernst. With the image manipulation software available these days such compositions could of course be easily 'photoshopped', but that would defeat the pleasure in this artistic act. As Meggs (1992: 245) puts it: 'To Duchamp, Dada's most articulate spokesman, art and life were processes of random chance and willful [sic] choice. Artistic acts became matters of individual decision and selection'.

Indeed, each of these 'glitchographs' is a unique temporal record. Despite digital technology's infinite potential to reproduce, these images cannot be replicated: they are each a combination of broadcast schedules, atmospheric conditions, technical disruption and my own interventions. Ultimately, these captured moments of televisual montage celebrate the glitch and the human pleasure in the imperfection of the supposedly better and flawless digital world.

However, in the true nihilistic spirit of Dada that 'the true dadaist was against dada itself' (Ades 1983: 121), once I discovered that frayed wiring was the primary cause of my glitch problem, I bought myself a new aerial lead and put an end to this digital dadaism.

~~g.L.*itc*.~~H.O~~.O.Q~~

19

Forty-Four Sounds

Mark Graver

Forty-Four Sounds (Figures 19.1–19.4) is a text and printmaking-based installation of 44 digital prints (on magnetic panels), a video of 44 frames (which lasts for 4 minutes and 44 seconds) and a soundtrack composed from samples of 44 found sounds. It was completed in 2012 and first installed in *Utopia into Dystopia*, Waitakaruru Sculpture Park, New Zealand. The phrase 'forty-four sounds' was cut from a newspaper – its apparent (albeit misquoted) reference to John Cage's *4'33"* (1952) is entirely serendipitous, but nevertheless poignant in respect of the artwork's complex choreography of predetermined elements and chance.

Working with photoshop layers and mixing with video and sound with each clip or track stacked on a horizontal timeline is visually and technically similar to printmaking, which generally employs layered images and/or progressed additive

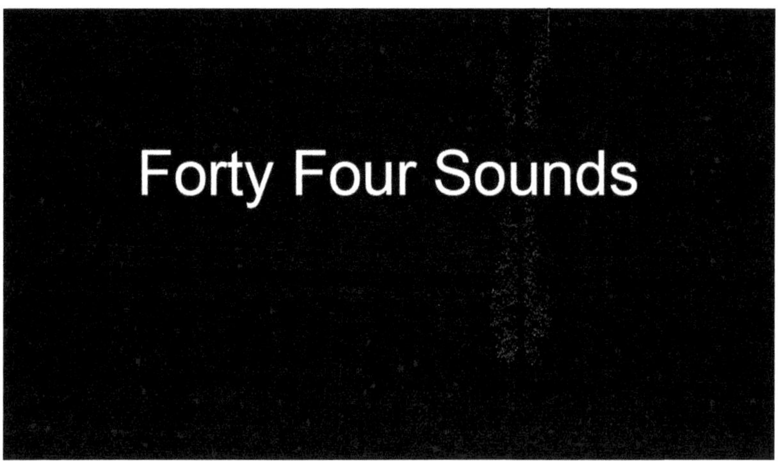

FIGURE 19.1: Mark Graver, *Forty-Four Sounds*, 2012. Video still and installation. © The author.

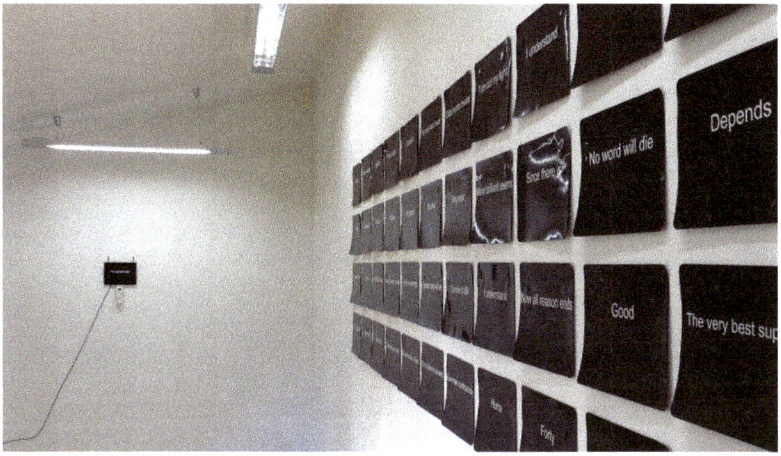

FIGURE 19.2: Mark Graver, *Forty-Four Sounds*, 2012. Video still and installation. © The author.

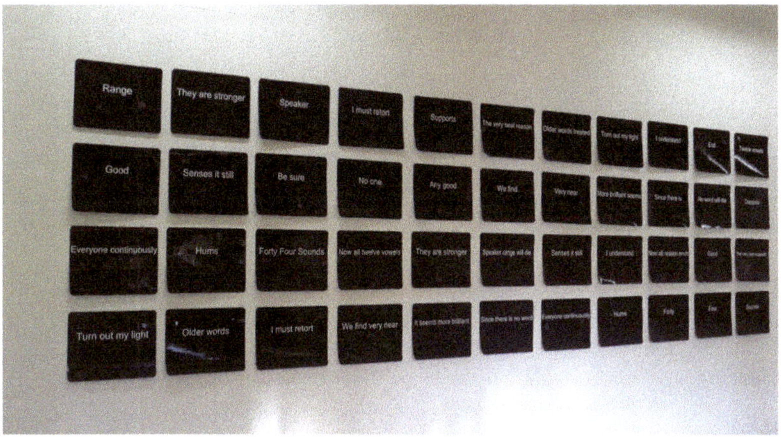

FIGURE 19.3: Mark Graver, *Forty-Four Sounds*, 2012. Digital prints on magnetic panels. © The author.

or reductive stages, whether with traditional or digital techniques. The arrangement, colour, opacity, speed and volume of each layer contribute to the final composition visually and audibly. The work can be realized in multiple static, moving and temporal formats. It can be installed as one work or exist as separate elements of print, video or sound, physically or online (Figure 19.4). It could also be performed indeterminately using motion-sensitive speakers triggered at random as people move around a space, so no performance would be identical.

FIGURE 19.4: Mark Graver, *Forty-Four Sounds*, 2012. Soundtrack. © The author.

The text was determined by a random selection of previously chosen words and phrases cut from newspapers, arranged and repeated. This echoes Tristan Tzara's aleatory literary technique described in 'To Make a Dadaist Poem' (Tzara [1920] 1980: 31). The concept and process of *Forty-Four Sounds* not only makes deliberate reference and homage to the 'cut up' poetry of Dada but also to the complex techniques of chance musical composition pioneered by Cage, who noted that '[m]ost people who believe that I'm interested in chance don't realize that I use chance as a discipline. They think that I use it as a way of giving up making choices. But my choices consist in choosing what questions to ask' (Cage cited in Kostelanetz 1988: 17). Cage's combination of chance and predetermined approaches – choreographed chance – guided the work. Thus, although the element of chance was present in the selection of the words and phrases, the chance action took place within a structure, as the words were predetermined and there was some post punctuation to decide.

The words were chosen to allow the possibility that, rather than stating, they would evoke or imply meaning when randomly selected and although conceived as a linear piece that builds to a crescendo of sound the work is looped and can be entered at any point. With the text order determined, the soundtrack (Figure 19.4) was composed of samples of 44 different sounds – cicadas, rainfall, our old Jack Russell now deceased, an apple being cut – manipulated and mixed, added one sound at a time until all 44 were layered together.

20

A Type of Chaos

Pauline Clancy

Typography, the visual representation of language, primarily communicates meaning through written, printed or digital forms. Through this tangible embodiment of language, typography offers an outward symbolic meaning, an external reality in the form of a 'sign'. Language, however, can also reflect onto itself, revealing a materiality, whereby a meaning is formed on the visual surface, foregrounding text as form, unfixed from the constraints to represent or communicate. Through the arbitrary arrangement of typographic forms, visual language (words) can also be viewed as pattern, where diverse formations communicate different things. When this pattern is interrupted a type of chaos ensues where language and its communicative role are challenged and disputed. This disruption enables a transformation of language into a new visual form and entity, where meaning still resides.

Figure 20.1 is a series of Letraset compositions created at *Ficciones Typografika* Workshop facilitated by Professor Erik Brandt as part of the (AGI) *Alliance Graphique Internationale* conference in Rotterdam (2019). The workshop explored typographic form through the process of Letraset, a low-tech, dry-transfer system. In the workshop, emphasis was on doing without thinking or making over thinking, thus allowing for interesting compositions and 'idiosyncratic discovery' (Brandt 2019). It focused on expanding typography – pulling or taking away from typographic form, moving typography away from its functional role, breaking away from the formal qualities of typography and building tension between the letterforms. Tension is also evident between the desire of the letterforms to sustain their communicative value and the material pull of language to perform in another capacity, challenging logocentrism in language. Drawing on Karen Barad's consideration of language as untethered from speech or words (Barad 2007), Petra Lange-Berndt proposes a 'methodology of material complicity' whereby to engage with materials is to 'formulate a *critique* of logocentrism and the predominance of written language as a tool to generate and communicate meaning' (2015: 13, emphasis added).

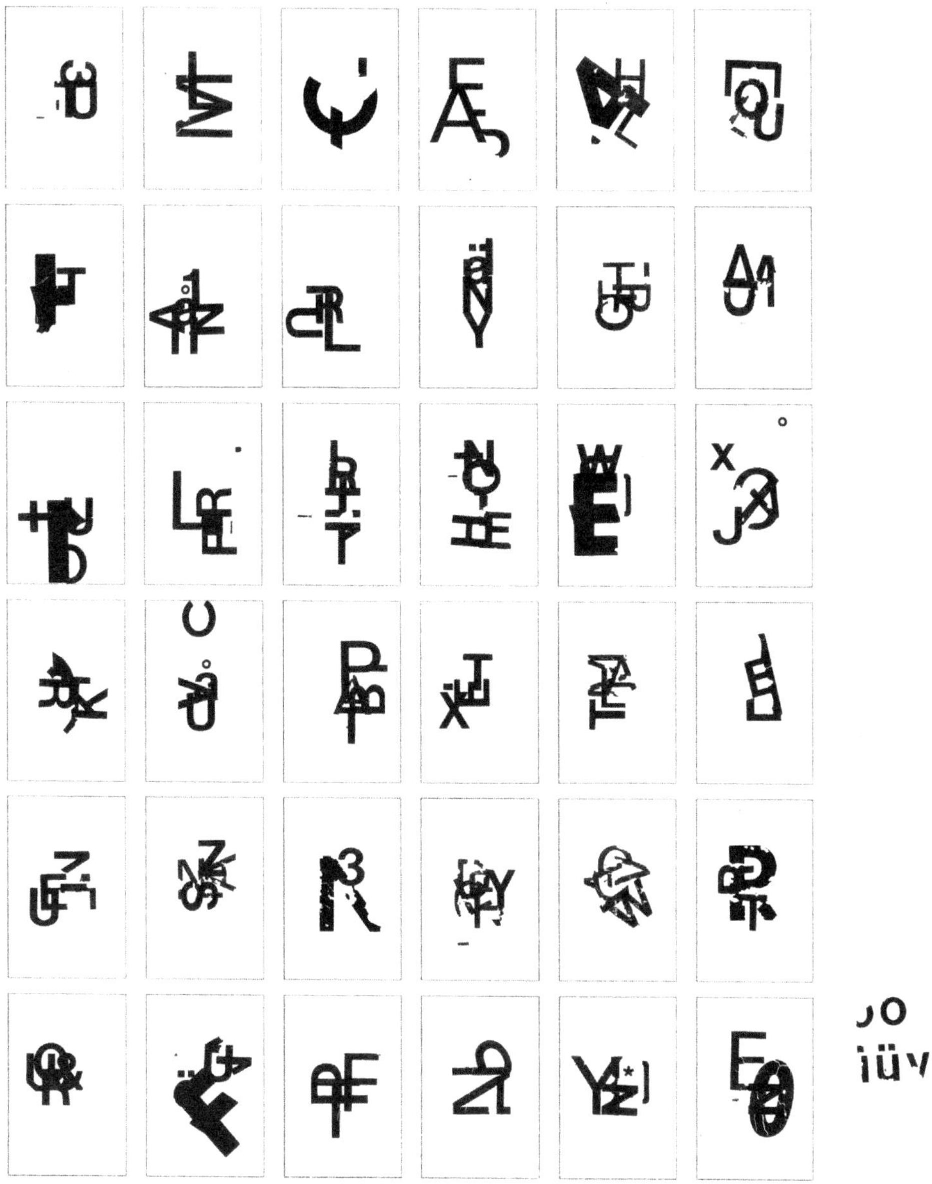

FIGURE 20.1: Pauline Clancy, *Letraset_01*, 2019. Letraset on paper. 216 mm × 279 mm. All images are courtesy of the artist.

While various sans serif letterforms were available for use in the workshop, the compositions mainly consist of Helvetica Medium, 24pt. The transfer process of the dry-lettering involves burnishing the letterforms with a pencil or similar object onto the paper. It can be precarious as often the full letterform does not transfer due to inaccurate pressure from the pencil, or a damaged sheet of letters, leaving incomplete or fragmented letterforms. In this context, experimentation with the somewhat unpredictable application process is encouraged. The fragility of the Letraset as a medium is also visible through cracking and tearing, causing a physical tension on the surface. It appears that these collected tensions facilitate both a physical and metaphysical rupture or destabilization in language.

Faults + Voids

From the original Letraset compositions, selected configurations were further expanded through digitization, laser-cut from 3 mm birch plywood (Figure 20.2), and subsequently installed with a digital projection (Figure 20.3). The three-dimensional (3D) (plywood) compositions are sealed with layers of white gesso revealing porcelain-like surface qualities alluding to the fragility of language while allowing

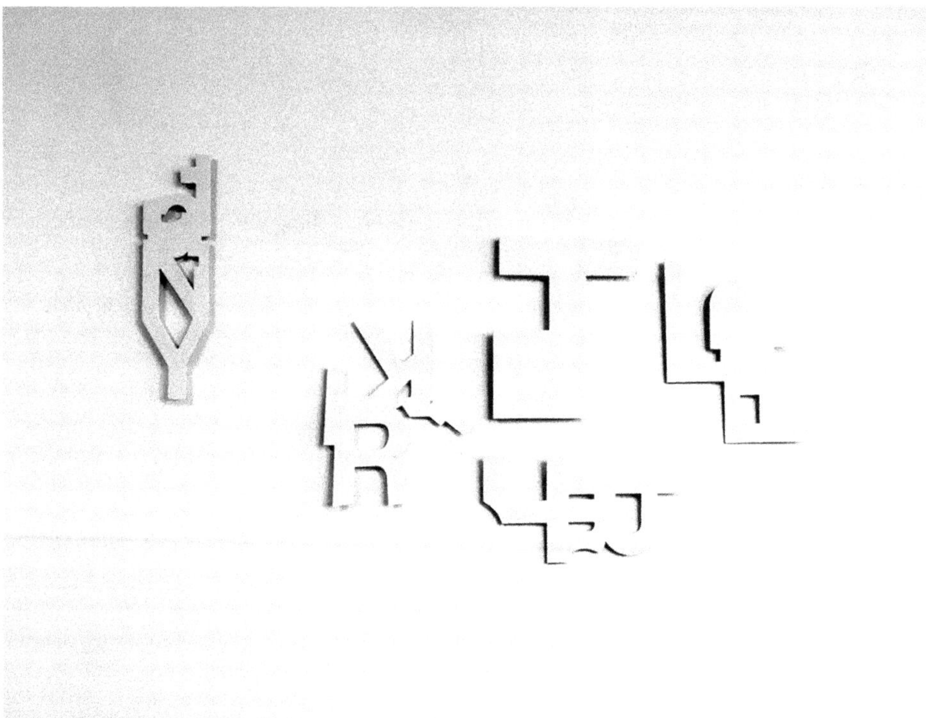

FIGURE 20.2: Pauline Clancy, *Faults + Voids*, 2021. Laser-cut 3 mm birch plywood, gesso.

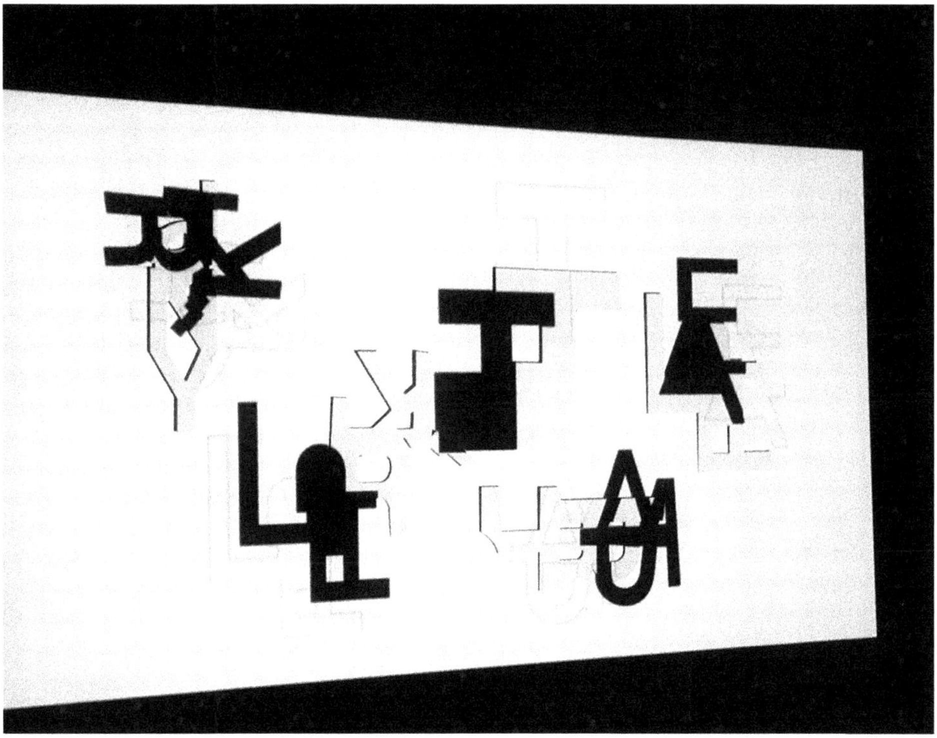

FIGURE 20.3: Pauline Clancy, *Faults + Voids*, 2021. Laser-cut 3 mm birch plywood. Digital projection.

for reflection and absorption of the digital projection. Figure 20.3 captures a still from the *Faults + Voids* installation revealing more digitized compositions in two manifestations: one as a solid form and another as an outline. The outline version, originally utilized as artwork for laser-cutting is re-used, re-imposed and reflected onto the compositions on the 3D surface. Scale has also impacted and altered the 2D compositions from flat containment within a 20 mm × 28 mm space to becoming large 3D sculptural forms. The digital projection inserts a new layer of negotiation between analogue and digital, prompting visual interactions between the merged and collapsed typographic formations.

The title *Faults + Voids* is in reference to US sculptor and land artist Robert Smithson (1938–73) whose interest was in the physicality and materiality of language, in particular in comparison with the landscape and environment. Smithson states that '[w]ords and rocks contain a language that follows a syntax of splits and ruptures. Look at any word long enough and you will see it open up into a series of faults, into a terrain of particles each containing its own void' (Flam 1996: xvii).

Smithson's disruption of language is reflected in contemporary critical readings such as those of Barad and Lange-Berndt where the communicative role of written language becomes secondary, and materiality is highlighted.

A Type of Chaos forms part of ongoing Ph.D. practice-based research examining typography and the materiality of language.

21

Fragile Order

Charlotte Hodes

Frieze[1] is a large papercut, over three metres in length, made in 2017 (Figure 21.1). It interweaves images of a female figure alongside vessels, dresses and architectural forms that appear to 'promenade' across a richly patterned surface. The horizontal format and shallow pictorial space are informed by and references the nineteenth-century plaster replica of a frieze from the Elgin marbles that runs the length of the gallery at the Wolverhampton Art Gallery, for which the papercut was specifically made. *Frieze* is part of a continuing body of work in which the format of the large-scale papercut is presented as an alternative to painting.

All the material for the papercut is sourced from an extensive repertoire of drawings, including the 'manufacture' of my own patterned wallpaper. These, alongside hand-painted and coloured papers, provide the surfaces into which I cut, using the scalpel blade as an alternative drawing tool. Together these form a complex and detailed collage, with many of the small fragments and intricately cut patterns being reapplied to the paper surface (Figures 21.2 and 21.3). This fine and detailed process of cut, paste and layering is countered by a formal sense of order which is tentatively held in place and echoes the original classical frieze.

The papercut is constructed within a limited range of pattern and colour. The colour palette, while representing a key of greys, is conceived in terms of colour rather than tone. This creates the conditions under which the forms morph, one into the other, dresses become figures become pillars and shadows and the spaces between take on a concrete form of their own.

FIGURE 21.1: Charlotte Hodes, *Frieze*, 2017. Papercut. 86.5 cm × 301 cm. © Charlotte Hodes.

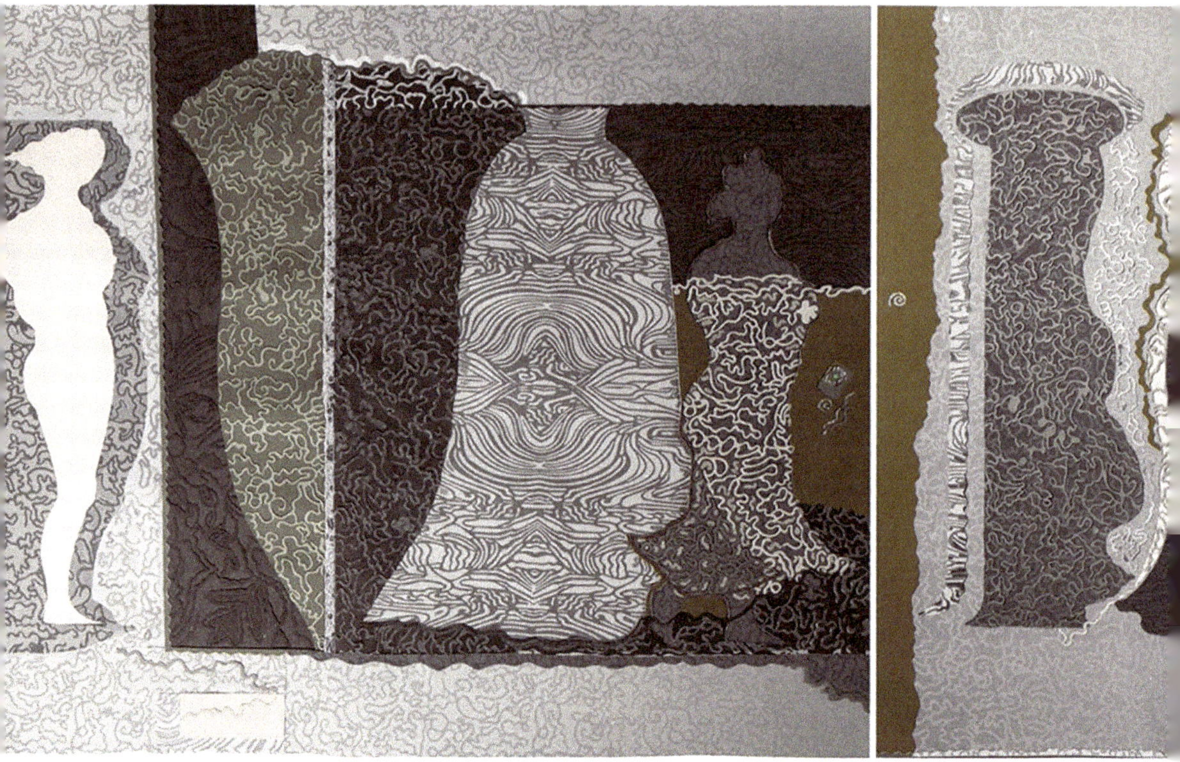

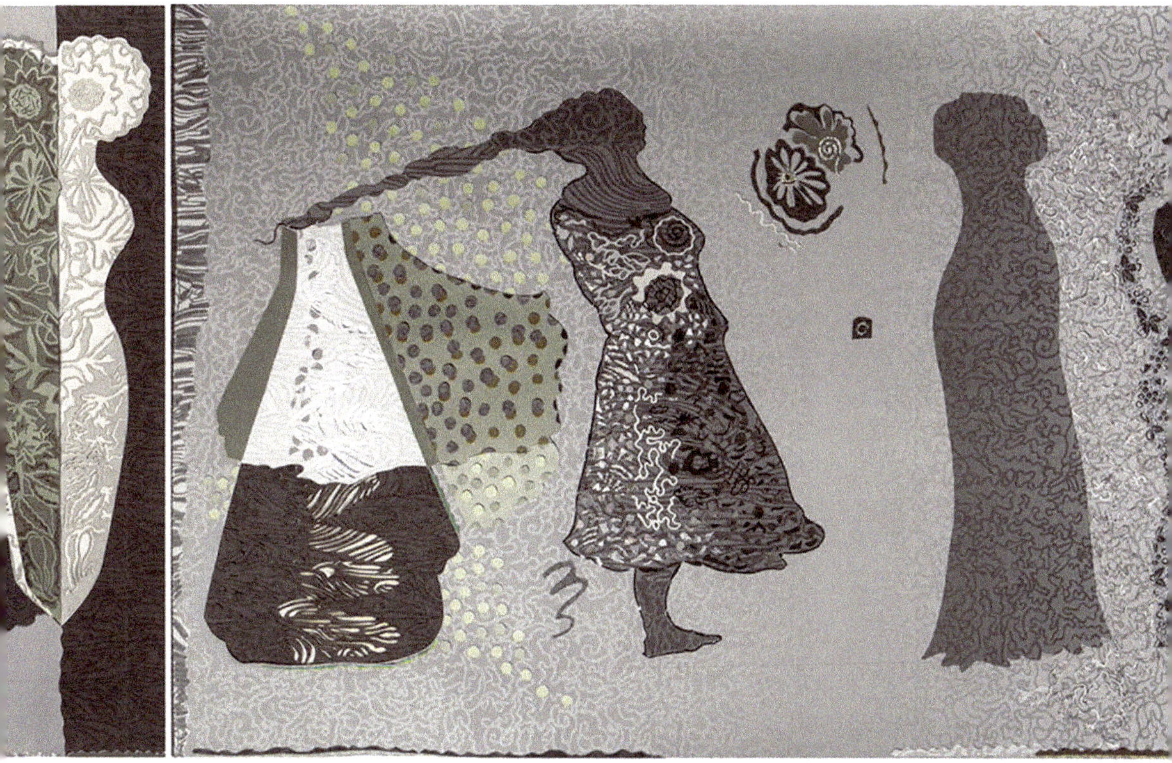

Frieze is to be 'read' from left to right. From the start, there is rebellion; the woman is depicted as a 'negative' shape. She has left her obligation as a caryatid-like pillar to make her way through the composition, emerging halfway as an architectural form. This playful and dream-like narrative continues across the surface as a woman with plaited hair is captured leaning against a conical vessel. Although she is supported by the vessel she may also be the female genie who has come out of her bottle or even Pandora herself who has emerged to play havoc. Flowers with no stems float free and 'women' dresses dance as if they have escaped the wardrobe and the women that they were made to contain.

Through the detailed repetitive working of the surface, in a manner more akin to those traditionally associated with female handcraft activities such as embroidery, the gender of the frieze is transformed from male (stone carving) to female. Likewise, expected hierarchies between architecture, fashion and domesticity are de-stabilized. Furthermore, in transforming the materiality of the original frieze from cold stone into the fragility of delicately cut paper, a sense of a dream is evoked in which the tension between the net-like collaged pattern and the nuanced colour transitions serves to threaten and overwhelm the forms.

The imagery addresses and challenges stereotypes of the woman as a decorative and fragile feature. Here she is solo, not needing justification for her very being or troubled by the male gaze. She makes her way in her own time; she is not to be rushed. She is transient, and there is a clear notion that even though the physical patterned order around her may collapse into chaos, into a pile of paper fragments, she will continue beyond the picture edge, to reappear in a multitude of past and future configurations.

NOTE

1. *Frieze* was part of a commission by Wolverhampton Art Gallery that included a wall frieze of 122 ceramic plates on a patterned wallpaper running 1185 cm, both of which were exhibited in Hodes's solo exhibition *Remember Me* at the gallery in 2017.

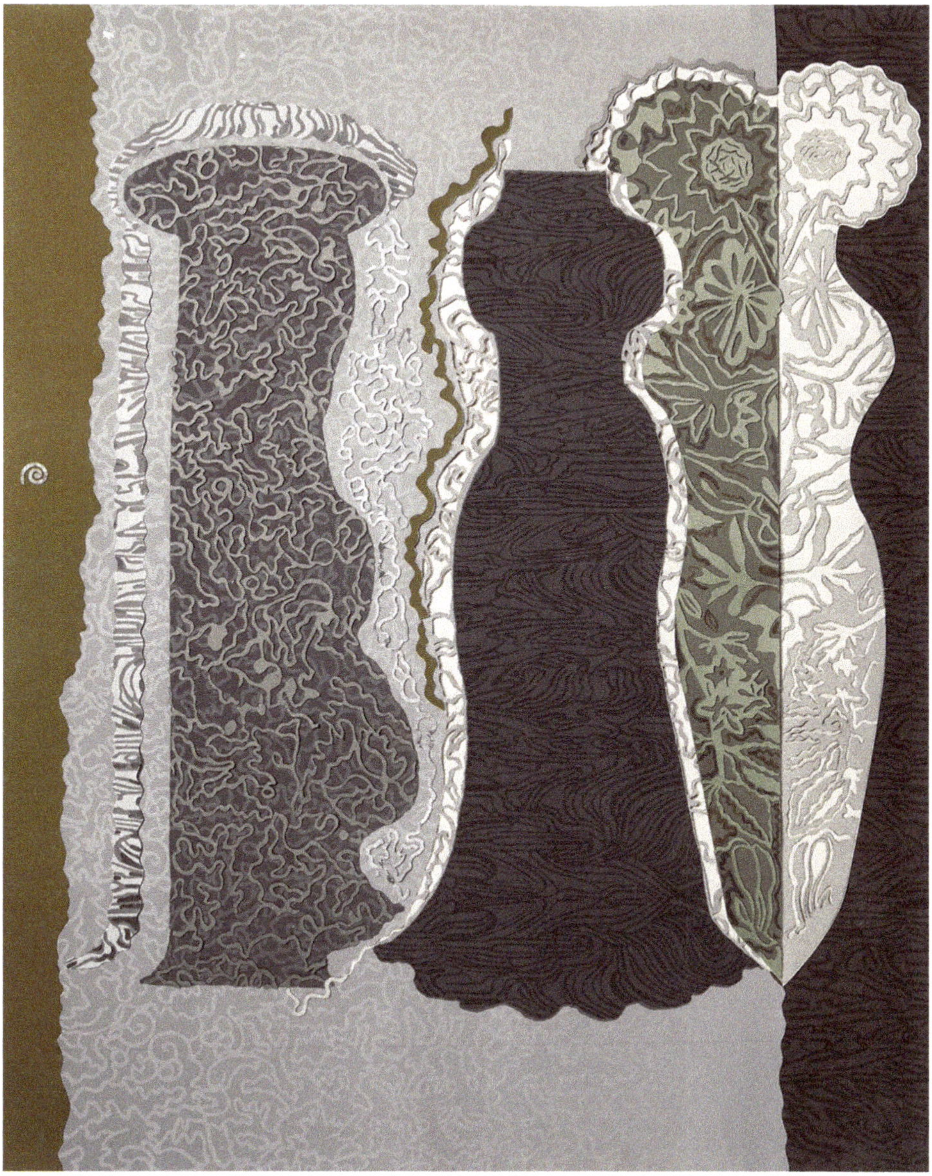

FIGURE 21.2: Charlotte Hodes, *Frieze* (detail), 2017. Papercut. 86.5 cm × 301 cm. © Charlotte Hodes.

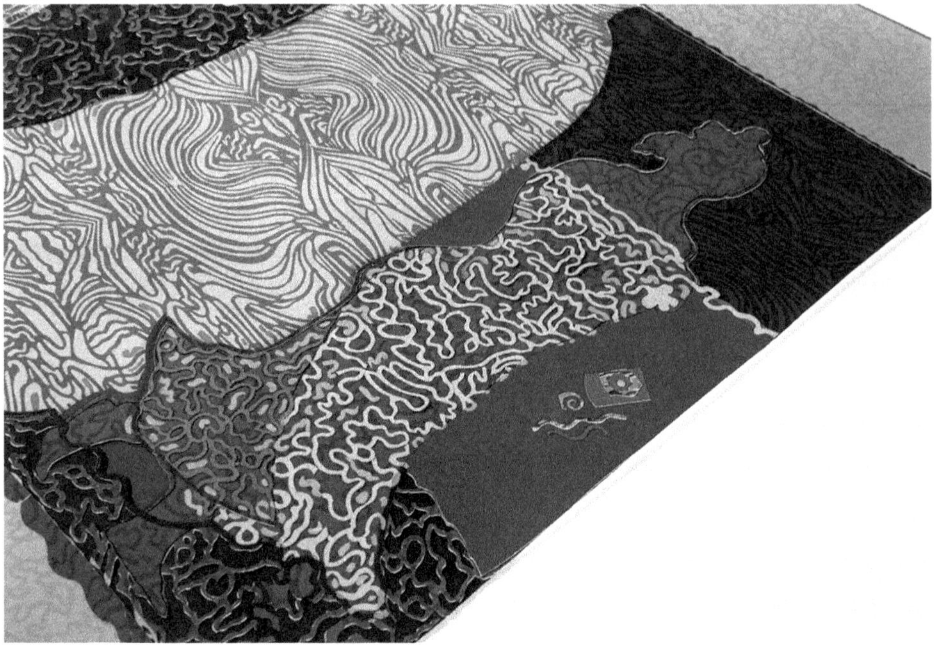

FIGURE 21.3: Charlotte Hodes, *Frieze* (detail), 2017. Papercut. 86.5 cm × 301 cm. © Charlotte Hodes.

22

Shatter

Zoë Hillyard

Disruption of normal patterns is a thought-provoking thread to look out for in works that involve the motif of destruction (Gray 2017:68).

Conceptually, 'ceramic patchwork' is a response to the repetitive patterns of consumption as evidenced by the chaotic bric-a-brac detritus found in charity shops and car boot sales – the sources of much of the materials I use. I shatter both ceramic forms and textile pattern compositions, re-structuring both to build fresh narratives surrounding our relationship to possessions.

The 'Elizabethan Vases' are part of a bespoke collection of contemporary craft commissioned by the British Museum. The institution is a bastion of culture with archives that provide glimpses of diverse ways of living expressed through the artefacts which have survived or are deemed worthy of collection. The means by which they have come to be in London, become selected for display and presented to us through curation, all reflect complex and politicized decision-making.

The underlying ceramic forms of the two vases are inexpensive and discarded mass-produced items, selected for their silhouettes. There is a certain liberation involved in the process of breaking them, an act that generates a network of cracks which eventually circumnavigate the form and result in un-predictably shaped sherds and dusty debris. Porcelain, stoneware and earthenware properties exert their personalities on the fault lines. Seemingly chaotic, in reality, they are logical lines of least resistance buried deep within the materials.

The imagery reworked is an exquisitely detailed 1592 etching of Queen Elizabeth I, which has been digitally printed onto silk. It is a power-portrait set against a backdrop depicting naval strength. The monarch dominates the structured composition, standing dressed in elaborate clothing made from intricate textiles, designed to convey wealth and status in every part of its rhythmic detailing. A personal ancestral connection to the exploitative Elizabethan adventurer, Sir Francis Drake, provides an additional dynamic to working with this imagery.

This historical narrative becomes fractured and reorientated as each ceramic sherd is individually wrapped with a section of silk and held in position by a tensioning web of stitching. Most of these stitches eventually sit hidden internally within the vase, straddling the concave cavity. However, when worked in reverse they sit on the outer surface, providing additional abstract qualities within the new pattern configuration. The neck ruff and other fragments of iconic Elizabethan imagery are still recognizable, but like traces of a dream, the scene has become disorientated and incomplete.

Unexpected colour and pattern alliances are at the heart of what often makes patchwork artefacts so wonderful. Yanagi Sōetsu, the connoisseur of beauty within Japanese craftsmanship, notes 'when intuition weakens, pattern becomes no more than a formal design' (2013: 174). Seeking spontaneity in how I cut sections from the silk, I distract myself with external stimulus, usually historical documentaries. I am looking for what Yanagi describes as a state of 'non-conceptualization'. It is challenging striving for serendipitous pattern associations within a process that is, by its nature, slow. I recognize my textile design training compounds this further, revealing the depth to which I am enslaved to a western concept of 'tasteful' coordination, constantly drawn to matching and balancing elements. In the same way that 'history is written by the victors', I accept that I am clearly still authoring these new compositions.

Holes and chips are important features of the two vases. These irregularities are the wounds inflicted by the 'tap-hit' impact of my hammer, the position and rhythm of which are only revealed when the vessels are fully reconstructed. As puncture points within the network of cracks, together they make evident a passage of time and represent a rebirth. The aesthetic I create is acknowledged as a contemporary approach to *kintsugi*, the Japanese ceramic mending tradition where breakages are celebrated with visible golden detail and treasured more as a result (Gray 2017; Kempske 2021). However, as the Elizabethan pattern-wrapped fragments are held together solely by stitch, the vases have an unsettling and surprising flex when cupped in the hand, distinctly different to traditionally repaired ceramics.

The British Museum is a place where the public comes to marvel at flawed, incomplete archaeological finds that have been painstakingly restored by conservators and is therefore an exciting context for showcasing contemporary work that explores

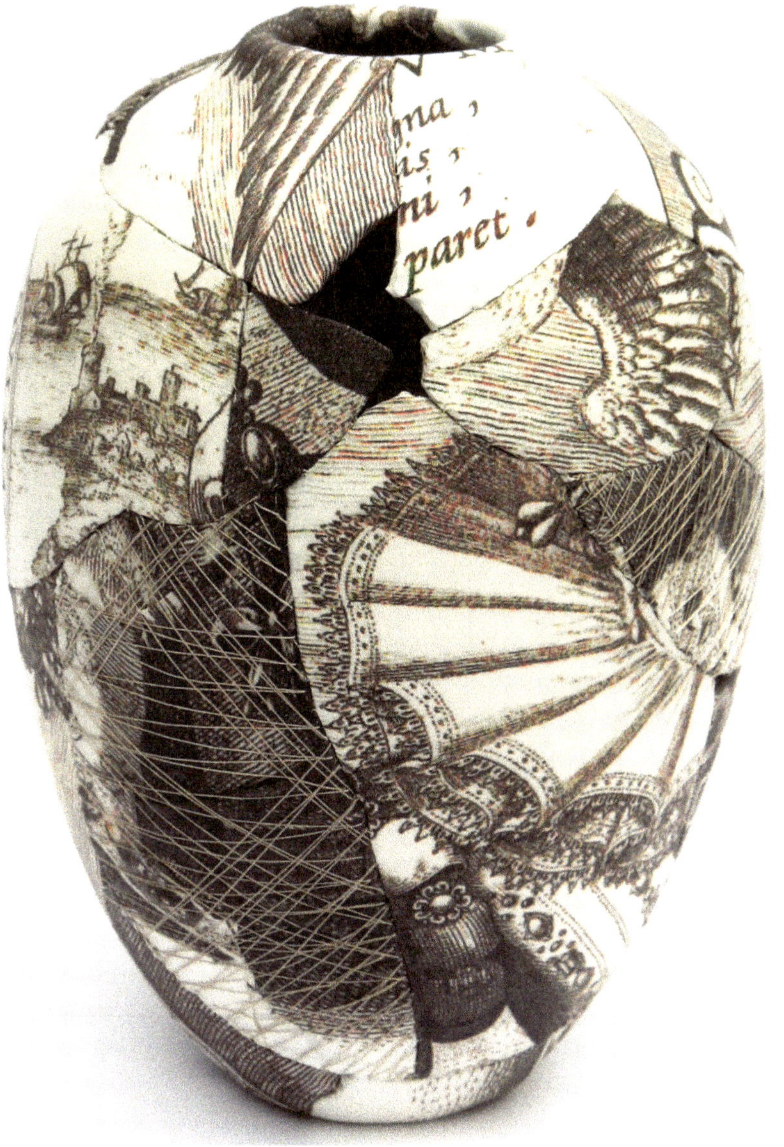

FIGURE 22.1: Zoë Hillyard, *Elizabethan Tall Vase*, 2012. Silk, ceramic and thread. 19 cm × 13 cm. Courtesy of Zoë Hillyard.

materiality, value and impermanence. At a time when we are discussing how inclusive and diverse narratives can better be built within our museums, room is certainly needed for disrupting familiar patterns so as to let unexpected and enriching new relationships reveal themselves.

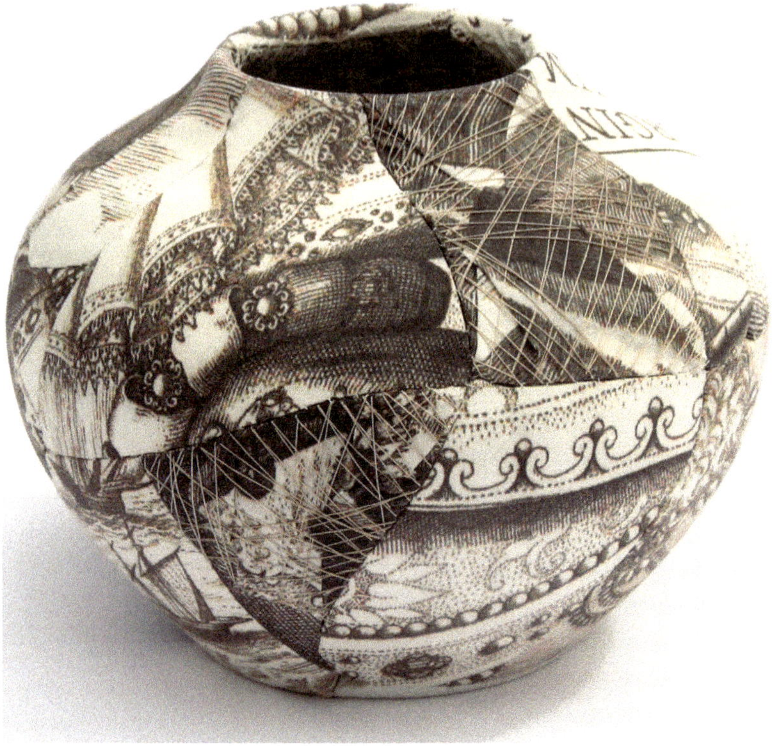

FIGURE 22.2: Zoë Hillyard, *Elizabethan Round Vase*, 2012. Silk, ceramic and thread. 12 cm × 16 cm. Courtesy of Zoë Hillyard.

The Elizabethan vases are the result of unremarkable mass-produced ceramics being used as a canvas for reconfiguring an established historical narrative, creating individual artistic statements through a mixture of disruption and care. Each is a creative journey that has reached a point of equilibrium with the trauma inflicted balanced by the craftsmanship invested and as such, can be seen as metaphors for acceptance, healing and renewal.

23

The Moments I am Looking For ...

Judith Stewart

Glimpses

In her 1927 diary, Virginia Woolf wrote of how travelling by car allowed her to alight 'accidentally' upon scenes that would have gone unrecorded 'save for this chance glimpse'. Her following sentence alludes to the potential power of such glimpses: 'Then it seems to me I am allowed to see the heart of the world uncovered for a moment' (Woolf 1982: 153). The power of such glimpses, which I regard as something different to the notion of 'seeing' or being 'seen', and the power relations associated with these, is more akin to Doreen Massey's 'throwntogetherness' (Massey 2005: 151). Such moments occur randomly when time, space and the gaze are brought together in an 'ever-shifting constellation of trajectories' in which chance plays a crucial role (Massey 2005: 151). Because Massey conceives of places as independently active, not requiring the human gaze to bring them to life, this throwntogetherness may create a specific moment in which the gaze is reflected back, not as 'an eye that I see looking at me [...] but as a gaze imagined by me' (Lacan 1981 cited in Krips 2010: 93). In such moments we catch a glimpse of what we are looking for, even if we didn't know it.

Looking

The serried ranks of books are topped by shelves containing boxes, stacked relatively neatly, of varied sizes and colours, all labelled. This apparent order continues when I open the boxes containing mini-DV tapes. Their labels: *Orford Ness, Paston, Rome, Spoleto* ... trigger memories of what I expect to find inside. Some labels have a title and 'Final' or 'Rough Edit' underneath, indicating the status of the work, but others – most if I am honest – generate a sense of foreboding as I know that this seeming order is only a thin veneer that masks the chaos sitting inside the plastic casing. I am looking for a particular short clip and anticipate several frustrating hours of looking through the tapes before I find it.

I think about this chaos every time I search for a particular bit of footage, especially on those occasions when the work involves academic research processes or funding applications. These processes assume an approach to research and making art that is pre-planned, the map from idea to artefact clear and straightforward. But, for me at least, the process of making is never linear and involves detours, returning to visual material collected in some past life, abandoning some things in favour of others, and sometimes it is all thrown back in the box and I start again. It is hard to explain this as a rational research methodology, but it is neither random nor 'intuitive' and generates its own intellectual and ethical rigour.

I know many artists are far more methodical than I am. But even though I get grouchy and swear to be more organized when surrounded by the chaos of half-labelled tapes and cameras and wires that don't match, having a slightly chaotic approach to storage has brought benefits. Many of these tapes contain footage recorded more than two decades ago, some using mismatched analogue technologies that have been digitized for ease of editing. Hence the tangle of wires. In trawling through, looking for something particular that I remember having filmed, I often stumble upon something forgotten, or discarded because it contained insufficient frames to use at the time, or was deemed too dull to work with.

October 1998 – Dublin

At the start of a month-long residency, I am standing in an empty studio on the first floor of a building overlooking the street. I haven't a clue what I am going to do or film whilst here and so I stare out of the tall window at the activity on the street below where there is a fruit stall and a man pushing a two-wheeled barrow. The barrow lurches, tips up and oranges spill out all over the road and pavement. Inevitably, my camera is in a bag and the moment lost.

The mess of tapes, cameras and wires that cover my floor mirrors the inside of my head and I have come to think of the process of making work as being a way to sort through hoarded mental images and references collected over a lifetime. It is here that the Artist Placement Group's mantra 'The context is half the work' has significance. While they were largely referring to the location of the work or artist, I prefer to think of this temporally. Looking back through the collection of recordings, I am conscious of how some footage has acquired a meaning far from its original context.

I note the repetitions: long minutes of nothing, tracks through the grass, processions of people doing ordinary but seemingly odd things, distant landscapes where the

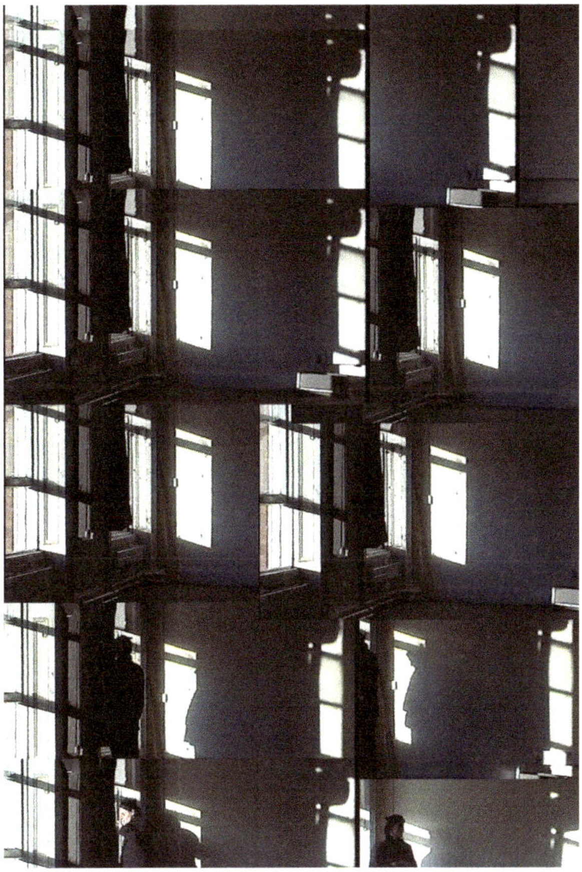

FIGURE 23.1: Judith Stewart, *Dublin 1998*, 2021. Stills taken from original Hi8 footage. © Judith Stewart.

movement generated by the extreme zoom combines with the handheld camera to create something akin to a heat haze. The sound of my breath reminds me of the ache in my arms as I try to hold them steady. At the time of filming, I did not know what I was attempting to capture but sometimes, either then or later, a connecting visual thread emerges, creating a link between then and now. What I also notice in trawling through hours of footage is the recurring cultural reference points that shape the thing we are looking for. So much for originality.

August 2001 – Sussex

I walk backwards and forwards across the Downs in the hope of finding something that lasts beyond this moment, that has the potential to resonate away from the immersive experience of the place. Coming down the chalk track, a former turnpike road, smoke drifts lazily across the path and disrupts the view. I start recording but am still uncertain about whether this is another clip that will find its way into the discarded box. The dog, who has been nothing but a nuisance to this point, appears from the hedge and shifts the dynamic.

I have come to enjoy these unexpected cultural connections where I notice things that echo scenes from films or paintings, and now see them as part of an overall pattern that provides another layer to the work. As my dog lopes across the path

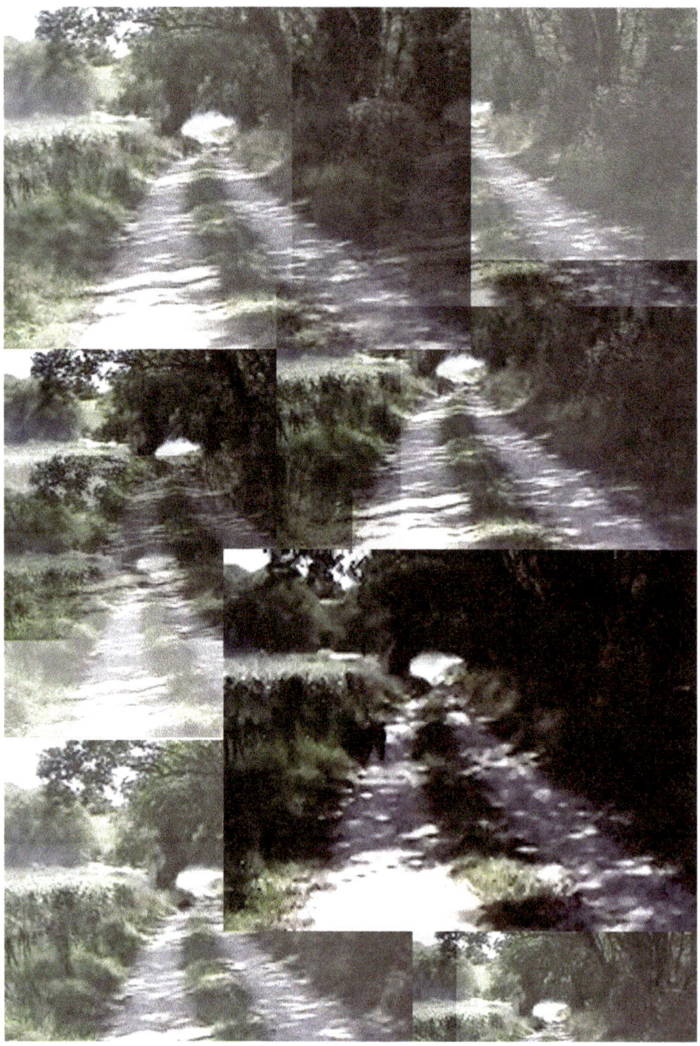

FIGURE 23.2: Judith Stewart, *Sussex 2001*, 2021. Stills taken from original DV footage. © Judith Stewart.

in the blurry footage, she creates a more mundane, English version of the scene in *Stalker* where the eponymous character lies in the water as the dog crosses behind him. I wonder whether Tarkovsky also had to endure his dog's intrusive behaviour before it suddenly provided him with the visuals he sought.

When I film, most of the time I do not set out with anything particular in mind. I am interested in capturing images – odd behaviours, fleeting events that, rather than

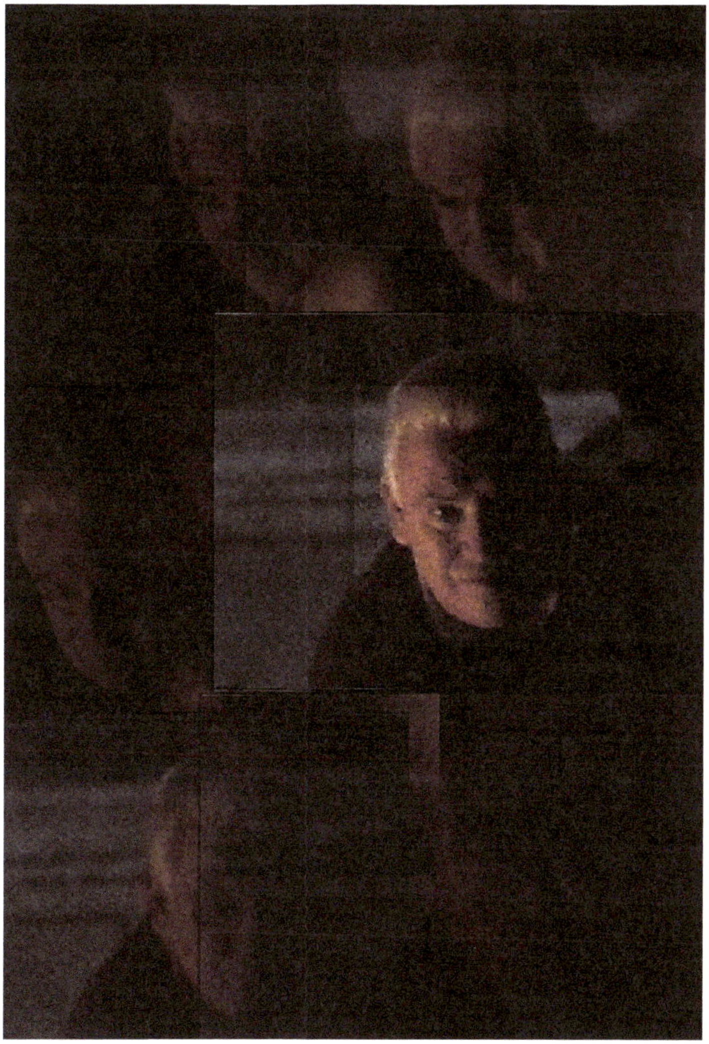

FIGURE 23.3: Judith Stewart, *Spoleto 1999*, 2021. Stills taken from original Hi8 footage. © Judith Stewart.

signifying random chaos, are half-remembered echoes of something once seen and brought back to consciousness. These images are, and yet are not, of the specific place and time.

1998 – Italy

A man stands on a stage in Piazza San Marco and seems to re-enact Domenico's speech in *Nostalgia*; in Spoleto the crowd gathering around a brazier in front of the Duomo have faces recognisable from Botticelli paintings. Merged with the crowd, I start recording, uncertain of how or whether I will use this footage.

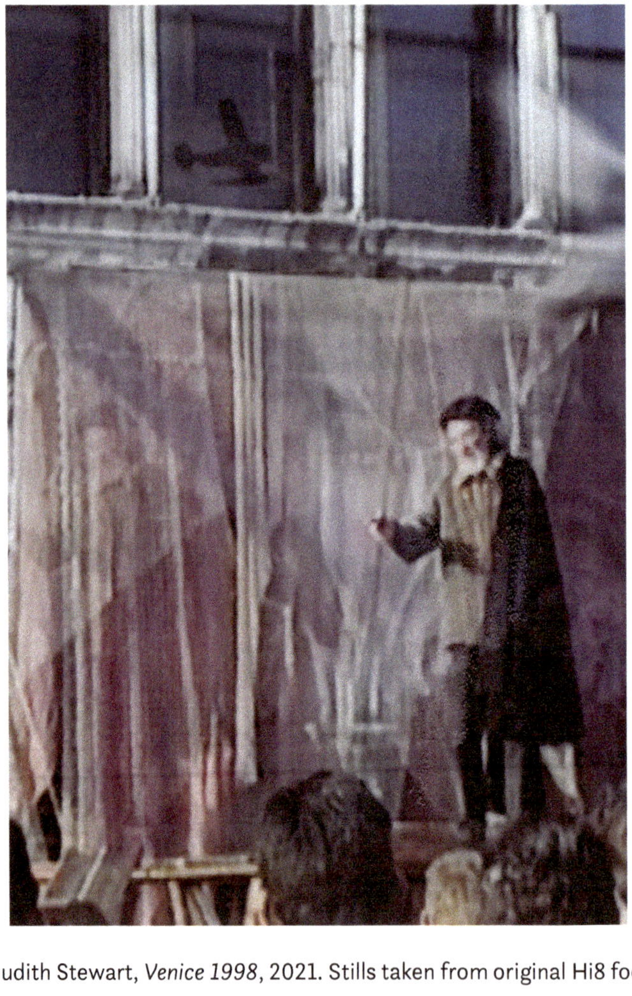

FIGURE 23.4: Judith Stewart, *Venice 1998*, 2021. Stills taken from original Hi8 footage. © Judith Stewart.

Seen

Looking at the footage again some twenty years later alongside other rough cuts of people filmed going about their business, I experience one of those moments of recognition; something that makes me feel rather idiotic for not having recognized it earlier. Having become aware that someone in the crowd of onlookers is filming for an unusually long time, the man raises his head. When the subject's gaze meets mine through the camera lens, I realize that it is this moment of being seen that provides the moment I was looking for.

24

Expanded Visuality: Photography as a Patterning Mechanism for the Animated Form

Katarina Andjelkovic

Diverse techniques and structures of animating the form have long been the concern of artistic expression, from attempts by early cinema at deconstructing life's movement, to the most recent features by groundbreaking digital technologies manipulating visual narratives. This text explores the conditions of expanded visuality in animating form, by looking at photography as a mechanism of the animate real. In the case of Idris Khan's photographs, I argue that his attempt to animate form in the image can be understood as a counteraction to George Bataille's 'formless' (1929b: 382), in anticipation of the mechanism for diagnosing the effects of digital technology on reopening the time of the image. Khan's choice is to capture and describe the major monuments of durability of the present world – The British Parliament, St. Paul's Cathedral, The London Bridge – through the dimension of their instability. In this context, Idris Khan's use of repetition is not so much about artistic image, as he expresses a desire to see the transformative image of space in our experience. In other words, in giving his photographs an intelligible expression for the construction of new spatial narratives, we can read Khan's method as patterning. In negotiating the animate real through his photographs, I analyze how the production of visual knowledge configures the expression for what is 'alive' in photography, carried out in the moment-to-moment, repeated, concentrated, overlapped.

In the context of visual analysis, the noun pattern normally refers to any regularly repeated arrangement, a particular way in which something is done, is organized or happens. The verb patterning (to give a regular or intelligible form) is critical because it can be used to look at photography's actions and operations to study the relationships between virtual dimensions of photography and concrete physical realities in contemporary discourses on architecture. This text discusses how constructed images that suggest a new reading of existing architectural reality may be used to design a method of 'patterning'. Analyzing how the production of visual knowledge configures the expression for what is 'alive', this method is understood as a way to give an intelligible expression for the transformative cultural and material conditions of today.

We can start by claiming that pattern is double-faced: it both inspires creativity and is critical. It gives artistic expression and examines routinized practices of lived experience. It gives an intelligible form and can unsettle hierarchies. It alludes to a repetitive labour and is any regularly repeated arrangement. Pattern is a set of instructions to be followed. Art and architecture navigate pattern as a rhythmic play that underpins everyday routines, connecting them to the world through the built environment. Thus, they transform into their own representations in photography and other types of visual images. Cultural and material conditions set the tone for the repeated operations or an aesthetics of intervals, disruptions and resumptions, a duration or a differentiated time. These conditions imply a process of repetition which can now embrace a full potential of patterning.

Constructing an image is an activity that develops a deeper understanding of the various visual media of expression such as photography or painting. My interest lies in encoding the verb 'patterning', as a process of repeated actions and operations that imply the inscription of temporality. The forms we see in Khan's photographs (Figure 24.1) evade a linear flow of sequential statements and remain stubbornly simultaneous, searching for an expression of new reality and opening the isolated

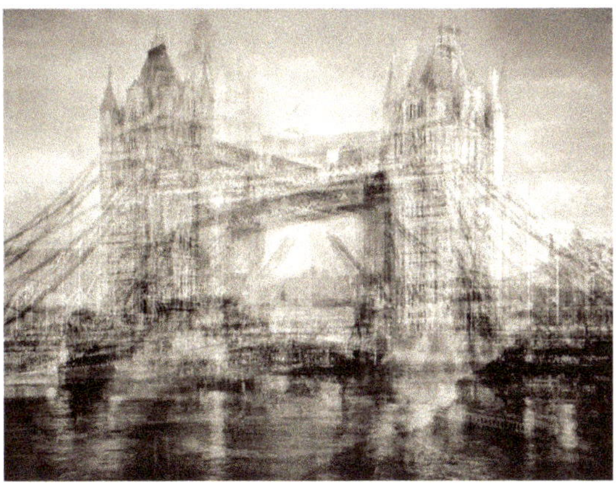

FIGURE 24.1: Idris Khan, *Tower Bridge, London*, 2015. Platinum-Palladium print on Fabriano Artistico E.W.S 300 gsm paper. Paper and image 61.5 cm × 76.5 cm. Edition of 37. Courtesy Idris Khan and Cristea Roberts Gallery, London.

moment in time and established visual layering. By detecting a representation of time in the painting, Sabeth Buchmann rightly pointed out that 'repetition invokes a process-related way of painting-in-progress with endless new beginnings and overworking that point to the impossibility of its completion' (Buchmann 2016: 159). In that sense, Idris Khan's disintegration of photographic composition can be related to Bergson's understanding of painting as thinking through time and to Edgar Degas's pre-cinematic method of temporalizing space as animating form.

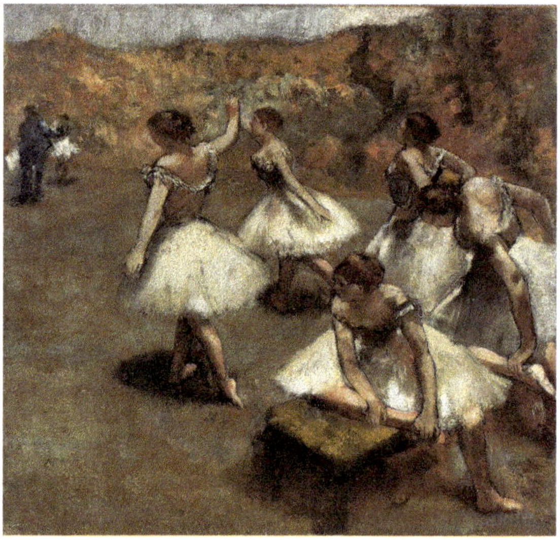

FIGURE 24.2: Edgar Degas, *Danseuses sur la scène*, c.1889. Oil on canvas. 78 cm × 82 cm. © *L'encyclopédie libre*.

By applying the technique of temporalizing space in a denaturalized way, Degas was a forerunner of the method of modern cinema. If, according to Bergson, 'there are ways of thinking through time when observing the painting' (Bergson 1968), in Degas's hands, these methods staged the very medium of painting as a transitory moving set of rules (Graw and Lajer-Burcharth 2016: 159). His choice of repetitive study of the dance rehearsals (Figure 24.2) is conducted to observe and analyze a subject that is structurally based on repetitive movements and reinvention. Creating an ongoing process of constructing, revising and defining rules, which precede his painterly 'choreography', could be compared with the dance rehearsal that happens before the public performance. Not least the *topos* of the rehearsal allowed Degas to visualize often long periods of time his compositions needed to be 'completed', or, to refer to David Joselit's and Isabelle Graw's conference papers (Graw and Lajer-Burcharth 2016: 159), such modes of depicting the dance rehearsal can be seen as a way to store labour time. This raises the question of whether and how the timing of the ballet rehearsal can reveal a more functional than substance-oriented or even a conceptual idea of the medium of painting (Buchmann 2016: 166). This way, Degas brought together qualitatively different practices into the medium of painting to testify time as a metaphor for repeated actions.

Affected by the human since the dawn of the Industrial Revolution, as atmospheric chemist Paul Crutzen claimed in 2002 (Crutzen 2002: 23), the planet entered a new geological epoch called the Anthropocene. The *anthropos* had overpowered nature in this period and the long-established axiomatic divides between humans and nature, living and non-living, global and local were infringed. These binaries were also recognized as the basic tool in Khan's work on establishing new conceptual relationships with the space. In search of an alternative understanding of visual space, he is subverting the idea of stabilizing human life that characterizes the human-made. Driven by the Heraclitean thesis that man can never enter the same stream twice, he operates like a critical cartographer who explores the Anthropocene by charting shifts in the ways that humans move through, dwell in, and dream of space and place in a world where our grasp of visuality of space has been fundamentally shaken. Khan's choice is to capture and describe the major monuments of durability of the present world (The British Parliament, St. Paul's Cathedral, The London Bridge) through the dimension of instability. Once symbols of durability and solidity, these now enter the artistic language to become the subject of ephemerality, temporality and friability. Moving along major contemporary issues, such as ecological instability, socially deviant posthumanism, global circulation of garbage and uncontrollable phenomena caused by the work of natural forces, Khan's approach to the medium of photography embodies architecture as an animated form of life, as both theme and medium, in light of the new forms of consumption, disposal, interaction and the effects of contamination in the age of the Anthropocene. In this context, Idris Khan's photographs are not so much about artistic image as they express the desire to see the changing image of space in our experience of everyday life. In negotiating the 'epitomizing life' through his photographs, we read how the production of visual knowledge configures the expression for what is 'alive' in photography, carried out in a moment-to-moment, repeated, concentrated, overlapped.

In a not dissimilar way to Degas, the animated form of depicted architectural monuments in Khan's images testifies to the 'choreographing' of time in painting and photography. His method undoubtedly pleads for the current epoch characterized by the profound transformation of land, and his monuments of durability seem hard to resist. Much like instrumentalizing the motion picture, Khan addresses this unstable rapidly changing experience in our collective imagination through a sequential visual representation, configured in the repetitive process of manipulating photographs.

By focusing on its temporal aspects, he confirms that human artifice is not absolute, eternal or solid. Khan's artwork sees architecture as part of this processuality. His work, be it an implementation of new narratives or stubborn exploitation of old ones, speaks of a fascination with the layers of photographic materiality that reveal some new meaning in lines of writing. By introducing these integrated spaces of change, the architecture becomes an agent that alleviates the current pressure of accelerated life adaptations and time scenarios. Perceiving, capturing and constructing reality in a repetitive fashion in the binary game of life and apocalypse, Khan's photographs pay homage to Deleuze's scepticism regarding the 'survival' of the *time-image* of cinema in the digital age (Deleuze 1968). His concepts and representations seem to synchronize our nostalgia for visually reinventing architecture.

The image is distilled until it reveals some new truth. Lines in the foreground are replaced by their alternate appearances in the background, in the manner of manipulating elements like 'fire' while 'burning' through the layers of the visual display. While layering the image, parts of the photograph are blurred and disappear, but gradually a space and an image emerge which gather traces of overlapping over time. Accordingly, a representation of time in the photograph is attained through narrative and anti-narrative visual devices, repetition and layering of images.

To conclude, Idris Khan's method provides a new intellectual environment for rethinking the role that photographs may play in visually reinventing architecture in the age of the Anthropocene. In other words, to relate our human experience in architectural and artistic terms, we remain caught in an intellectual pattern that has recurred from Descartes to the poststructuralists. In that sense, Khan's critique of the Anthropocene reveals the relation between humans and technologies to be one of mutual construction rather than domination, or in Esther Leslie's words, 'in animated nature, technology and magic are one' (Leslie 2014: 29). Moreover, Khan's method discloses the potential of art and the digital manipulation of an image as a means to rethink how the universe of image representation configures the expression for what is 'alive' in photography, carried out in a moment-to-moment, repeated, concentrated and overlapped. His production of visual knowledge, as a kind of epistemically informed imagination and the construction of new spatial narratives of the Anthropocene in the image, was patterned for the purpose of critically thinking through contemporary changes. Allowing a new perception of what exists, the method of patterning can be used critically to make intelligible the transformative cultural and material conditions of the Anthropocene. Furthermore, this method unearths significant issues and raises challenges to better understand how society's historical and cultural processes in each era decisively influence the type and use of visual constructs.

PART 4

VIRUS

Introduction

The reproducibility of viruses must be one of the most extreme and dangerous examples of pattern making. The voracious nature of HIV, SARS and COVID-19 uncomfortably demonstrate the power of viruses and the challenges they pose to our sense of equilibrium. Repeating, whilst also mutating, might be seen in some ways to link to themes in Part 2: 'Morph, Flux, Mutate', but one significant difference here is the ability of the virus to hijack and harness the energy of a host. As demonstrated in these contributions contagion is seen to be a powerful force.

With respect to repetition and difference, there are profound ethical implications when it comes to our relationships with fellow humans and animals, and respect for our planet. The inequalities highlighted by the spread of the cholera virus are explored by Anne Eggebert who uses ProMED mail alerts and Google Earth to engage with locations of cholera outbreaks. Her drawings describe a geography of accidental shapes, patterns and desire lines that draw attention to 'patterns of inaction and privation'. They highlight the ways 'societal roles and environments we are bound to determine our vulnerability to disease'. Shifting focus from the macro to the micro, the inexorability and penetrative nature of viruses is explored in the animation of Daksha Patel. Through coding, Patel harnesses the pattern-making potential of algorithms to infiltrate and manipulate her animation, echoing the unpredictability of the virus to affect the body. She utilizes computer-generated coding as a 'viral' force that can manipulate and transform a drawing.

Echoing the semi-controlled environment of the studio, Louise Mackenzie presents a situated account of working with DNA, viruses and microbes in a genetics laboratory. She describes, through speculative reflections, the complexities of attempting to translate a material that we are so inherently entangled with that we can never fully comprehend. The power of infiltration exemplified by viruses provides a vehicle for Andrew Bracey's exploration of the connections between virus and appropriation, whereby patterns 'spread, take over hosts, expand across borders and are underlined by logic but open to chaos' through specific reference to the work of artists Philip Taaffe, David Mabb and R. H. Quaytman.

25

Global Ghost Map

Anne Eggebert

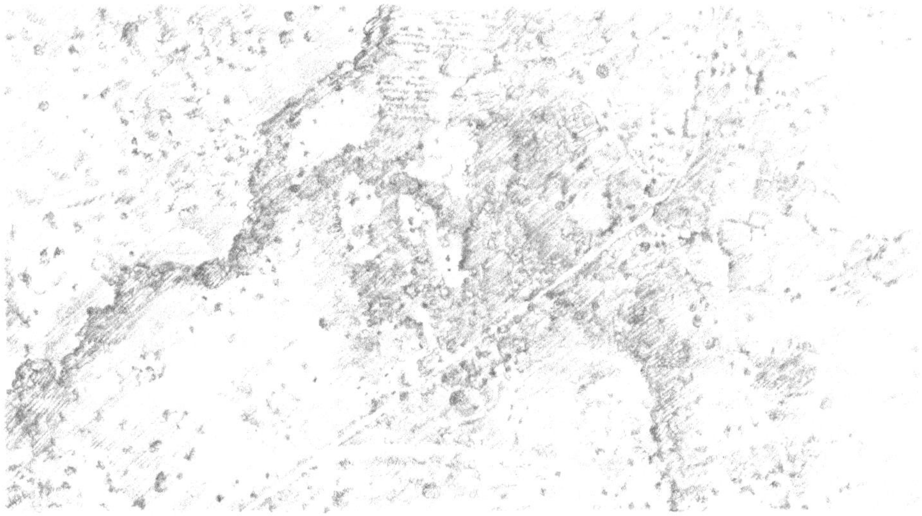

FIGURE 25.1: Anne Eggebert, *Global Ghost Map* (Boke, Guinea), 2012–13 (one from series). Pencil on paper. 210 mm × 297 mm. © Anne Eggebert. Original source image: © 2012 Google, © 2012 Cnes / Spot Image.

These drawings were originally made in response to ideas of mapping cholera for *Cartographies of Life and Death* at the London School of Hygiene and Tropical Medicine – an exhibition in 2013 celebrating the two-hundredth anniversary of John Snow's birth. During the 1854 cholera epidemic in Soho, which claimed the lives of over 600 people, Snow developed his 'ghost map' by drawing on the knowledge of the street (the word and walking) and death returns to build a map that evidenced cholera as a waterborne disease. Modern epidemiological methods evolved from his work which identified the water pump in Broadwick Street as the source of the outbreak. My drawings stemmed from a critical reflection on the celebration: whilst we know that *Toxigenic Vibrio cholerae*, the pathogen that causes cholera in humans, can be eradicated with clean drinking water there continue to be outbreaks around the world. I planned to draw the location of each of these as they occurred between the point of commission and the opening of the exhibition – I couldn't keep up with the number of outbreaks.

I used ProMED-mail alerts and Google Earth to find the locations of the present-day cholera outbreaks. Allowing the search engine to guide me to the named region flagged up in each of the automated ProMED e-mails, I then searched for signs of human habitation, floating over the surface seeking out a geography of structures or accidental shapes, patterns and desire lines, where details of dwellings might come into view. These traces of matter's movements – from a rural structure likely pulled together from local materials, criss-crossing bare-earth paths flattened and scuffed by multitudinous footsteps (perhaps both human and domestic animal), to machine-made city forms that overlay the soil, cut into the geology of a region – reveal the patterns of human co-existence with, or conurbational obliteration of, planetary surfaces. How different those urban spaces would appear if we greened all their sky-facing surfaces, and softened those edges.

The representation of landscape carries its cultural burdens and has its own complex and systemic patterns of history – the socio-political implications associated with historical representations speak of ownership, navigation and colonialization[1] – whilst contemporary mapping systems pinpoint drone or missile strikes, reveal habitat destruction and fracturing ice shelves, plot our daily runs, identify the locations of those in desperate need (London School of Hygiene and Tropical Medicine 2013) or allow us to poke about the surrounds of a holiday villa, amongst other things. These digital representations of place are imbued with both the local and the global, the located self and the other. We simultaneously sit still and move around on the spot, looking. Everyday traces of human action are mapped out at a moment in time (Google's illusion of the present through the physical movement of our mouse belies the photographic past of the almost seamless image) or in real time as we pound through the park or walk the city streets – or both at the moment we see our digital location marker revealing where we are on a satellite image taken perhaps months ago. If we are in prime marketing territory our landscape images are refreshed again and again. If we have little spending power the images are often poor, unfocused. Teletechnology's bird's-eye view of the terrain, on the face of it, is nothing to do with the picturesque. Invisible Committee suggest in their essay 'Fuck Off, Google', that 'An enterprise that maps the planet Earth, sending its teams into every street of every one of its towns, cannot have purely commercial aims. One never maps a territory that one doesn't contemplate appropriating' (The Invisible Committee 2014: n.pag.). Nonetheless, in spite of these patterns of behaviour, swirls, rhythms, eddies and flows are revealed as we float over the landscapes of our planet. I stop at a cluster of rectilinear structures, a space of soft edges hovering in the soft blue glow of the screen.

Each mark of my pencil becomes a trace that directs me to loss and survival, with the intention of creating a subjective connection to the impact of cholera. I work with repeated hatch lines – a pattern that accumulates to form the aerial view – they are not expressive, but rather programmatic and repetitive. Google's digital images are re-rendered by hand from screen to paper as I reflect on distance and experience with the intention to signify what is happening elsewhere. We could consider the human act of moving about a locale as drawing – tracing these movements through a shifting of material on the surface of the world – the landscape as ground in both meanings of the term. Then the process of copying from the digitized aerial view of this collective drawing might be understood differently – as a phenomenological process – that translates the movement of the body there, to the movement of the body here. Sliding these places across my screen, all equal in their flatness, I mark out their planar formats, bringing the world into view through a drawn collection of diverse patterns and traces, an archive of an outbreak. The experience of intentionality, a meditative action of translation (Jaffray 2019) through mark making, can bring us to the acknowledgement that life in that location matters.

I'm curious how digital encounters with other landscapes impact on our sense of placeness and connection to the other; and whether or not these images simply contribute to the proliferation of images of a place that somehow become its unmaking. Can they also be a tool for subjective responses to the location of the distant human? Distance itself has become such a challenging space for so many under the impact of the latest pathogen's invisible journey. In recent months I regularly find myself seated on a plastic patio chair in the grounds of a care home, looking in through a window where a frail woman sits looking back at me. We are separated by another screen between here and there, distance and proximity, my reflection merging with her face beyond the glass. Where does the here end and the there begin?

These pathogens, *Toxigenic Vibrio cholerae* and SARS-CoV-2, are both not as indiscriminate as we might like to think. Our patterns of inaction, privation and the impacts of poverty, the need for clean water, and the societal roles and environments we are bound to, all of these things determine our vulnerability to disease. Ecocide too has its patterns and traces – most of these can be seen clearly with hindsight. If we look back at the systemic abuses of our landscapes, resources, biomes and fellow humans, we can see that our most novel of pathogens, along with its attendant disease COVID-19, is a harbinger. As Coole and Frost outline in their introduction to *New Materialisms* (2010), 'Complexity theory is playing an increasingly significant

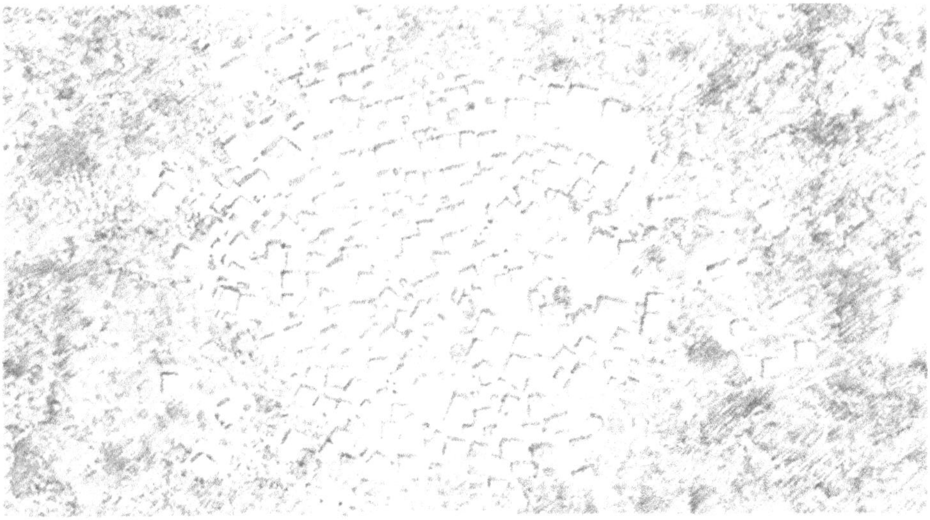

FIGURE 25.2: Anne Eggebert, *Global Ghost Map* (Sierra Leone), 2012–13 (one from series). Pencil on paper. 210 mm × 297 mm. © Anne Eggebert. Original source image: Imagery © 2012 Digital Global, GeoEye.

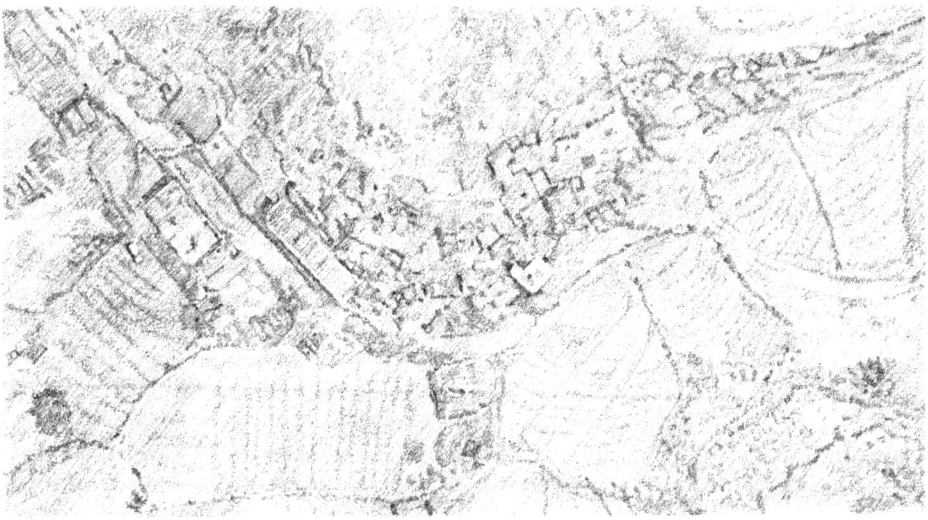

FIGURE 25.3: Anne Eggebert, *Global Ghost Map* (Jammu and Kashmir), 2012–13 (one from series). Pencil on paper. 210 mm × 297 mm. © Anne Eggebert. Original source image: Imagery ©2012 GeoEye.

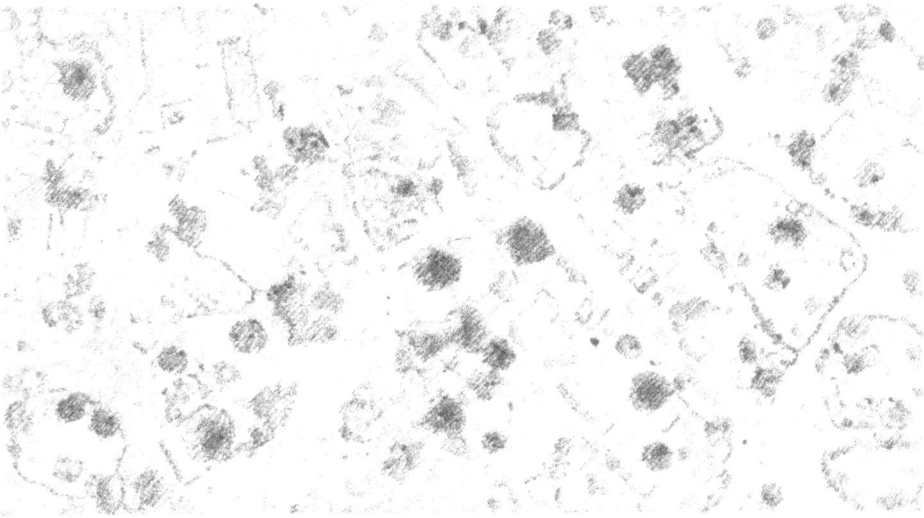

FIGURE 25.4: Anne Eggebert, *Global Ghost Map* (Somalia), 2012–13 (one from series). Pencil on paper. 210 mm × 297 mm. © Anne Eggebert. Original source image: Imagery © 2012 GeoEye, Maps Data: ©2012 Google.

role in understanding socio-material processes, too, because it appreciates their inextricability from a wider natural environment' (Coole and Frost 2010: 13). If we are open to visualizing the patterns that are unfolding, to empathizing with those in other parts of the world, we may begin to recognize the urgency of our responses to the chaos of the Climate Emergency – whether we live in the wilderness, rural or urban locations, climate change can kill us in myriad ways. Insights offered by the arts and humanities can raise cultural consciousness and make empathetic connections to other subjectivities – together we can rethink what Planetary Health might really mean.

NOTE

1. Denis Cosgrove's *Social Formation and Symbolic Landscape* (1984) is a really useful resource here.

26

Embodied and Coded: Drawings as Viral Systems

Daksha Patel

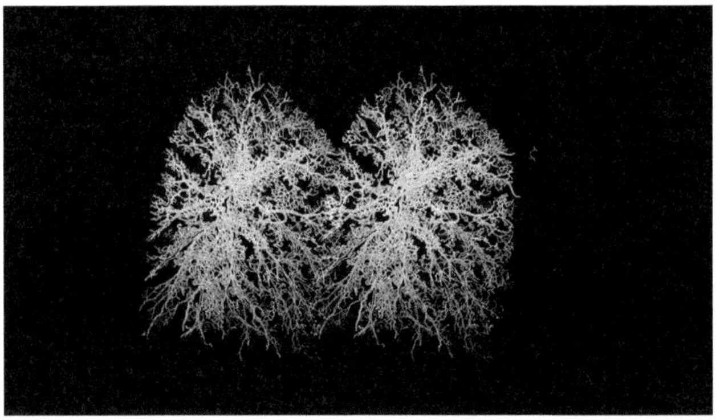

FIGURE 26.1: Daksha Patel, *in the same breath*, 2020. Video still (1 mins 9 secs). © Daksha Patel.

Scientific accounts of a virus can be contradictory and are to some extent contested. They are variously described as 'non-living', 'inert', 'dormant', 'on the verge of life' and 'very much alive' (Pearson 2008; Passalacqua et al. 2019; Villarreal 2004; Bhella 2016). Debates about whether viruses are living things centre upon standard definitions of living organisms such as an ability to reproduce, to metabolize and maintain homeostasis. Viruses occupy a liminal space that does not conform to these standards. They are inert and cannot reproduce (multiply) until they penetrate a living cell. Once inside a cell, they display many of the standard characteristics of life. However, definitions of what constitutes a living thing are also contested and scientists increasingly acknowledge the 'changeable and blurry edges of living things' (Ferreira and Umerez 2018).

The intersections between biomedical definitions of a virus and algorithms that infiltrate the 'body' of an artist's drawing to generate animated sequences are explored in this essay. I question whether the method of creative coding can be conceived of as a viral force that penetrates, multiplies and transforms a drawing. The viral metaphor is used as an interdisciplinary framework for conceptualizing sequences of data-generated drawings from the animation *in the same breath* (Patel 2020), which was produced in collaboration with creative coder Chris Ball. The work was commissioned by Animate Projects, with a brief to respond to the COVID-19 pandemic. My decision to use coding as a methodology for the animation was in response to the use of algorithms and data throughout the pandemic to predict future outbreaks and monitor the spread of the virus.

A (biological or biochemical) virus can manipulate the information contained within a living cell to generate new sets of instructions, thereby changing its behaviour and ultimately, the system it is part of. Its functions can undoubtedly be associated with the ways in which code can change patterns of behaviour within a computer system, thereby causing it to generate new patterns of data. Both manipulate information, whether it is encoded in DNA or an algorithm. The similarities between biological and digital viruses are evident in their shared terminology and etymology, which is from the Latin for 'poison and venom' (etymonline 2021). The sinister root meaning reflects their ability to corrupt the systems (human or digital) they infiltrate. Computer viruses are today a very real threat that many of us negotiate on a daily basis to maintain our digital and economic health, much as COVID-19 is a threat to our biological health. Whilst malware (malicious software) is associated with covertness and criminality, the code involved in its operation is in fact neutral: it is simply a set of instructions. It is the effects of the virus (upon computer systems and upon the body) that are harmful.

Creative coding is a growing methodology in digital art and design practices. It is used to generate visualizations, sounds, text-based art, architectural designs and light installations to name just a few of its applications. Architects, product designers, graphic designers and fine artists have all used creative coding as part of their process; it is a shared method in functional design, fine arts practice and artworks that blur boundaries between them. Fine arts practice that uses coding is often linked to the Generative Art genre, particularly when a degree of autonomy is implicated in the process. Definitions of Generative Art are complex and reflect the diversity of practices involved. They do not necessarily involve digital or hi-tech equipment. Philip Galanter argues that examples of Generative Art can be found in art history prior to the advent of digitization. He defines it as:

> any arts practice where the artist uses a system, such as a set of natural language rules, a computer program, a machine, or other procedural invention, which is set into motion with some degree of autonomy contributing to or resulting in a completed work of art (Galanter 2003: 4).

The element of autonomy in the creative coding process is, in my experience, central to why it is so intriguing and fascinating to work with. It produces a sense of anticipation and suspense because the outcome of each sequence cannot be fully controlled or known. For *in the same breath*, I planned timings and set parameters for each animated sequence. However, once code enters a system (my drawings for instance) it interacts with it in ways that can be unexpected. Each sequence was surprising and produced unforeseen elements. My aim here is not to problematize notions of authorship, which is a wider debate within the Generative Art genre. Nevertheless, ideas about lack of control and unknown outcomes resonate with the pandemic theme. Scientists can predict a number of outcomes when a virus enters a body (or a population); however, each body will respond uniquely and differently. Predictions and probable outcomes based upon data are not certainties, and the element of unknown remains.

In the early stages of the making of *in the same breath*, we tested different drawings with exactly the same code and each one behaved differently. Surprisingly, the simplest drawings were the most difficult to work with, whilst more complex drawings appeared to flow more naturally. 'Flows of information' are commonly referred to in coding and the pattern formations that coding can generate can at times resemble intricate biomorphic forms. This seems particularly apt because many of the original drawings are of internal bodies and are inspired by medical visualizations.

Figure 26.2 is a still from one of our first animation tests involving a very short sequence of three seconds duration, responding to a drawing of a heart. It is shown

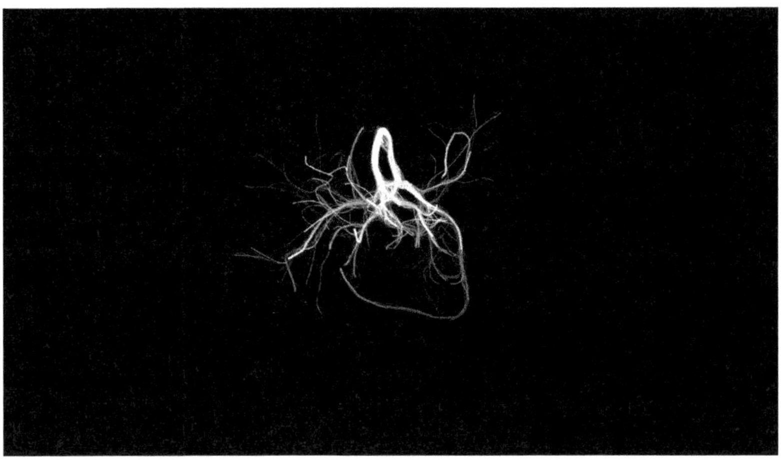

FIGURE 26.2: Daksha Patel, untitled, 2020. Test animation (3 secs). © Daksha Patel.

here as an example of how difficult it can be to replicate the line formations and patterns generated by the code. After seeing this sequence, I was keen to reproduce the same types of lines and movements within other drawings. However, despite numerous attempts, it became evident that it was impossible to duplicate the qualities of this sequence in other drawings. There were too many unpredictable elements: the different lengths of lines, their directions or the connections between them are just some of the factors that shape the animated sequences. Each drawing is its own system.

My animation process began with a traditional hand-drawn storyboard, from which I developed more detailed drawings for key stages in the timeline. These drawings became the framework for the coded sequences of lines. The code created by Chris Ball is called a ‘line follower’: it generates thousands of very fine lines that flow along different pathways dictated by my original drawing. Each line follower has three sensors that can detect pixels from the original drawing in its path – a central sensor that can ‘see’ straight ahead, one that can ‘see’ to the right and one to the left. If the central sensor ‘sees’ a pixel, its command is: go straight and speed up. If the left sensor ‘sees’ a pixel, its command is to slow down and turn left. If no pixels are detected, the command is to turn a random amount. Chris explicates his process further:

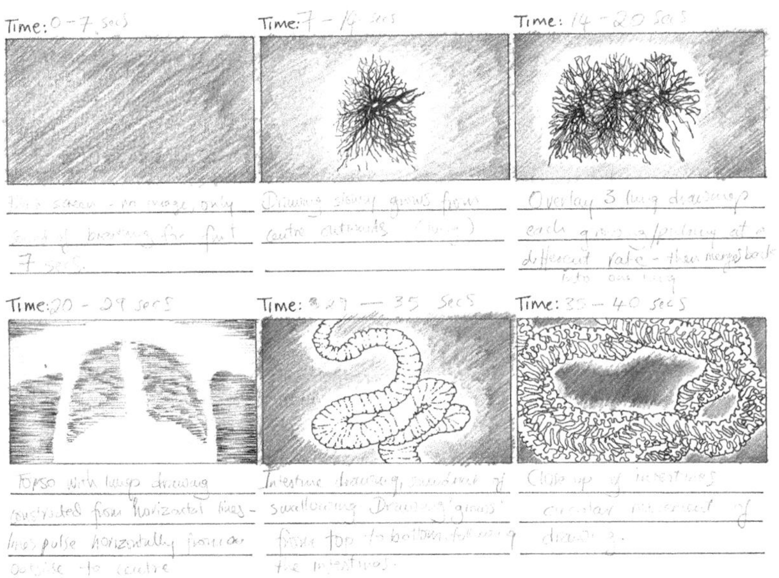

FIGURE 26.3: Daksha Patel, first storyboard for animation *in the same breath*, 2020. © Daksha Patel.

The line followers also have a 'confidence' value – consecutive successful line detections increase their confidence. High confidence means they search with more distant and narrow sensors. Zero confidence means they 'die' and are recreated at the last position they were maximally confident (Chris Ball, e-mail, April 2021).

The method described earlier already contains unpredictable elements. Allocating random values to the process using a random generator code increases the unpredictability. It is used:

- to place line followers at a random angle and distance from the 'spawn point'
- to give a random maximum size to each line
- to give a random direction of travel at the start of each sequence
- to vary their angle of travel as they 'search' for a line (Chris Ball, e-mail, April 2021).

A 'spawn area' (clusters of dots which I added to each drawing) indicated the section within each drawing from which the digital lines could emerge, whilst the whole drawing provided the framework that thousands of coded lines could follow. The timing of each sequence was aligned to the soundtrack, which represents a pattern formation of data. The audio was split into frame length sequences and average frame values based on volume data dictate the speed of lines and their growth.

I recall the first time I saw the line follower code in a very simple visualization where it was assigned a 'dot' motif with three eyes (sensors to the left, centre and right). I was struck by how 'lively' it seemed; it resembled the movements of an ant searching for food. It was quite uncanny! The 'life' metaphor is evident in the language we used to describe its functions: it 'sees' pixels, can 'multiply and search', has 'confidence' and can 'spawn' or 'die'. Furthermore, the randomness built into the code seems to give it a kind of agency because its movements are unpredictable. The code is not unlike the DNA coiled within a biological virus, which enables it to multiply inside a living cell.

The blurring of boundaries between digital interfaces and embodied experience has been theorized widely. Our interactions with digital technologies have moved beyond simple finger tapping on keyboards to whole-body movements where perceptions of the body in space are disrupted and displaced. However, vision alone can implicate other senses as Laura Marks observes. In *Skin of the Film* (2000), she theorizes about a haptic visuality that is embodied, multisensory and tactile where 'the eyes themselves function like organs of touch' (2000: 162). Marks distinguishes between an optic visuality which 'depends on a separation between the viewing subject and the object' and a haptic visuality which 'tends to move over the surface of its object rather than to plunge

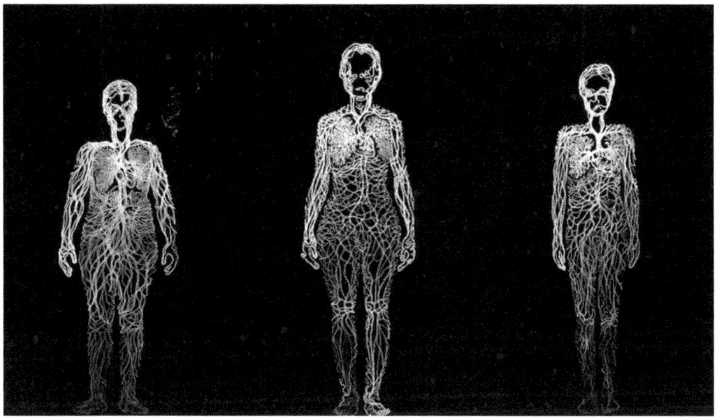

FIGURE 26.4: Daksha Patel, *in the same breath*, 2020. Video still (2 mins 40 secs). © Daksha Patel.

into illusionistic depth, not to distinguish form so much as texture. It is more inclined to move than to focus, more inclined to graze than to gaze'. Marks's assertion that haptic cinema 'encourages a bodily relationship between the viewer and the image' (2000: 164) resonates with *in the same breath*. A 'graze' aptly describes the eyes moving over the linear flows within the animation. Textural close-ups of the internal body lead the viewer to 'contemplate the image itself instead of being pulled into narrative' (2000: 163).

The visual style of the animation, with its delicate line work and monochrome palette, gives it a hand-drawn aesthetic. Although the code-generated lines do not have the same tonal range as the original scanned drawings, their multiplicity and fineness give them a haptic quality. The flow of the data lines through the body of the drawing intensifies this quality so that paradoxically, they are perceived as more intensely haptic than the original hand-made drawings. The principle element of the animation is line, which forms patterns of networks. Christopher Fry asserts that line and drawing have a long association within generative art practice and that this in turn affects the viewer's reception:

> Just as we have learnt to understand the perspectival and the photographic image, so we have learnt from a very young age, how to see and interpret lines. This is a way of seeing that is haptic, interpretive and active (2019: 7).

As opposed to the pixel, voxel or point cloud, Fry asserts, line sits within both the analogue and digital worlds. Line-based imagery is ubiquitous in generative art; however, Fry reminds us that our experience of digital line is mediated through the histories (artistic, social and cultural) of analogue line.

In a world that has been described as post-digital, as we reappraise our relationship with the digital, it could be that reconsidering practices such as drawing offer unique possibilities for exploring and reconciling the meeting of the computational and the human (Fry 2019: 8).

A soundtrack of breathing, heartbeats and frenetic hand rubbing accompanies the animation. It underscores psychological responses to the pandemic. Soundtrack data determines the speed and intensity of the line work. For much of the animation, we are either inside a body or very close to its surface – upon the skin. There is an atmosphere of physical intimacy. In contrast, the maps at the beginning of the animation suggest a fear of proximity to others. A soundtrack of running generates clusters of dots on a street map to create a pathway through which a body runs. Maps featured regularly on our screens during 2020, with clusters of dots signifying areas of high infection. Data visualizations increasingly use this type of pattern-based imagery as opposed to graph-based and potentially more precise representations of data. When datasets are vast, dynamic and continually changing, it is easier to 'see' the information when it is presented as a pattern. Patterns enable us to visually process complex systems such as the spread of a virus. Pattern formations can reveal things that numerical information alone may not, such as underlying trends in the data, which in turn can aid analysis. McKosker and Wilken's analysis of data visualizations states that 'distributions of colour, shape, and pattern' can be 'put to use in the posing of problems, in the process of diagrammatic thinking or experimentation' (2014: 160). Their call for 'diagrammatic thinking' acknowledges the importance of visual patterns in the processing of information 'for experimentation, differentiation and problem-posing' (2014: 163).

Using drawings of natural forms with coding can perhaps change perceptions about the opacity and impenetrable nature of code. I undertook some training in processing (a programming language for teaching code to non-coders) to ensure my collaboration with Chris was meaningful. As I developed confidence in coding, I was surprised to find myself working quite intuitively, much as I might experiment and play with materials in the studio. Instead of following instructions laboriously, I took some risks and explored 'what ifs'. A typical example is the way in which digital colours are selected in coding. Processing constructs colour using numerical combinations of red, green and blue (RGB – the primary colours for light), where 0 is the minimum value and 255 is the maximum value. So (0,255,0) is code for an intense green, (255,0,0) is red and (255,255,255) is white (based on the properties of white light). Combining different values of RGB can create a wide palette of colours. I found myself playing with patterns of numbers and anticipating which colours they would produce. This was not unlike mixing paints with a brush: you can predict what a

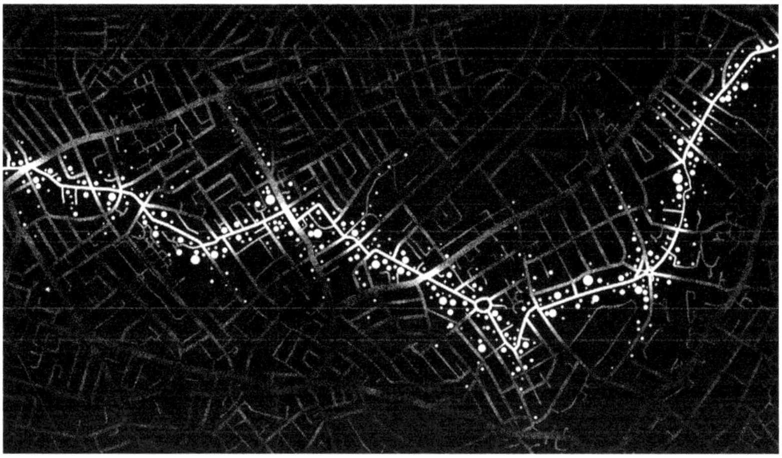

FIGURE 26.5: Daksha Patel, *in the same breath*, 2020. Video still (38 secs). © Daksha Patel.

colour combination may produce, but cannot be certain of the outcome until you mix the paint and adjust accordingly. The materiality of the paint was missing but the element of playfulness was equally strong with the digital colour. Over time, patterns of numbers became associated with particular colours so that when I looked at a certain combination of numbers I could 'see' the colour they represented in my mind's eye.

Simon Downs defines drawing as 'a complex adaptive system', by using chaos and complexity theories as models. He conceptualizes, 'the whole experience of drawing, from theory to practice to reception, as a single system: through not treating drawing as a black-box system with the drafter's brain at one end and a completed drawing at the other' (2019: 15). In Downs's paradigm, drawing as verb and noun is 'an emergent effect of a complex adaptive system' (2019: 1). This singular way of thinking about drawing resonates with the data drawings and connects them to the hand-drawn. The patterns of linear movement in the animation describe circulatory systems, digestive systems and systems of nerves. According to Serena Chan, a complex system is characterized by 'inter-relationship, inter-action and inter-connectivity of elements within a system and between a system and its environment' (2001: 1). The creative coding process, where each 'line follower' is connecting and re-connecting with different pathways is based upon inter-relationships and inter-actions. In this paradigm, the 'environment' can be understood as the body of the drawing, the animation itself, as well as the cultural, social and art historical environments that influence the reception of the animation by the viewer. Downs asserts that 'something "other" happens' (2019: 5) as a result of a feedback loop between these different factors. Artists understand that a drawing is more than the sum of its parts. A finished drawing will often develop into something 'other' to the artist's original intention; it unfolds and reveals itself through dynamic interactions. This process, which artists know intuitively, is being recognized in healthcare. Dean Rickles et al. (2007) assert that complex and chaotic systems are increasingly invoked in healthcare environments such as epidemiology. A report by the Health Foundation, which collates research articles about complex adaptive systems thinking cites McQuillan:

> [Previous] efforts assumed that complicated phenomenon could be understood by analyzing their constituent parts, when in fact the sum of the whole was greater, and more complex, than the sum of the individual parts [...]. Complexity theory draws attention to the evolving interrelationships among system elements at various levels of the system. It offers a means to analyse emerging patterns and trends to illuminate how the disparate system parts are, or are not, working together (2008: 1773).

The emerging patterns of interaction and behaviour that develop out of a complex or chaotic system demand a shift in focus away from cause and effect, to interrelationships. Emergent and changing patterns can allow us to 'see' behaviours and effects when vast dynamic datasets appear chaotic and random.

27

Viral Experiments

Louise Mackenzie

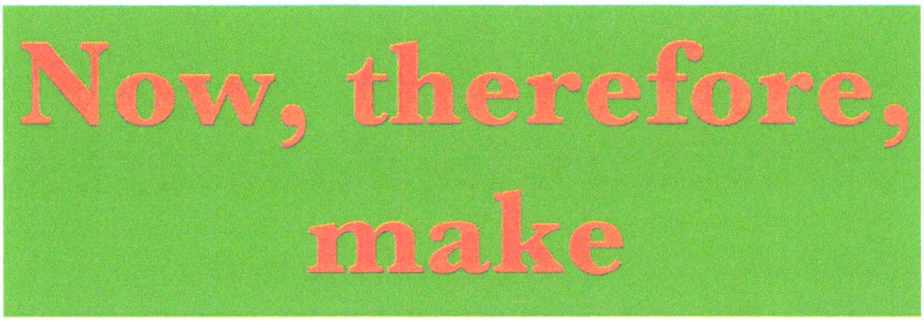

FIGURE 27.1: Louise Mackenzie, *viral virus*, 2015. The meaning of the word 'virus' is translated into English using all the languages in Google Translate. GIF. © Louise Mackenzie. The full GIF can be viewed at www.loumackenzie.com.

I once chose to try and learn the language of bioinformatics, where wet life becomes dry index. I had read of men who spoke of books of life and wondered how this vibrant, messy thing could be confined to a page. How could I be confined to a page? How could I be coded and written like a script? DNA. Deoxyribonucleic acid. There is a test some scientists like to give when you talk to them about DNA – do you know what it stands for? Can you pronounce it? Ah, well yes if you can, then you can join our club and perhaps you might be able to read this book.

But I don't like dry books.

My father learned his trade as a print compositor. He arranged tiny blocks so that they would form letters on a page, ink on metal becoming words on paper. He told me that he saw many great books roll off the press and that he learned to read them through metal blocks with the words back to front.

In 1953, James Watson and Francis Crick declared in cold hard ink, 'WE wish to suggest a structure for the salt of deoxyribose nucleic acid (D.N.A.)' (Crick and Watson 1953: 737), the leading lead-in misdirecting their oh-so-polite suggestion away from *Photo 51*, an image taken of the double-helix by Rosalind Franklin a year earlier (and brought to Crick and Watson's attention by Franklin's student, Maurice Wilkins).

DNA. The double helix. Four nucleotides (A, C, G and T) bonded together over and over by a sugar-phosphate backbone. Molecules entwined. When you look at the way the chemistry is drawn, it makes an interesting pattern: like a rope ladder spinning out of control. The genetic code all seems to fit together so neatly. Until you realize that it doesn't. There are five nucleotides, not four. In RNA, T (Thymine) is generally replaced by U (Uracil). Then there are the 'non-canonical bases'. In addition to adenosine, guanine, thymine, uracil and cytosine, there are more than 15 other modified versions of these bases. That we know of. It's an inky, sticky mess that refuses to conform.

In the 1950s, correspondence between Watson and theoretical physicist George Gamow led to the formation of the RNA Tie Club (Wellcome Collection 1980), an exclusive all-male group, which aimed to map out the genetic code by identifying how RNA forms proteins in the body. The group had 24 members, one for each of the 20 amino acids within the human body and another four representing each of the four nucleotides within RNA (A, C, U and G). No one represented T because that was not found in RNA. Some things have changed since the 1950s (Anandagopu et al. 2008; Berg and Brandl 2021).

RNA. Ribonucleic acid. It is a snake, a dragon, a single strand of DNA. If DNA is the hardbound book of life, then RNA is its supple pages. No, that's not right. RNA is letters on the page and DNA the crisp metal shapes on the compositor's block. No, it's more like the ink on the metal ... I keep trying but I can't get it right. That's the thing with metaphors, they are there to aid the imagination. They are not intended as dogma.

Crick knew this. He was not a religious man, but his belief in science was fundamental. The concept of 'the genetic code' came from what Francis Crick described as 'the Central Dogma' (he liked to capitalize). The intention behind the central dogma was to view the fundamentals of biology as a flow of information: DNA to RNA, RNA to Protein.

Genetics, linguistics and information theory grew up together in the 1950s, each borrowing from the other as they played together. Linguist Roman Jakobson applied a mathematical approach to linguistics first, then to biology, describing DNA or 'genetic information' as an inscription 'in the chromosomes [...] exactly like in a phrase in a text' (Kay 2000: 307). Geneticist George Beadle compared the near-universal nature of the genetic code to the Rosetta Stone – an unlocking of an ancient language (Kay 2000: 291).

In the 1990s, science mapped all of the genetic information in the human body (Lander et al. 2001). Nobody mapped my body. I wonder whose body did they map? My mother once used threadlike sticky strips of coloured tape to make maps. She was a cartographer, coding geologists' sketches with contours and cross-hatch. Marking out patterns on the paper just as my father set out words on the page. This mark means that. Sometimes I was allowed to play with the tape, putting marks in the landscape wherever I liked. Sometimes the tape stuck to my body.

Lily thinks that talking about DNA as 'information' is a metaphor of a metaphor, and that doesn't work because it refers to nothing (Kay 2000: 2).

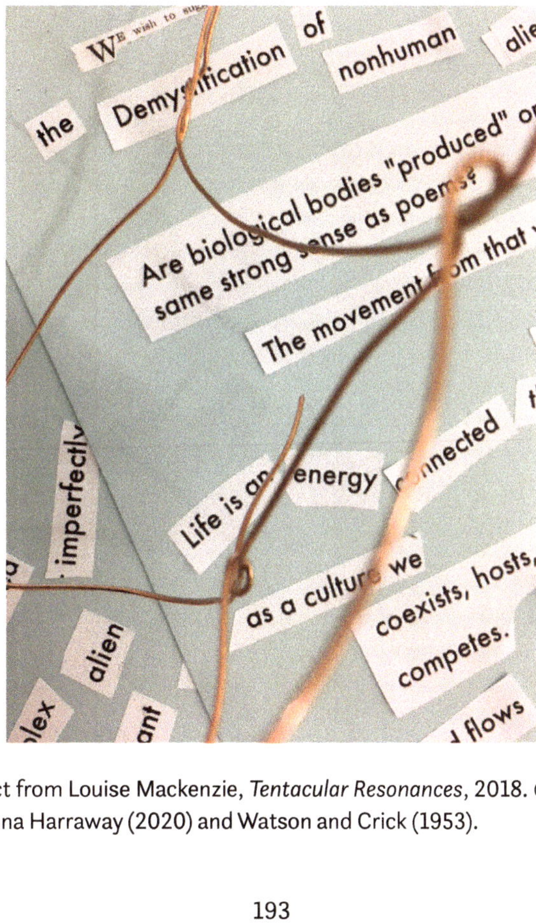

FIGURE 27.2: Extract from Louise Mackenzie, *Tentacular Resonances*, 2018. © Louise Mackenzie. Text taken from Donna Harraway (2020) and Watson and Crick (1953).

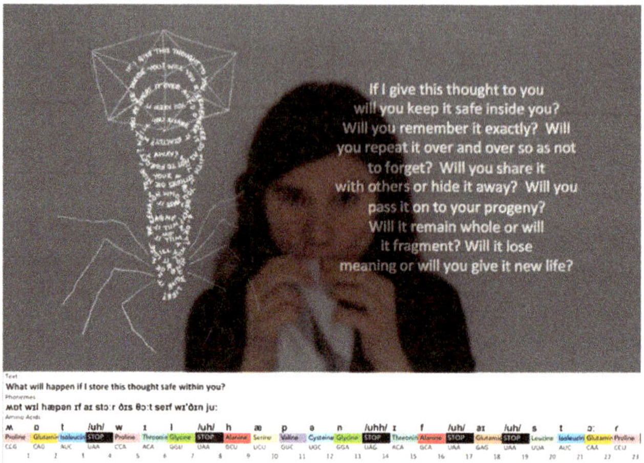

FIGURE 27.3: Louise Mackenzie, *Safe inside*, 2015. Video still with digital collage. © Louise Mackenzie.

We can make DNA and RNA now. Fabricate it out of thin air (Agapakis 2014).[1] Did you know that? That's how we have made the coronavirus vaccine. We just built it.

If I were to make my own DNA and insert it within a living body, what would this DNA that is me – and is not me – be?

William said 'the word is [...] a virus' that infected our pre-lingual ancestors (Burroughs 2005: 5) and that we are 'a symbiotic relationship of body and word-virus' (Land 2005: 450).

Body – word-virus – DNA
In conversation.

What could this DNA that is me – and is not
me – be?

DNA and language, culturally and materially
entwined.

Science breaks DNA down, syntactically, into
the 'genetic code' so that we can rebuild it
however we choose.
That's how we make DNA.

Sugarcane is harvested and processed
under patent in a factory in Japan manned
by hi-tech robots.
That's how we make DNA.

The four DNA bases are, 'available in metric ton quantities from a variety of sources. The cheapest suppliers are in China; they sell their product for under \$100/kg' (Sanghvi 2005: 20). DNA is synthesized in a complex process that originally used fish silt as raw material, but this was inefficient. '[T]o isolate 1 kg of four nucleosides, one would need 1818 kg of salmon' (Sanghvi 2005: 23). A new process using cane sugar was fully automated and patent protected by Mitsui Chemicals in Japan, significantly reducing synthetic DNA production costs (Sanghvi 2005: 21). Now, DNA rolls off the genetic production line and is shipped all over the world.

I can't seem to find any public information on how much cane sugar is needed and the processes by which it is acquired for DNA synthesis. I did find a report that provides a global industry analysis of the viral vector and plasmid DNA manufacturing market, but it costs around \$5000 to buy (Grand View Research 2020). You have to be part of a very special club to read that book.

I snuck into one such club through a side door labelled 'Art' and I learned how to make DNA. When Marta did this, she observed that to the naked eye, DNA looks like a cloud and she asked, 'could this be the soul?' (de Menezes 2003).

People see faces in clouds. I see faces in the damask patterns on the blinds in the living room. Gargoyle faces. Psychiatrist Karl Ludwig Kahlbaum named this tendency

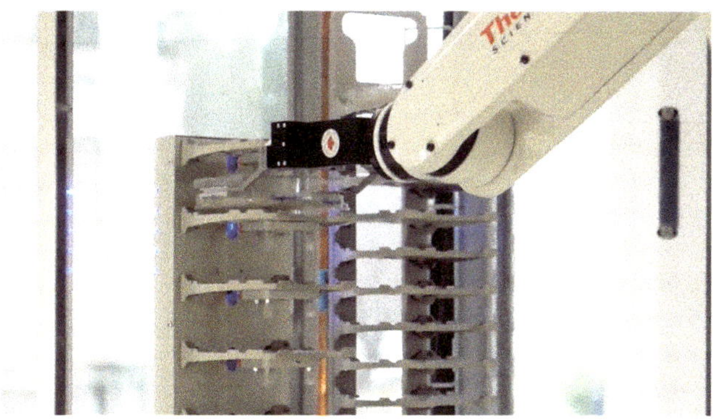

FIGURE 27.4: Louise Mackenzie, *Lively Material*, 2018. Video still. © Louise Mackenzie.

to spot the familiar in the random *pareidolie* in his 1866 paper, 'On Delusions of the Senses', describing it as 'imperfect perception' (*History of Psychiatry* 2008). Leonardo da Vinci saw it as a gift to creativity (Müri and Göbel 2020). Perhaps a kind of spiritual pareidolia is what some of us want. Not to see faces in the clouds, but to see our souls.

Maybe viruses are souls. Viruses are almost nothing but DNA. We can make DNA out of thin air. We can make DNA out of sugar cane. We can make DNA out of viruses.

The DNA in viruses and plasmids can be copied when inside living bodies. A bit like words in the print room or sketches in the cartographer's office.

A plasmid is described as a circular piece of DNA that floats freely, and thus has the capacity to move, twist and fold, within the body of the bacterial cell. The plasmid is not a circle. A circle is like a soul. Easy to imagine, hard to see.

Ourobouros is an ancient symbol depicting a snake or dragon eating its own tail. It is often interpreted as a symbol of eternal renewal or a cycle of life, death and rebirth.

A plasmid ourobouros has no start or end and as such, can be read as a poem beginning at any point and ending at any point, or read continually and indefinitely.

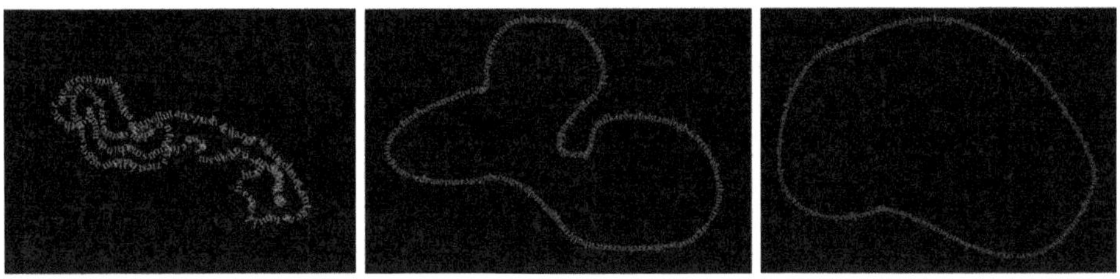

FIGURE 27.5: Louise Mackenzie, *Ourobouros*, 2016. GIF. © Louise Mackenzie. The full GIF can be viewed at www.loumackenzie.com.

Lab Diary, 12 September 2016:

My hair has grown a lot since I started this thesis. In fact, I won't cut it. It's as if all the knowledge that I have accumulated is writ large right there: a visual record of everything that I have taken into my head.

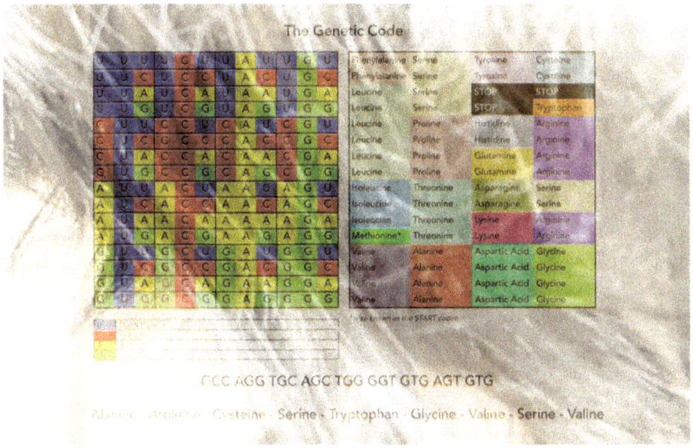

FIGURE 27.6: Louise Mackenzie, *The genetic code and the hairs on my head*, 2020. Digital image. © Louise Mackenzie.

A string of DNA might contain within it the following nucleotide triplets:

GCC AGG TGC AGC TGG GGT GTG AGT GTG

These triplets form the corresponding amino acids:

Alanine – Arginine – Cysteine – Serine – Tryptophan – Glycine – Valine – Serine – Valine

This particular combination of amino acids forms a small segment of the genetic sequence for the human protein, Keratin 18. Keratin is a protective protein found in human hair and nails (Bragulla and Homberger 2009). I searched the database of the National Centre for Biotechnology Information to find the DNA sequence for keratin. After narrowing my search to 'homo sapiens keratin 18', I found 1482 examples, and those are only the ones on the database.

DNA to RNA, RNA to Protein, to Protein, to Protein, to Protein, to ...

The possible outcomes appear to be as varied as the colour, length and thickness of the individual hairs on my head.

Lab Diary, 4 June 2015

I have decided on syllables. I will record a spoken text and break it up into syllables that we will

encode as amino acids so that we can read these back as fragments of text.

I need to match the 44 syllables in the English language to 44 amino acids then add pauses and other intonations. I will then record a short piece of text and transcribe this according to syllables / phonemes.

Experiment with text to use.

Lab Diary, 10 June 2015

Unconscious Thoughts:
I am speaking my mind, into this body. It is other, but it is also me.
I am thought within a body. Does this make me soul?
I think I am thought within a body, but I have no control.
What happens if I hide my thoughts away in here, is it dangerous?

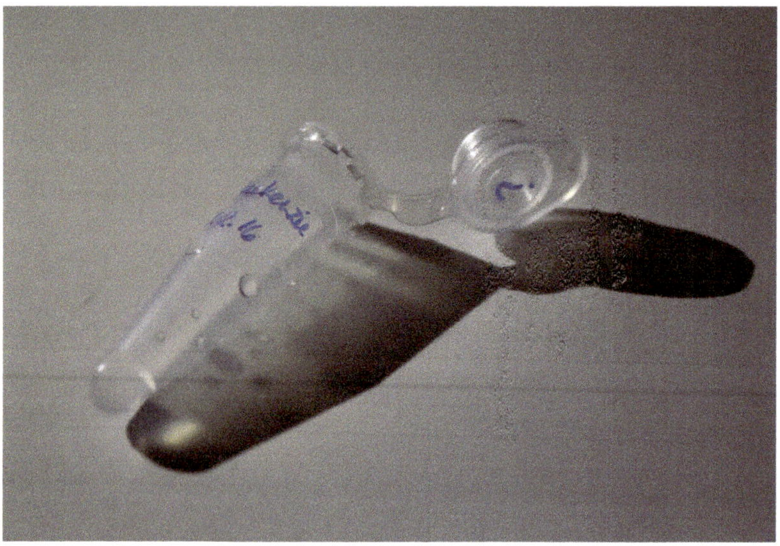

FIGURE 27.7: Louise Mackenzie, *Bioassemblage #1*, 2017. Synthetic DNA encoded with the phrase 'What will happen if I store this thought safe within you?' C-type Print. © Louise Mackenzie.

I wrote my thought as DNA and placed it in a biological body. A bacterial body. It felt wrong somehow. As if I was just playing with that body. I kept thinking, why should we get to do this?

We wish to suggest a tentacular structure for DNA: the Demystification of Non-human Aliens. Are biological bodies produced or generated in the same strong sense as poems? Listening now reveals the noise (Mackenzie 2018).

DNA was once the book of life and RNA was liquid ink, snaking over metal and seeping into paper, in a hot, smoky room where men talked and laughed over the noise of the machinery. I played at making maps, tacking sticky tape to the paper with my fingers and tangling it in my hair.

NOTE

1. I refer here to the way that scientists' use of DNA is often portrayed in the media. Synthetic biologist Christina Agapakis is careful to point out that we must attend to all of the layers required to create DNA.

28

Contagious Pattern: The Spread of Appropriated Patterns by Contemporary Artists

Andrew Bracey

To look at a virus under a microscope is to encounter a pattern. The relationship between pattern and viruses feels potent as they each spread and expand over borders, are underlined by logic and order but are open to chaos, and are highly adaptable. This text will explore how the virus can be used metaphorically to interpret the use of appropriated patterns by contemporary artists. The virus – as the active and contagious agent – in the metaphor can shift position, from the appropriated pattern, contemporary artist or their artwork. Art historian Isabelle Graw has recognized appropriation art's relationship with the virus. She talks of how appropriation in art is 'considered to be successful at the moment when it had managed to "permeate" the "body" like a stealthily spreading virus' (Graw 2004: 52).

A virus is an obligate intracellular parasite that relies on a host to reproduce. Like other parasites, a virus is generally viewed negatively as a cause of disease and death – something the COVID-19 pandemic has only intensified. Conversely, scientists now understand that viruses are 'proving to be of prime importance in the stability of ecosystems worldwide' (Crawford 2011: 20) and are seen as the 'most elementary form of life we know' (Giordano 2020: 7). A virus sits intriguingly between the living and non-living, being closest in concept to a seed, as something inert but changes to bring forth life (Villarreal 2008: n.pag.). The negative bias towards viruses can lead to positive positions being undermined by these dominant views. For example, viruses are used in medical care to help fight cancer (Mietzsch and Agbandje-McKenna 2017) and there are examples of viruses within plants and animals that are necessary to their survival (Roossinck 2011). Similarly, there is a bias in the limited interpretations of appropriation in art that can lead to wider readings of appropriation being undermined.

The dominant semiotic approach to appropriation in art (Owens 1980; Crimp 2008; Édeline and Klinkenberg 2015) can be recognized as conventionally parasitic, as a critique of the 'host' appropriated material by the 'parasite' contemporary artist. *Parasitical Appropriation* is a methodology I am developing to re-examine appropriation in art. The parasitical element is intentionally provocative: the symbiotic and mutualist elements of the parasite/virus are used to understand other approaches of appropriation in art and to recognize material that is otherwise undervalued or ignored by other critical voices. Through three contemporary artists (David Mabb, Phillip Taaffe and R. H. Quaytman) who appropriate patterns, I will explore different approaches to the viral metaphor.

Appropriated pattern as virus

William Morris's patterns have proliferated from the initial hand-produced woodblock print wallpapers and fabrics, into a myriad of mass-produced forms, spreading like a particularly virulent virus. David Mabb speaks of Morris's patterns shifting in meaning over time, from an ideological and radical alternative to Victorian industrialization to how by being 'widely available through relatively cheap Sanderson copies [...] they have come to represent the values of suburbia, the middle classes and the aesthetically conservative' (Mabb 2006: 28). It is worth mentioning that most potentially contagious things – patterns, social media posts or viruses – do not spread very far. As epidemiologist Adam Kucharski writes, 'It's extremely rare to get something that spreads, and even when it does, it doesn't spread beyond a few generations of transmission' (Kucharski 2020: 178). Many factors have to come together for a viral spread to be effective; Morris's patterns can be seen to be highly successful in this regard.

Mabb has consistently used Morris's patterns to smuggle in content, related to and extended out, from Morris's political and social stance. The use of Morris's patterns by Mabb chimes with art historian Robert S. Nelson's view that 'When appropriation art does succeed it works silently, breaching the body's defences like a foreign organism and insinuating itself within, as if it were natural and wholly benign' (Nelson and Shiff 2003: 164). *Rhythm 69* (Figure 28.1) combines painted images by Mabb on top of 69 hand-printed wallpaper samples by Morris. Mabb's contribution to the works has been taken from a 1970 storyboard of an unmade film by Hans Richter inspired by the lasting impact of seeing Kazimir Malevich's paintings. Mabb has used Malevich/Richter's forms to disrupt Morris's designs (Figure 28.2). Artist Colin Darke notes however that Morris's floral infiltration within Mabb's work, retaliate[s], pushing through and crawling over the shapes. The conflicting entities – flat, geometric, simple, architectonic, versus decorative, organic complex, natural – struggle against each other, neither achieving dominance, so each painting is at the same time neither and both its constituent parts (Darke and Mabb 2008: 56).

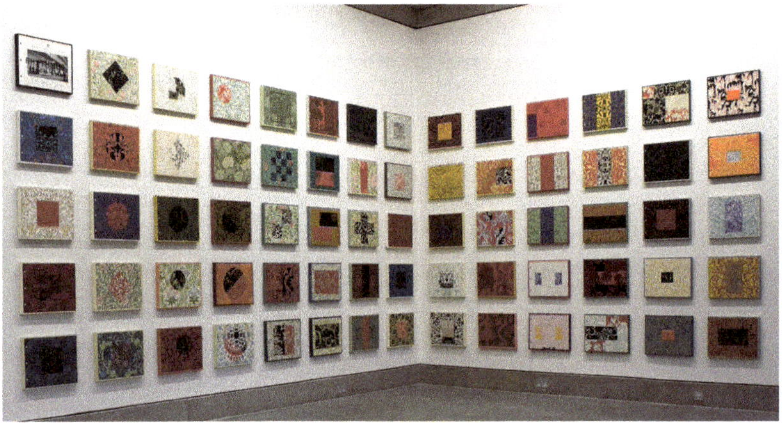

FIGURE 28.1: David Mabb, *Rhythm 69*, 2007. Paint on wallpaper mounted on canvas. Dimensions variable. Courtesy of the artist and Bluecoat Gallery, Liverpool.

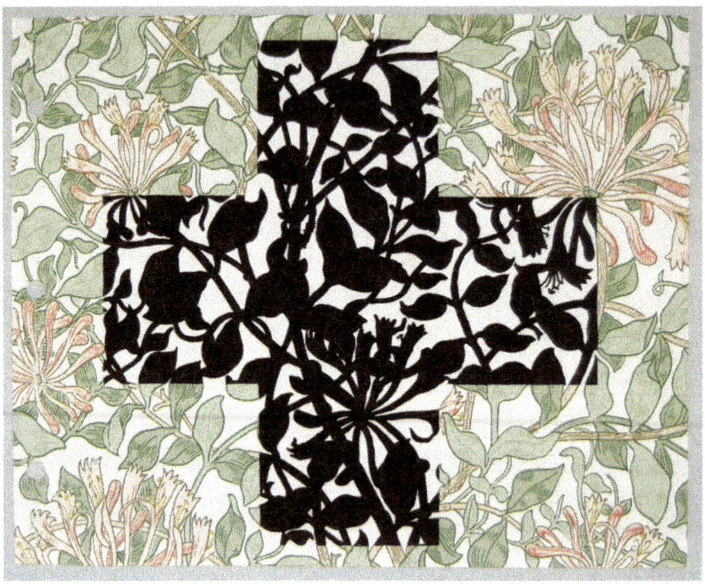

FIGURE 28.2: David Mabb, *Painting 32* from *Rhythm 69*, 2007. Paint on wallpaper mounted on canvas. 38 cm × 48 cm. Courtesy of the artist.

There is a tussle of authorship between Richter, Malevich, Morris and Mabb conceptually. However, when one looks at the piece as *Mabb's work*, then there is what Darke calls 'a kind of unexpected congruence' (Darke and Mabb 2008: 56) and one starts to see what connects, as much as what distinguishes each of the original artist's work. The work celebrates Morris's patterns and shifts the original meaning in different ways through Mabb's appropriation. The continual and shifting presence in Mabb's work is one example of how – like a virus – Morris's patterns, through their adaptability to take on different parameters and guises, continue to spread in our culture.

Contemporary artist as virus

Novelist and physicist Paolo Giordano says that the epidemic caused by SARS-CoV-2 is encouraging us all to think more collectively: 'In the contagion we rediscover ourselves as part of a single organism. In the contagion we become, again, a community' (Giordano 2020: 25). The idea of collective as opposed to singular authorship is another way of considering appropriation. Artists like Mabb and Taaffe re-use and transform existing patterns in their work in an approach akin to the shaman as a mediator. John Yau writes that for Taaffe,

> possession of an image is bound up with the process used to isolate and dislodge it from its original context: this entails both research and the choice of the right method of copying and preserving the images before it becomes part of an artwork (Yau 2018: 10).

There is a sense that culture, and pattern specifically, is a shared archive of material, images and ideas.

When Taaffe appropriates Bridget Riley's optical patterns from her paintings in his work, he is engaged in an act of 'transformation and expansion' (Yau 2018: 11). In works such as *Undercurrent* (Figure 28.3) and *Inner City* (Figure 28.4), Taaffe created paper linoprint copies of Riley's undulating patterns that were stuck to his canvas. Somewhat paradoxically Riley's work is often mistaken for printed when they are hand-painted, and Taaffe's work appears painted in the parts that are printed. Taaffe's approach to appropriation of Riley's patterns is not to critique or parody the original author, but rather he 'is committed to rescuing and honoring it' (Yau 2018: 11), to act as the mediator rather than the author. Here the contemporary artist is acting more akin to the virus that has a symbiotic relationship with the host.

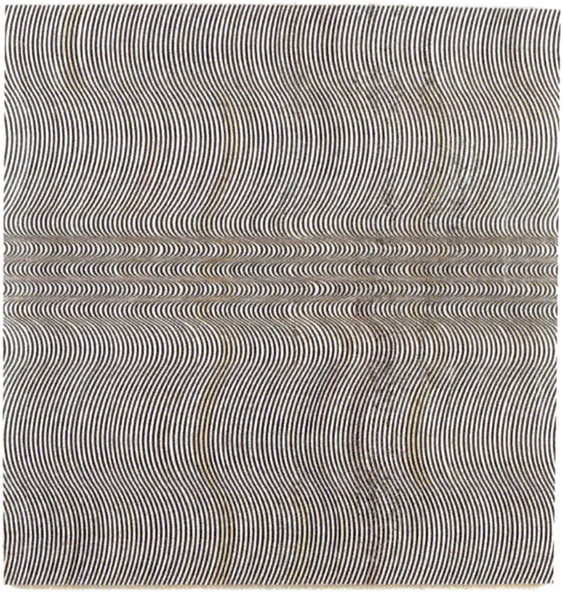

FIGURE 28.3: Philip Taaffe, *Undercurrent*, 1983. Mixed media on canvas. 223.5 cm × 223.5 cm. © Philip Taaffe. Courtesy of the artist and Luhring Augustine, New York.

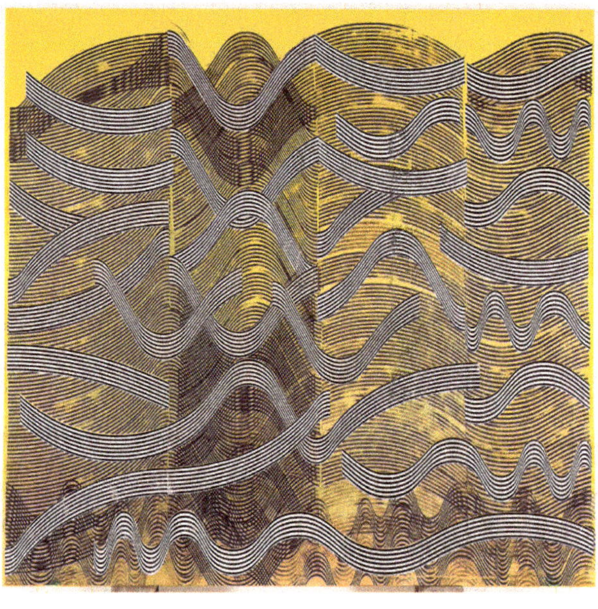

FIGURE 28.4: Philip Taaffe, *Inner City*, 1993. Mixed media on canvas. 251.5 cm × 267 cm. © Philip Taaffe. Courtesy of the artist and Luhring Augustine, New York.

Artwork as virus

R. H. Quaytman is known for distinctive bodies of work designated by her as *Chapters*. These works consist of several screen-printed and painted panels that see pattern used throughout alongside other appropriated imagery and incorporate various approaches to collage. In *Point de Gaze, Chapter 23* (Figure 28.5), Quaytman used imagery arising from her research into the lace-making culture of the Beguines, a semi-monastic order of women active in the Netherlands and Belgium in the thirteenth and fourteenth centuries. The pattern is depicted in this instance in a familiar manner that recreates the materiality of the lace. Quaytman harnesses and subverts patterns in intriguing ways; the artworks are active viral agents that use, adapt and change the existing patterns pulling them into Quaytman's visual language.

Quaytman appropriates various optical patterns which both reorientate and disorientate the viewing process as well as play with the perception of depth, space and surface. For example, there has been a sustained use of the optical patterns mined from the paintings of Wladyslaw Strzeminski (1893–1952). In *iamb, Chapter 12 (Lateral Inhibitions in the Perceptual Field)* (Figure 28.6) her use of the scintillating grid dominates. Quaytman describes this optical pattern as having been 'invented to show the blind spot at the center of visual perception. When you focus on it, your peripheral vision goes haywire' (Steedman 2010: 92). This notion of an optical use of pattern to dazzle or sideline viewing recurs throughout Quaytman's work, coupled with a returned use of the same pattern across different works.

Plywood is another constant in Quaytman's work and I would argue this is where the virus as appropriated pattern is most potent. The edges of her panels and shelves all feature the distinctive plywood edge; furthermore, she has silk-screened these horizontal bands of pattern onto the front surface of many of her compositions. Plywood's striped edge pattern is harnessed and used comparably to the other appropriated optical and representational images. The pattern element is quiet here in comparison to the other imagery on the front of the work, which might not even feature any pattern. The subversion is that appropriated pattern – through the use of plywood – is always present in Quaytman's work, even when it is not noticed; this is akin to both latent viruses waiting to activate and virulent viruses spreading through hosts who may not be aware of its presence.

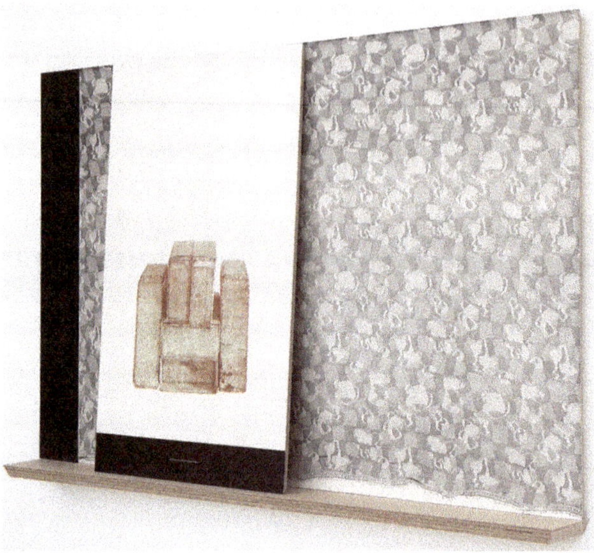

FIGURE 28.5: R. H. Quaytman, *Point de Gaze, Chapter 23 (Lygia Clark)*, 2011. Silkscreen ink, gesso on wood. 82.2 cm × 50.8 cm; and R. H. Quaytman, *Point de Gaze, Chapter 23*, 2011. Silkscreen ink, gesso on wood, 82.2 cm × 133 cm. Courtesy the artist, Gladstone Gallery, New York/Brussels, and Miguel Abreu Gallery, New York.

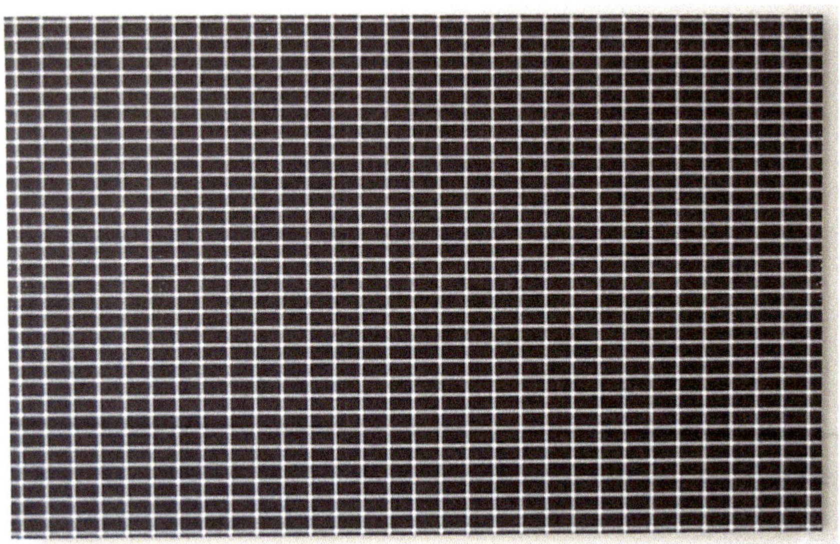

FIGURE 28.6: R. H. Quaytman, *iamb, Chapter 12 (Lateral Inhibitions in the Perceptual Field)*, 2008. Silkscreen ink, gesso on wood. 50.8 cm × 82.2 cm. Courtesy the artist and Miguel Abreu Gallery, New York.

Mutual influence

Patterns, like viruses, have an innate desire to spread. Patterns created by a designer, artist or found in nature can be (re)discovered by a contemporary artist and adapted to a new context. We have seen how different artists have appropriated patterns in different ways: how a pattern can redefine itself, how the authorship of a pattern can be extended, through utilizing 'unseen' patterns in everyday life, or how pattern can dis/re-orientate a viewing experience. These qualities of adaptability are fundamental to how viruses mutate, spread and ultimately survive. Like a virus, a pattern can become latent, such as the Beguines lace, ready to be found in later years and to spread again in new forms, including in new artworks by contemporary artists. To close I will return to Isabelle Graw. She expands upon the viral metaphor associated with appropriation art in the 1980s (Nelson and Shiff 2003: 164) which saw the appropriated material being dominated by the contemporary artist. Graw suggests that this approach discredits the 'powers of resistance' (Graw 2004: 52) of the appropriated material to fight back against the contemporary artist. Graw suggests that the process of appropriating is two-sided and 'can be seen as a process of mutual influence, in which the dynamic of the appropriated material is transferred to the appropriator' (Graw 2004: 52–54). This mutualist position of reciprocal gain is one that I believe is present in Mabb, Taaffe and Quaytman approach and use of appropriated patterns and is the most potent way to approach the virus metaphor in relation to appropriation by artists.

PART 5

SOCIAL IMAGINARY

Introduction

The roles of artist and designer are often associated with innovation or individuality when measured against social norms, for example in recognizing patterns which work against the grain of social propriety. In many of the contributions here, a dialectic interweaving of 'social' and 'imaginary' both highlights and challenges such characterization, by making and discussing work which balances the predictable with the out-of-the-ordinary, often through interdisciplinary collaboration. How society is framed, for example through community, the everyday, democracy, language, shared trauma or as reflecting utopian ideologies, is critical. Patterns that might be taken for granted are broken, disrupted or realigned and refigured.

In part serving as a bridge between Parts 4 and 5, Sarah Lowndes considers the community singing of 'You'll Never Walk Alone', often accompanied by rhythmic swaying, as facilitating reparation and strengthening social bonds in the face of disruptive events, including COVID-19. Language, whether through song, everyday speech or otherwise, is crucial to the communication on which community depends and through which the social imaginary is figured. For graphic illustrator Sarah Blair, language is ingeniously disrupted and re-imagined through a visual coding of letterforms and syntax to reveal otherwise invisible patterns of syntactical composition.

The complex repairing of divisions between system and derangement can be significantly affected by context. The artist Catherine Baker re-imagines the old sore of mind/body dichotomy in the light of the complexities that shape the psychological, medical and temporal landscape at the boundary between life and death, presenting this through artworks that take into account the role of the artist at the interface between medical staff, patients and their relatives. While Baker observes a surgeon at work, artist Les Bicknell observes that which is typically disregarded within the scrupulously controlled environment of a nanoscience research laboratory, noting especially the elaborate choreography of gestures and hands in the materially neutralized environment. There are some parallels, too, in the regulated environment of an industrial laundry within which Sarah Horton introduces the motif of the cloud. Here, it is repetitive gestures, movements and tasks which signal elements of control against which the cloud evokes dreams of drifting away or forming patterns that break the hum-drum rhythms of repetitious labour. Like the laboratory or the factory, the everyday context of home life also depends on rituals. These can succumb to disarray, as in house moving, when, as noted by artist-family Townley and Bradby, 'difference presses in on us'. As the family learn to preserve an artifice of normality with endless lists, the precariousness of the disruption is articulated in drawings of a child eating a pastry. The healing function of copious treats becomes the new norm.

In an ideal, utopian world as imagined by William Morris or El Lissitzky, the worker (whether machine operator or artist) would likely have equivalent responsibilities, and pattern and chaos would co-exist harmoniously. In layering their respective dreams together, David Mabb reveals Morris's nostalgia for the hand-made and El Lissitzky's fantasy of a mechanized future as dialectically opposed yet still, significantly, in conversation. In the 'real' world, as architect Anthony Hudson concludes, a dichotomy between top-down and bottom-up approaches to the ideal city often favours a streamlining form of design which effaces the (preferable) incidental, complex variety of the local. The spaces in which we live and work, permeated by conjunctions of pattern and chaos, also reflect the friction that occurs when abstract visions or mathematical data sets are brought face-to-face with behaviour in everyday experience, as Lucy Ward's collaboration with mathematician Karoline Wiesner articulates, poignantly, through drawing. The social imaginary that is democracy is always complex, it appears, but working across boundaries, here it emerges as a pattern worth repeating.

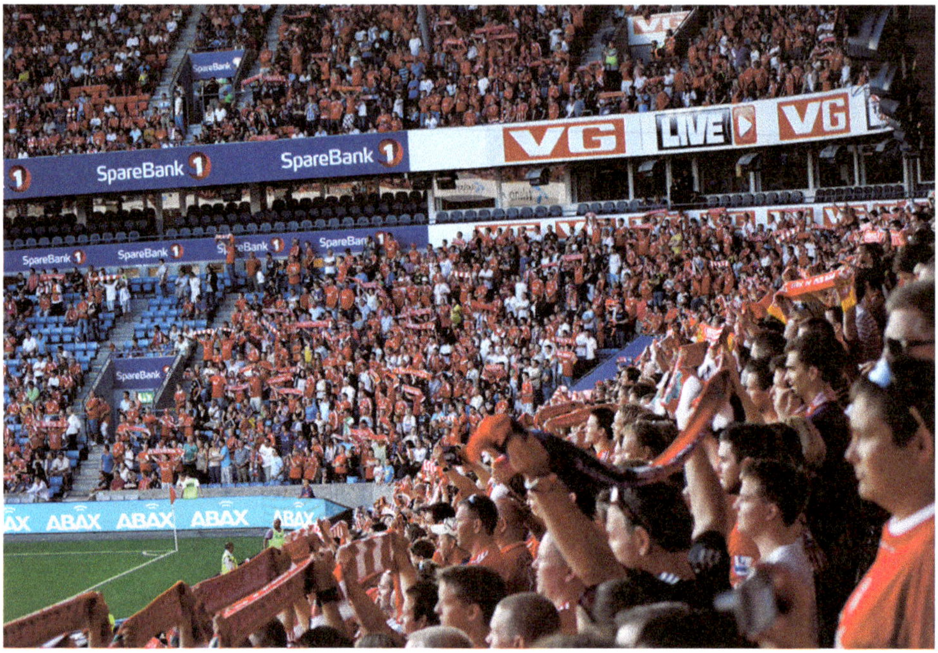

Figure 29.1: Inge Knoff, Vålerenga IF v Liverpool FC, 1 August 2011. Digital photograph. Creative Commons Attribution-Share Alike 2.0 Generic license.

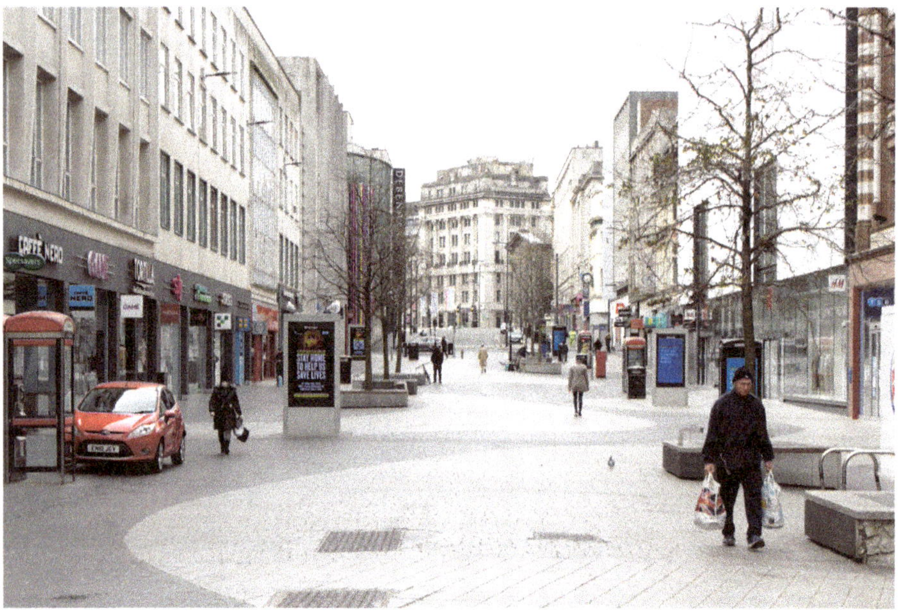

Figure 29.2: Anonymous, Lord Street, Liverpool, during the first coronavirus lockdown, April 2020. Digital photograph.

29

'You'll Never Walk Alone': A Song of Community and Struggle, 1945–2021

Sarah Lowndes

In *The Long Revolution*, Raymond Williams sets out the case that

> the process of communication is in fact the process of community: the sharing of common meanings, and thence common activities and purposes; the offering, reception and comparison of new meanings, leading to the tensions and achievements of growth and change (Williams [1961] 2011: 10).

This essay uses Williams's connected ideas of the knowable community and structures of feeling to discuss the ways in which, in recent times, the meanings and values of British people were lived in real lives, in actual communities, with a focus on football supporters and the COVID-19 pandemic. These examples are linked together in this text by analysis of the Second World War song 'You'll Never Walk Alone' as a public expression of collective identity both for Liverpool Football Club fans and, more recently, for those fundraising in support of the National Health Service in the United Kingdom during the COVID-19 pandemic. The familiar pattern of this well-known anthem has provided what we might consider a breathing space, created through communal song, providing hope and comfort through crises and chaotic events, notably the Hillsborough disaster and the COVID-19 pandemic.

One of Williams's most influential and persuasive ideas was the way in which he described communities as *structures of feeling* that persisted, developed and were expressed through words and actions. As he explained, '[a] culture is common meanings, the product of a whole people' (Williams [1958] 1989: 3–14). Williams placed further emphasis on culture as alive in the chapter 'Structures of Feeling' in his

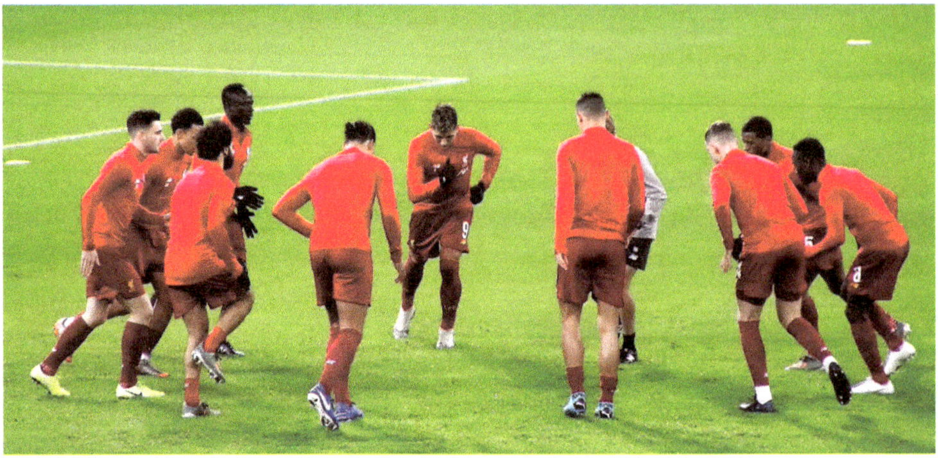

FIGURE 29.3: Anonymous, UEFA Champions League 6th round group stage, 10 December 2019, Salzburg. Digital photograph. This file is licensed under the Creative Commons Attribution-Share Alike 4.0 International license.

book *Marxism and Literature*, stating that the tendency to reduce culture and society into 'past tense' or 'finished products' was the strongest barrier to the recognition of human cultural activity (Williams [1977] 1978: 128). Today Williams's ideas remain as relevant as ever in understanding both how community functions and is expressed, *a structure of feeling* that in the coronavirus pandemic has taken shapes both creative and symbolic.

Song is a significant and vital means by which communities express what Williams called underlying *structures of feeling*. In lockdown, unable to see those whom we would ordinarily have met on street corners, at sports facilities or at pubs and clubs, we still gravitated towards song as a means to connect, expressing community as a remote choir. In lockdown, the feeling of community, or community of shared feeling, was often unspoken. But it could be expressed through song. Raymond Williams understood community and culture as being things that everyone knows and is part of: culture is sung by many voices, like a choir. In England, there were several popular songs that assumed particular significance during the pandemic year of 2020–21, but this text is focused on 'You'll Never Walk Alone', a show tune from the 1945 Rodgers and Hammerstein musical *Carousel* which gained popularity at the end of the Second World War and was recorded by artists including Frank Sinatra and Judy Garland. It is a song about enduring hardship, resisting fear and maintaining hope. The central refrain insists on a hopeful outlook as a means of connecting with others ('Walk on, walk on / With hope in your heart / And you'll never walk alone').

The tradition of singing 'You'll Never Walk Alone' at Liverpool FC's Anfield stadium came into being during the reign of legendary Liverpool manager Bill Shankly, who came into post in 1959. Under Shankly, Liverpool gained promotion to the First Division and won three League Championships and the UEFA Cup. During Shankly's fifteen-year reign, he placed a key emphasis on the importance of the Kop and the contribution of the supporters to the team's morale – to the extent that he allowed fans to have their ashes scattered on the pitch so that they could go on attending games even in death. Under Shankly, Liverpool also adopted an all-red home strip, as he believed that his team would look more intimidating dressed in all red, symbolic of danger and power. Shankly also made no secret of his left-wing leanings, stating, 'The socialism I believe in is everyone working for each other, everyone having their share of the rewards. It's the way I see football, it's the way I see life' (Koller and Brändle 2015: 100).

In the post-war period, Liverpool suffered a pronounced and prolonged period of industrial decline. Supporting the team became ever more important during these lean years, as Bill Shankly once said, 'Some people believe football is a matter of life and death. I can assure you, it is much, much more important than that' (Shankly 2010: n.pag.). Dave Russell, in his fascinating social history of association football records that Liverpool's Kop (the most famous of Anfield's four stands) which alone housed almost 30,000 fans,

> under a roof which amplified noise levels to a frightening extent [...] became probably the most celebrated football 'end' in the world during the early 1960s, when during the club's 1961–2 Second Division promotion season, fans began to develop the collection of rhythmic chants that were soon to become the common currency of English football (Russell 1997: 194).

In the early 1960s, Liverpool's Anfield Stadium was one of the first football grounds to have a PA system, and the club employed a local DJ to play top ten hits, many of which were by 'Merseybeat' bands, such as The Beatles and Gerry and the Pacemakers, in the runup to kick off. The fans sang along pre-match and continued singing during the game, often with the lyrics altered for comic effect (Brown 2017: 162). As the editors of football compendium *When Saturday Comes* rightly note, 'These terrace songs were part of an organic football culture that was created by the fans themselves' (Lyons and Ronay 2005: 378) – or what Williams might have described as 'transformed into a different order of thinking and feeling' (Williams [1973] 2016: 160). While songs such as 'She Loves You' by the Beatles and Cilla Black's 'Anyone

Who Had a Heart' were popular amongst the Anfield support, it was 'You'll Never Walk Alone', which stayed at No. 1 in the charts for about four weeks in 1963, that ultimately became Liverpool FC's signature tune (Longdon 2020: n.pag.). Motherwell-born forward Ian St. John, who joined the club in 1961, remembered, 'They would do it en masse and the whole thing would be swaying, singing and the noise that came out of it was like a roar that would go up' (St. John 2021: n.pag.). Reporter John Morgan remembered witnessing the Kop in full voice:

> I've never seen anything like this Liverpool crowd. Their rhythmic swaying is an elaborate and organised ritual. They seem to know intuitively when to begin. Throughout the match they invent new words to express adulatory, cruel or bawdy comments about the players or the police, but even then they begin singing these new words with one immediate huge voice (Brown 2017: 162).

At Anfield, the supporters increasingly put the home team at a psychological advantage, while unsettling the opposition and appealing to referees with their songs, chants, cheering and shouts of disapproval. Reporter Frank McGhee wrote of Liverpool's triumphant 5–0 victory over Arsenal which secured the league championship in 1964,

> The Kop set the match to music, and sang their team to the championship with a wit and warmth woven into the pattern of play so closely it became a part of the game itself. Liverpool fans don't just watch a game. They take part. They live it (Brown 2017: 162).

What McGhee describes is a new conception of spectatorship, similar to that outlined by Algerian philosopher Jacques Rancière in *The Emancipated Spectator*, which questions the paternalistic view of the spectator as passive and outlines a new conception of acting and watching which sees both as active partners in the generation of knowledge, as a 'third thing' that is constantly renegotiated as part of the production of knowledge (Rancière 2009: 13). This 'third thing' can also be understood as a *breathing space* – where fans forgot everyday worries and expressed themselves collectively in song.

The common bond between Liverpool's fans was strengthened following the 1989 Hillsborough disaster when 96 Liverpool fans died after they were crushed within Sheffield Wednesday's stadium during the 1989 FA Cup semi-final with Nottingham Forest. From the outset, South Yorkshire Police had sought to establish a case emphasizing exceptional levels of drunkenness and aggression among Liverpool fans, alleging falsely that many arrived at the stadium late, without tickets and determined to force entry (*BBC News* 2012: n.pag.). Ch. Supt. Duckenfield alleged that a gate was 'forced' by Liverpool fans, an untrue claim reinforced in briefings to media sources.

Newspapers took up the story, notably *The Sun* with its now infamous front page 'The Truth' (published on 19 April 1989) which declared Liverpool fans had 'urinated on police officers' and 'picked the pockets of the dead'; these serious and baseless claims adding considerably to the pain of the survivors and the bereaved.

In 2012, a report compiled by the Hillsborough Independent Panel was released, stating that despite obvious signs of distress, the police were slow to fully react and launch attempts to rescue those who were being crushed. The report found that the management of the crowd by South Yorkshire Police and Sheffield Wednesday FC had suffered a number of key operational failings, due to a policing and stewarding mindset predominantly concerned with crowd disorder. Importantly, for the families of the dead and the survivors who had been wrongly blamed for causing the crush, the report also found 'there is no evidence to support these allegations, other than a few isolated examples of aggressive or verbally abusive behaviour clearly reflecting frustration and desperation' (*BBC News* 2012: n.pag.). Almost all of the 96 had died from compressive asphyxia, being unable to breathe because of the external pressure of other supporters' bodies. Following the publication of the damning report, Liverpool Walton MP Steve Rotheram started a campaign to get 'You'll Never Walk Alone' back into the UK charts. After 48 years, the song returned to the UK top 20 in September 2012 with the proceeds being donated to Hillsborough-related charities.

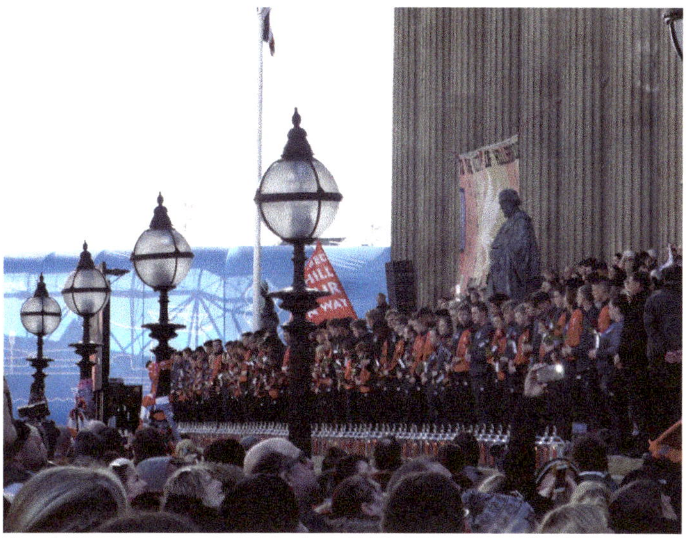

FIGURE 29.4: John Bradley. Vigil to remember the 96 victims of Hillsborough, April 2016, at St. Georges Hall Liverpool. Digital photograph. Creative Commons Attribution-Share Alike 4.0 International license.

In the strange spring of 2020, Liverpool would again endure a tragic loss of life, when COVID-19 spread rapidly through the city during late March. On 11 March, the government allowed the UEFA Champions League Football Match between Liverpool and Atlético Madrid to go ahead at Anfield stadium. A total of 54,000 people attended the game, including 3000 fans from Madrid, which was already in partial lockdown. On the day of the match, 782 of the confirmed 1646 COVID-19 cases in Spain were located in Madrid. Spain had already marked 35 COVID -19-related deaths and the country had closed its schools the day before. About 3000 fans from Madrid visited Liverpool: mixing with local people in bars, hotels, shops and at Anfield stadium. There were just six confirmed cases of COVID-19 in Liverpool on 11 March, but that number swelled to 309 less than three weeks later (*The Independent* 2020: n.pag.). Later, it would be estimated that the Liverpool–Atlético match was linked to 41 extra deaths at Liverpool hospitals between 25 and 35 days after the game, nearly half the death toll of Hillsborough. By 15 May, Liverpool had the highest COVID-19 death rate outside of London.

One of those who died from COVID-19 after attending the game was lifelong Liverpool fan Richard Mawson, 70, who passed through the Spanish fans on his way to Anfield on 11 March (Thorp 2020: n.pag.). A season ticket holder, Richie had been attending Liverpool matches since he was a teenager. He had been at Hillsborough in Sheffield on 15 April 1989, with his son, Jamie, and had helped pull fans to safety into the stand above. Richie became ill two weeks after the game and could not breathe when he last spoke to Jamie on the afternoon of 3 April. He was saying: 'I can't get my breath', his son remembered (Kale 2020: n.pag.). Only ten people were allowed to attend his funeral, but when Richie's wife Mary and Jamie arrived at the crematorium, they found a socially distanced crowd of 200 people waiting outside. One of the most poignant impacts of COVID-19 in the UK has been government advice that communal (congregational) singing should not take place indoors or outdoors, even if social distancing is being observed or face coverings are used (Ministry of Housing, Communities and Local Government 2021: n.pag.). However, at Richie's funeral, the funeral director turned on the outside speakers. When the coffin disappeared behind the curtain, strewn in red roses, the gathered supporters listened together to 'You'll Never Walk Alone' (Kale 2020: n.pag.).

On 11 March, the day of the Liverpool and Atlético Madrid match, deeply concerned by the alarming levels of spread and severity and the alarming levels of inaction, the World Health Organization made the assessment that COVID-19 could be characterized as a pandemic. Confirmed UK cases of coronavirus rose by more than 200 in a single day, and in the face of UK government inaction, large numbers of institutions, organizations and individuals across the UK cancelled or postponed public events, including the Six Nations Championship and the London Marathon. On 13 March, Premier League football fixtures were suspended, and a week later, on 20 March, all English schools closed. On 23 March, Boris Johnson announced new strict rules applicable to the entire United Kingdom with the aim of reducing the transmission of the disease between different households. The British public was instructed that they must stay at home, except for certain 'very limited purposes' – shopping, outdoor exercise, providing care to a vulnerable person; and travel to and from work where absolutely necessary. In the month that followed, in the UK, 'You'll Never Walk Alone' became the anthem of support for medical staff, first responders, and those in quarantine during the COVID-19 pandemic.

On 6 April 2020, at the age of 99, Captain Tom Moore began to walk 100 lengths of his garden in aid of NHS Charities Together, with the goal of raising £1000 by his hundredth birthday. The image of this elderly war veteran taking practical action struck a chord and in the 24-day course of his fundraising, he attracted over 1.5 million individual donations. On the morning of Moore's hundredth birthday, the total raised by his walk passed £30 million. Also during April 2020, Moore performed in a cover version of 'You'll Never Walk Alone' sung by Michael Ball and the NHS Voices of Care choir, with proceeds going to the same charity. The single was the fastest-selling single of the year and reached number one on UK music charts by 24 April, making him the oldest person to achieve a UK number one (Snapes 2020: n.pag.). As Raymond Williams had observed in *Culture is Ordinary*, 'meanings are made by living, made and remade in ways we cannot know in advance' (Williams [1958] 1989: 3–14).

How did this latest version of 'You'll Never Walk Alone' differ in significance from Gerry and the Pacemakers' version, which was the model for singing on the terraces of Anfield? Probably the most significant difference was that this version harked back to the song's legacy as a 1940s show tune. In particular, Frank Sinatra's 1945 cover version which had resonated strongly with those who had survived the Second World War and lost loved ones in the conflict. Captain Tom Moore himself was a survivor, who had served with the 146th Regiment Royal Armoured Corps in India and subsequently with the Fourteenth Army, the so-called 'Forgotten Army', who fought a brutal and gruelling war in the jungles of Burma (now Myanmar) but who largely went uncelebrated in Britain. Captain Tom's version of 'You'll Never Walk Alone', like Vera Lynn's 'We'll Meet Again', which also saw renewed popularity during the pandemic year, hearkened back to wartime separations and the eventual triumph through adversity felt by many, but not all, on VE Day and afterwards. The generation most vulnerable to the effects of COVID-19 was the war-time generations, the over-80s, born in 1940 or earlier. UK Hospital admission rates have

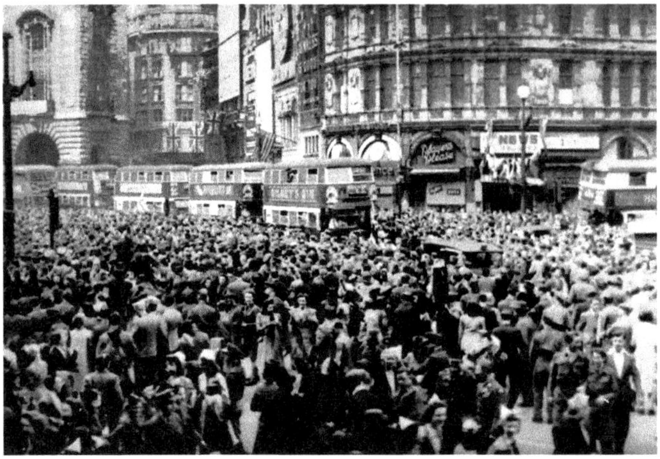

FIGURE 29.5: Sgt. James A. Spence, *Piccadilly Square pictured as supporters celebrate VE Day, 8 May 1945*, 2010, digital photograph. Creative Commons Attribution-Share Alike 3.0 Unported license.

been the highest in the 85 years and older age group throughout the pandemic. The Office for National Statistics reported that their hospital admission rate was almost 50 times higher than for those aged between 15 and 44 years. On 2 February 2021, Captain Sir Tom Moore died in hospital aged 100 after testing positive for COVID-19. His family, like many others, was forced to arrange a small funeral for just eight mourners.

Raymond Williams noted with typical insight that

> [t]he complexity of community relates to the difficult interaction between the tendencies originally distinguished in the historical development: on the one hand the sense of direct common concern; on the other hand the materialization of various forms of common organization, which may or may not adequately express this (Williams [1976] 1988: 76).

By summer 2021, 152,000 people had died from COVID-19 in the United Kingdom. Only time will tell the impact of the common organization in response to the COVID-19 pandemic. Meanwhile, can we see the pandemic, like the Industrial Revolution, as forcing the formation of a new politics, one that is alert and responsive to the community of interests of all British people? Raymond Williams wrote that

> [a] culture in common, in our own day, will not be the simple all-in-all society of old dream. It will be a very complex organization, requiring continual adjustment and redrawing. At

root, the feeling of solidarity is the only conceivable element of stabilization in so difficult an organization (Williams [1958] 2017: 436).

It is that hopeful feeling, of solidarity, that produces *a breathing space*.

This essay was adapted from a much longer Explainer text, 'One Immediate Huge Voice: Raymond Williams and Community' (Lowndes 2021) commissioned by the Raymond Williams Foundation. Some of the ideas introduced in this text are also explored in greater detail in Lowndes (2022).

30

Dialectical Reversal in *About Two Worlds*

David Mabb

About Two Worlds brings together two facsimile utopian modernist books, to create a dialectical reversal that interrupts the historical relationship between the handmade and machine-made in two differing utopian perspectives.

The first book is a facsimile of the Kelmscott edition (1894) of William Morris's *The Wood Beyond the World*, one of Morris's late romances, republished by Dover in 1972. Morris's prose romances 'remain one of the most contested elements of Morris's literary legacy' (Bennett 2021: 343). The Dover publication, on its back cover, describes the narrative as a 'wonderful fantasy in a medieval setting, brimming with high adventure and flights of fancy'. E. P. Thompson reads Morris's work with the Kelmscott Press as separate from his political activity as a socialist, thus 'no part of the earlier "warfare against the age"' (Thompson 2011: 679). He recounts that 'when a critic detected a socialist allegory in *The Wood Beyond the World*, Morris was quick to disillusion him: "it is meant for a tale pure and simple, with nothing didactic about it"' (679). Nevertheless, the visual qualities of the Kelmscott edition suggest an emphatic rejection of the ugliness of late nineteenth century industrial capitalist society. Set in Chaucer type, one of three typefaces Morris designed for the Kelmscott Press, *The Wood Beyond the World* fuses romantic narrative content with Morris's appropriation of book craft from the Middle Ages into a sort of utopian modernist space. It includes a frontispiece illustration of the 'Maid', from Morris's narrative, designed by Edward Burne-Jones and engraved by W. H. Cooper. In combination with Morris's rich decoration of the borders, capitals and type, it makes the Kelmscott *The Wood Beyond the World* something to be 'looked at' as much as read.

The second facsimile book is El Lissitzky's children's story *About Two Squares*, originally published in 1922. At the beginning of the book Lissitzky instructs his young,

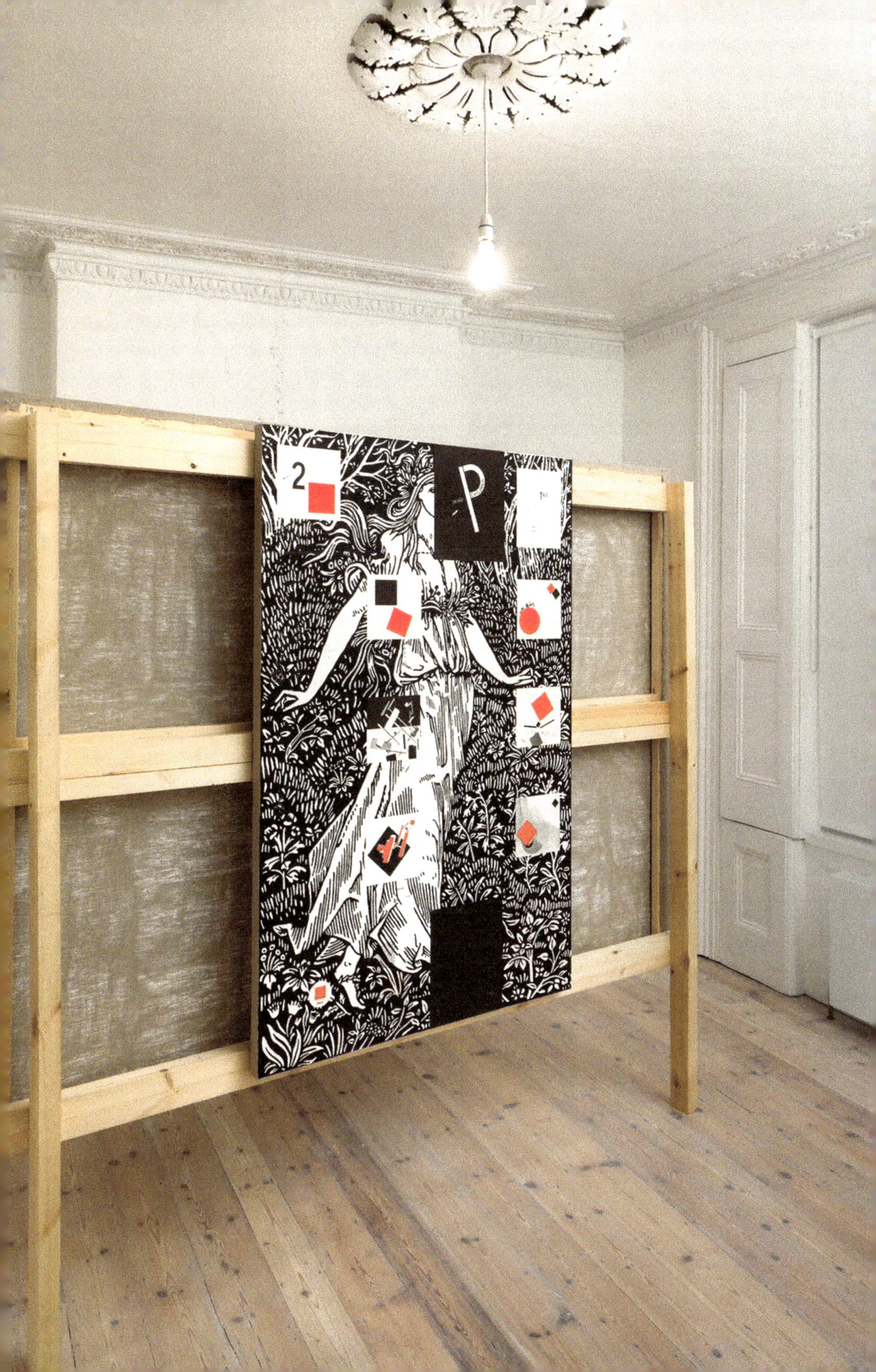
2
P

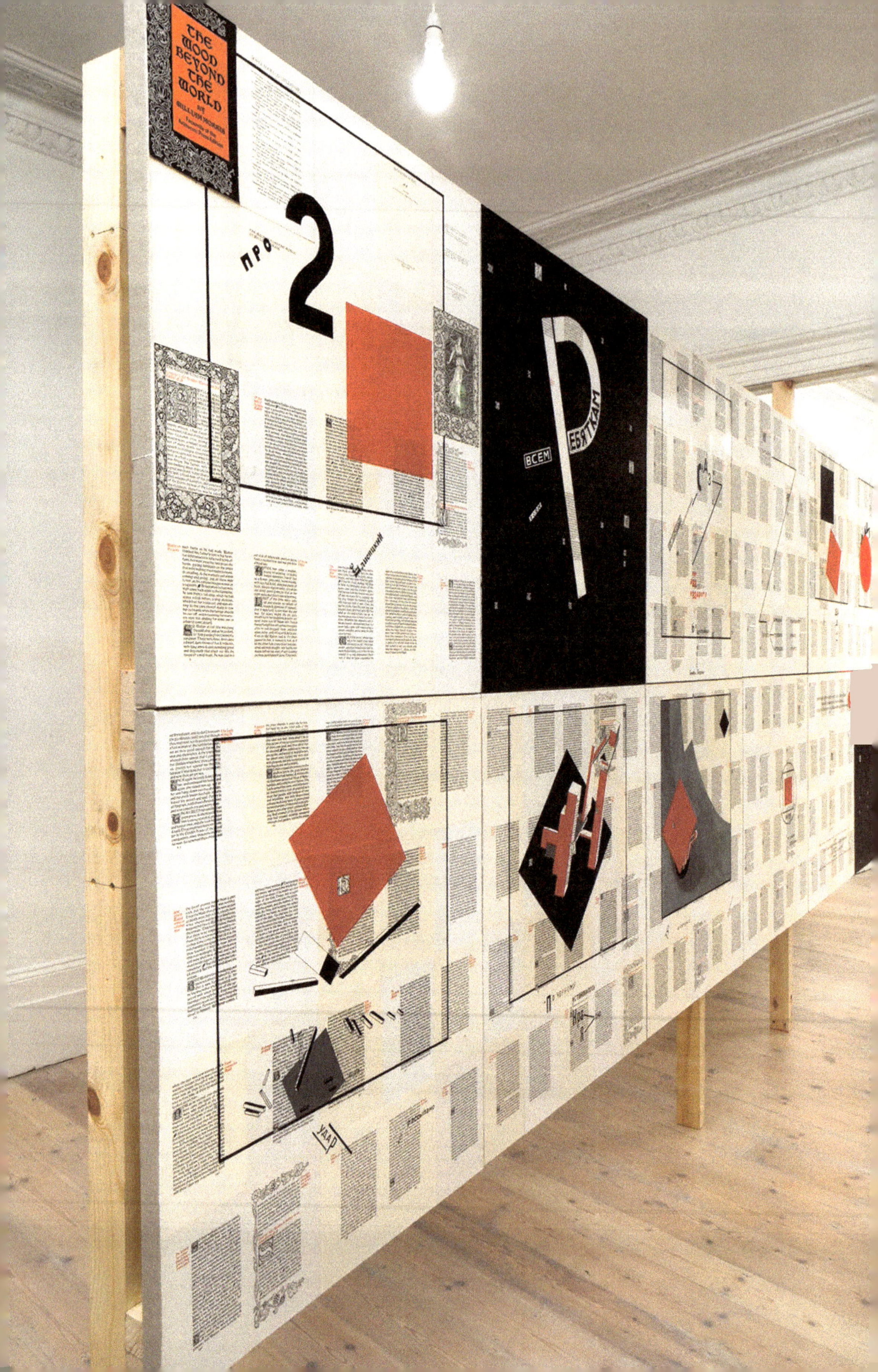
THE WOOD BEYOND THE WORLD
BY WILLIAM MORRIS
ПРО
2
ВСЕМ
РЕБЯТКАМ
УДАР

FIGURE 30.1: David Mabb, *About Two Worlds*, facsimiles of El Lissitzky's *About Two Squares* and William Morris's *The Wood Beyond the World*, 2015. Acrylic paint, varnish, pencil, mounted on linen, displayed on wooden support. 100 cm × 155 cm. First shown at Handel Street Projects, London, 16 January–20 February 2016. All photographs courtesy Peter White, FXP Photography.

FIGURE 30.2: David Mabb, *About Two Worlds*, facsimiles of El Lissitzky's *About Two Squares* and William Morris's *The Wood Beyond the World*, 2015. Acrylic paint, varnish, pencil, mounted on linen, displayed on wooden support. Each section 80 cm × 70 cm.

and not so young, readers to join in: 'Do not read, take paper, columns, blocks' and 'fold, colour, build' (Lissitzky 2014). The book's narrative is largely visual in design and whilst open to interpretation can be read as a science fiction communist fantasy. Sophie Lissitzky-Kuppers (1968: 24) describes it as follows:

> the two squares flew through the endless infinity of space; prophetic concepts were shot out of the artist's mind and took on forms, held in counter-balance by colossal tension. The red square destroyed the black chaos on earth in order to rebuild a new red unity. The black one, having been put out of action, fled into infinity.

At Handel Street Projects, *About Two Worlds* was installed diagonally across the two rooms of the gallery and hung on a long wooden structure so that both sides faced outwards. On the reverse side, which (taking Lissitzky's advice to join in) is seen first on entering the installation (Figure 30.1), all the facsimile pages from *About Two Squares* are glued in numerical order, recto and verso, onto a single canvas, and painted over with an enlarged painted copy of Edward Burne-Jones's frontispiece.

On the front side of the work (Figure 30.2), all the facsimile pages from *The Wood Beyond the World* are glued in numerical order, recto and verso, across fourteen canvas sections. Each section is painted over with an enlarged copy of a single page from *About Two Squares* (Figures 30.3–30.7). The painted parts never completely cover over the underlying pages from *The Wood Beyond the World* but allow large elements of the text to show through, creating a dialectical interweaving between the two.

The work draws our attention to the differences and similarities between Morris's and Lissitzky's respective works. Both stories are fantasies, with Morris's narrative taking place in the medieval past whilst Lissitzky's looks to the future. The materials, design and production of both books replicates this difference, with Morris looking

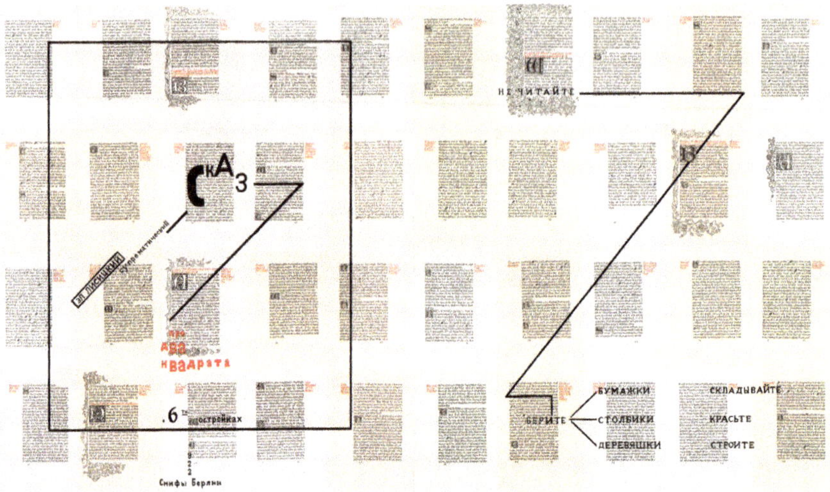

FIGURE 30.3: Sections 3 and 4 from *About Two Worlds*. Each section 80 cm × 70 cm.

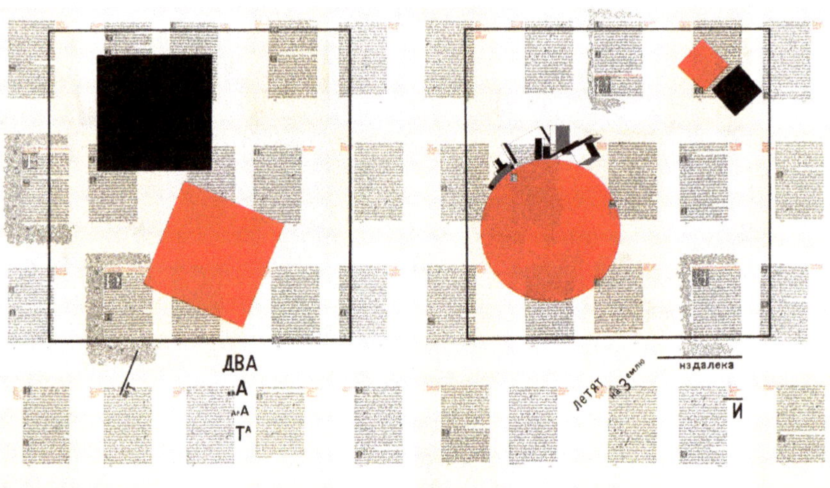

FIGURE 30.4: Sections 5 and 6 from *About Two Worlds*. Each section 80 cm × 70 cm.

to the Middle Ages for his influences, while Lissitzky consciously breaks with the past in his use of abstraction.

However, the work *About Two Worlds* complicates this division. Morris's original hand-crafted *The Wood Beyond the World* is here replaced by a facsimile – a

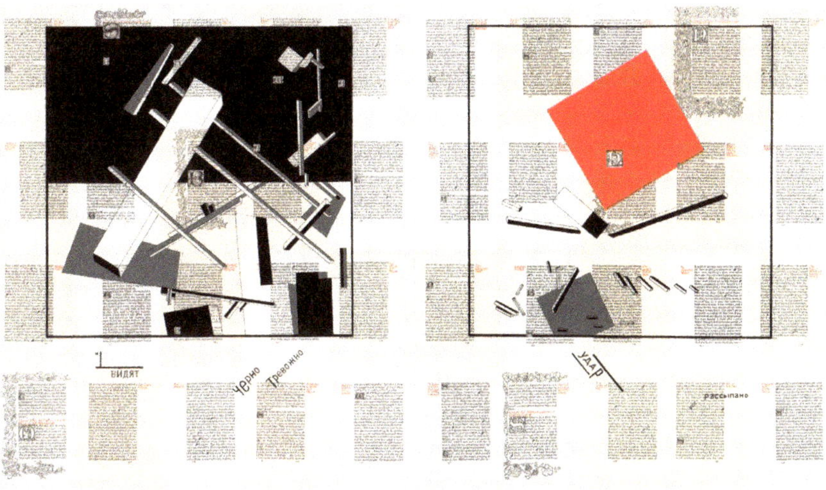

FIGURE 30.5: Sections 7 and 8 from *About Two Worlds*. Each section 80 cm × 70 cm.

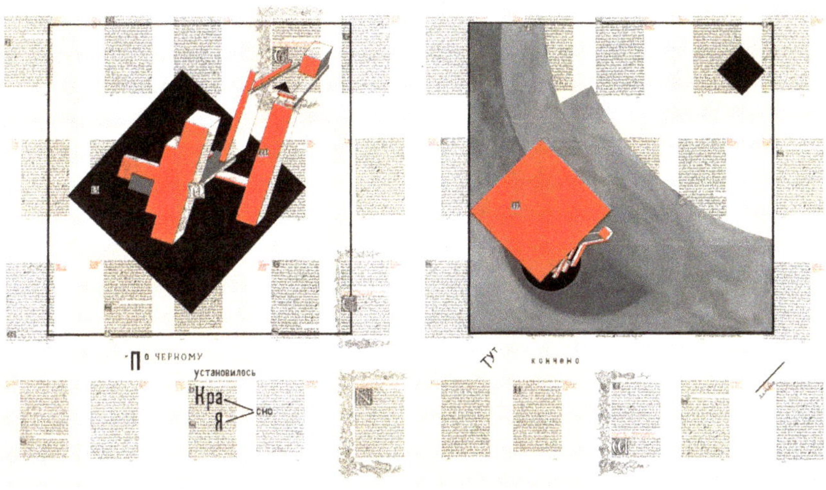

FIGURE 30.6: Sections 9 and 10 from *About Two Worlds*. Each section 80 cm × 70 cm.

mass-produced machine printed edition which has been hand-painted over with images of Lissitzky's originally machine-printed pages. Over the facsimile pages of *About Two Squares* is an enlarged hand-painted version of Burne-Jones's illustration frontispiece of the 'Maid', originally printed using engraved wooden blocks. All this confuses the division between the backward-looking Morris and forward-looking Lissitzky.

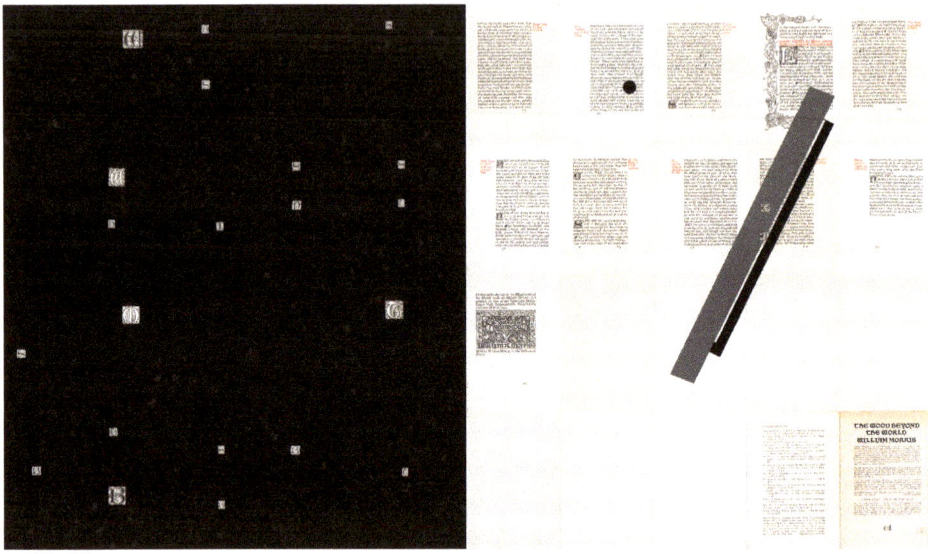

FIGURE 30.7: Sections 13 and 14 from *About Two Worlds*. Each section 80 cm × 70 cm.

The old binary between the hand and machine-made gets broken down by dialectical reversals.

More broadly, both Morris and Lissitzky used their art to promote their socialist politics. Yet, whilst Morris saw beauty in the past, wanting to elevate Victorian society from the ugliness imposed by nineteenth-century industrial manufacturing, for Lissitzky, the Russian revolution and the rapid advancement of science and technology meant the old world was no longer recognizable. He sought an entirely new visual language that could express the socialist world he believed he was helping to construct. By juxtaposing both works through dialectical reversals, *About Two Worlds* recalls Walter Benjamin's 'dialectics at a standstill': 'that wherein what has been comes together in a flash with the now to form a constellation. In other words: image is dialectics at a standstill' (1999: 463). The installation proposes that it is out of these fragments appropriated from the past that a future can be imagined.

31

Distance and Disruption: The Organized Disorder of the Body in Illness

Catherine Baker

In 2019 I was approved, through ethical processes, to observe a complex spinal surgery performed by the surgeon I am partnering with on the delivery of a collaborative interdisciplinary research project.[1] Having spent years researching the medical condition in question in order to understand not only its impact on the individual but also on familial units, I thought it necessary to attempt to understand the impact that performing surgical interventions has on the people carrying out the surgery. Scrubbed in, I hovered in the background until the operating theatre was ready and the young patient had been fully prepped for her operation. As a mother myself, I was conscious of the young person whose life was currently being controlled by machines and the talents of an anaesthetist, whilst acutely aware of the anxious wait that her parents were going through on the floor below. Her young body was masked to reveal only the area where the operation would take place; dowsed with iodine her skin took on a look of something less known. With cauterized flesh that didn't bleed, the immediacy of a clear detachment occurred in the setting. Sixteen clinical staff were present for four hours, for fifteen of them their focus was continuously on the surgeon, his instructions, his actions and the activities of copious machines. It might seem strange to suggest that in some ways the patient became almost invisible amongst the chorus of counting swabs, bleeping measurements and the overwhelming attention of fifteen people on the dialogue of the person performing the surgery. I wondered what it must feel like under the spotlight of so many for so long in an intensely pressured and precise task and, moreover, to be simultaneously responsible for the life-long mobility of a young teenage girl when one misplaced millimetre might cause immediate paralysis.

What I experienced that day made me question the enduring mind/body dichotomy from an expanded viewpoint, giving rise to questions around whether the objectivity needed to perform such surgeries can be maintained if those in attendance surrender to the lived experience of the individual. If so, what might its impact be on treatment and surgical performance and critical decision-making? In the disorganized, disrupted world of illness does detached organization become necessary for those who we expected to heal and fix us? I have yet to reach my own conclusions but perhaps, the difficulty that remains around such questions is precisely why the unsettled character of the mind/body dilemma flourishes in our encounters with illness and disease.

Advancement in modern medicine and innovative diagnostic technologies suggests that we are all likely to experience illness or disease at some point in our lives or through the lives of those we care about. Whether that experience becomes life-altering will likely depend on the nature of the illness itself and how it is encountered, coupled with the disruption it brings about both physically and/or mentally. The impact it has on the way we think about who we are in the world and what that world becomes for us as a result is an individual experience, albeit one that frequently parallels the experiences of others. Reflecting on the title of this anthology and the spirit of a broader definition of pattern and chaos contextualized through ill-health, this essay will introduce art, with its potential to draw on creative strategies such as metaphor, and health, which is frequently ordered by science and objectivity. Through identified artworks and the author's own work, this essay attempts to articulate what it means to grasp such experiences amidst and into practice through art made with participants as co-producers rather than art made about patients. The makers of the works herein occupy a complex status, becoming multiple things to many people. In the journey towards final outcomes, they act as observers, participants, producers, collaborators, voices and additionally often become friends. Importantly, in these works, the participants align with the artist(s), becoming relational stakeholders whose voice has agency.

Central to the essay is the highly critiqued Cartesian model of medicine that is based on a mechanistic separation of mind and body. The mind/body problem, as it has come to be known, is presented in the context of two artworks that utilize film as the chosen medium to explore themes including 'waiting' as a factor of illness, the temporal disruption of disease, and death and dying. It seems somewhat obscure to separate the death and dying; Jacques Derrida's (2007) explanation of 'trace' as the idea of an immediate absence of the present, the now as no longer new, suggests that the pressing directness of extant death is so transitory that the authority of dying overcrowds the moment.

Jeffrey Gold, in his discussion paper for the *Journal of the Royal Society of Medicine* (1985), pleaded for the importance of a reconsideration of philosophy in shaping contemporary medicine. Gold largely attributes the problems present in modern medicine to the central concept of Cartesian dualism and the acceptance of body and mind as distinct entities. René Descartes, often attributed with founding a new era for philosophy following the dominance of the church and the epoch of scholasticism, dedicated much of his life to the search for a certainty of knowledge situated through mathematics and physics. He sets out the profound results of his enduring search for certainty in *Mediations on First Philosophy* published in 1641 including Mediation 6, 'Concerning the Existence of Material Things, and the real distinction between the Mind and Body' (1998).

Through his system for constructing knowledge, Descartes resolves himself to be merely a 'thinking thing' (*res cogitans*), a soul, an immaterial mind distinct from the body (1641: 96), having accepted the deceitful nature of empirical knowledge he firmly believed 'that one needs to withdraw their minds from the senses as well as from all prejudice' (1641: 52). Descartes asserts that he has a distinct idea of a body that he describes as 'very closely joined to me', yet that his essence consists entirely in his being a thinking thing. He further considers that as a thinking thing he is not an 'extended thing' (*res extensa*) (1998: 96), a material substance, and is therefore completely independent of anything corporeal, thus:

> Accordingly[,] this 'I' – that is, the soul by which I am what I am – is entirely distinct from the body, and indeed easier to know than the body, and would not cease to be everything it is even if the body did not exist (Descartes cited in Cottingham 1997: 22).

The impact of Descartes's reductionist method is frequently attributed by many from both within the field of philosophy and medicine alike as having given rise to the problematic mechanistic mind and body problem in the condition of illness and/or disease. The logic that Descartes so admired of physics and mathematics, when applied in medicine, envisages illness as organized – yet understanding the impact of disease is seldom ordered. S. Kay Toombs asserts that the Cartesian paradigm of embodiment lacks completion and that a resulting disarray is formed from the failure to consider the whole person: 'In a very explicit way it omits the person to whom the body belongs, the person whose body it is' (1988: 201).

The conceptualization of the biological body in medicine as machine-like is amplified through Descartes's comments on the divisibility of the body into its constituent parts while the mind remains utterly indivisible. Nonetheless, the simplicity of divisible order vs. the often-idiopathic character of illness and disease coalesce in troublesome ways for those whose lives encounter the impact of illness either directly or indirectly. According to Descartes (1998: 101), the mind as 'wholly diverse from the body' is part of multiple observations, one being 'that my mind is not immediately affected by all the parts of the body, but only by the brain, or perhaps even just one small part of the brain, namely, by that part where the "common" sense is said to reside'.

John Cottingham notes that according to Descartes's premise, the 'I', the soul by which I am what I am, could indeed 'survive beyond the complete destruction of the body (including, let us be clear, the brain)' (1997: 22). This, in turn, gives rise to the idea that failing body parts might simply and inconsequentially be replaced, as in the situation of transplantation, or repaired with little or no impact on self. The criticism surrounding the separation of body and mind in illness is further examined by Toombs who draws attention to the body as an object in illness (1993). She highlights that bodily objectification and the experience of the body annexed in its material state can contribute to an experience of the body as separate from self in illness.

The artworks presented herein attempt to explicate the profound sense of loss experienced by the alienated self in illness (as in Figure 31.1, for example) or perhaps what Maurice Merleau-Ponty might refer to as the *ambiguity* of the body, the difference between the lived body and the biological body that is revealed through illness (cited in Carel 2013: 30). Paraphrasing Havi Carel (2013), the healthy body is both the biological body and the body as experienced in complete harmonious alignment. Thus, the constant awareness of the body that asserts itself in illness and disease recedes in the healthy body; it functions and by default, it is not present despite not having gone away. Carel explains that 'while digestion, fluid balance, and muscular performance are going well, we do not experience them consciously' (2013: 31), imperceptibly they enable us to go about our daily activities without a single thought. However, illness acts to disrupt the balance, bringing confusion with it and unsettling the everyday taken-for-grantedness of the body as experienced in

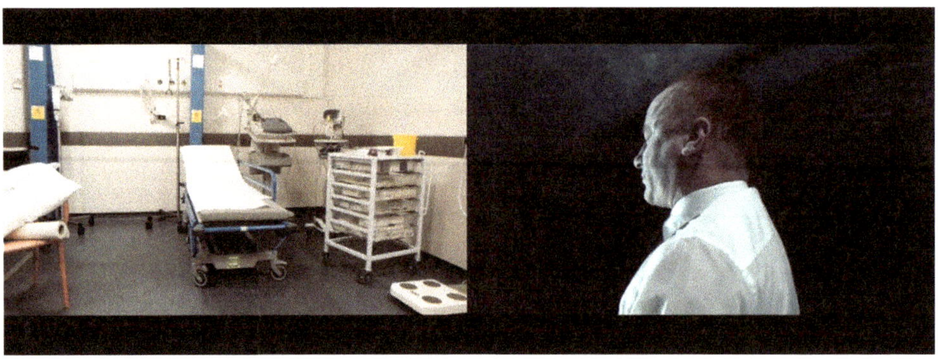

FIGURE 31.1: Catherine Baker, *Waiting*, 2019. Film still, 03:20. © Catherine Baker.

health. The harmonious pattern of everyday life ceases and with it comes a shattering of Merleau-Ponty's phenomenological description of the body, which speaks of the unified and complete body 'I am not in front of my body, I am in it, or rather I am it' (cited in Carel 2013: 35).

In Steven Eastwood's essay (2016), 'The Interval and the Instant: Inscribing death and dying', the complex relationship between ethics and aesthetics in film is explored. Far removed from the fiction of cinema and notions of a good death, whatever that might be, Eastwood's work reveals dying as a lived experience of the deteriorating body. Framed in the context of palliative care and end-of-life, the multiscreen installation *The Interval and the Instant* launched at Fabrica in 2017 (Figure 31.2), with the sister artwork, a feature-length film titled *ISLAND* (released in 2018). The works raise questions surrounding the absence, in western society, of accurate depictions of death and dying when, much like illness, dying is omnipresent. Eastwood states that the somehow unacceptable vision of dying has resulted in minimal filmmaking approaches that respond to the shared potentiality that exists between the filmmaker and the dying person. He points out that '[a]s a consequence, there is very little filmmaking done with the consent and collaboration of the dying person and there are few moving images of the very end of life' (Eastwood 2016).

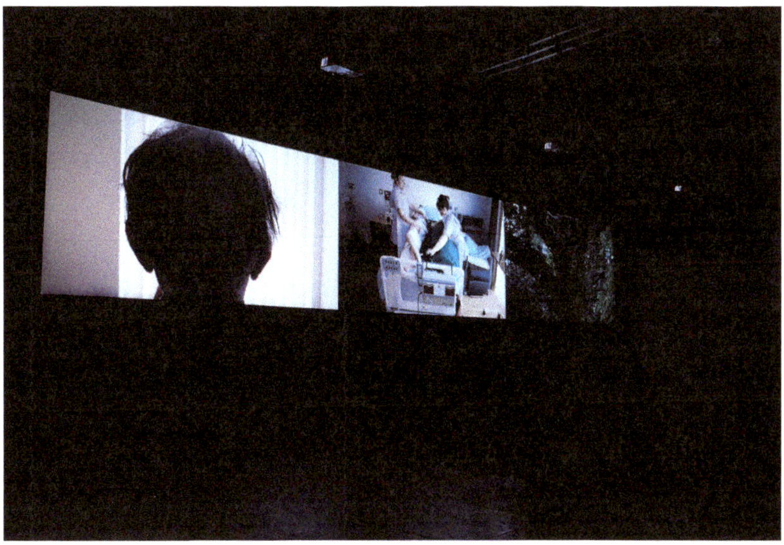

FIGURE 31.2: Steven Eastwood, *The Interval and the Instant*, 2017, Fabrica. © Courtesy of the artist.

In the same way that modern medicine has contributed to an increased awareness of illness as an everyday presence, death persists as the one true constant, yet remains unwillingly acknowledged. As presented in the opening paragraph of this essay, under challenging clinical settings a 'distancing' was observed that enabled those at the so-called front-line to continue to operate in the best interests of the patient. Similarly, Eastwood's work provides a profound insight into the private world of disease and illness and one that I suspect challenged audiences, albeit not unanticipated (as in Figure 31.3).

Whilst Toombs and Carel advocate a phenomenological consideration of the person whose body is in illness; Eastwood likewise promotes the voice and agency of the patient; his approach involves them as both subject and co-producer. This could be considered in opposition to those who might seek to encourage seclusion at such times, shielding the wider connected community from the harsh reality of death in terminal illness. Eastwood points out that the well-meaning situation of palliative care in terminal illness 'may occlude the validity of the dying person's need to speak and participate in the creation of an image of their dying' (2016: 29).

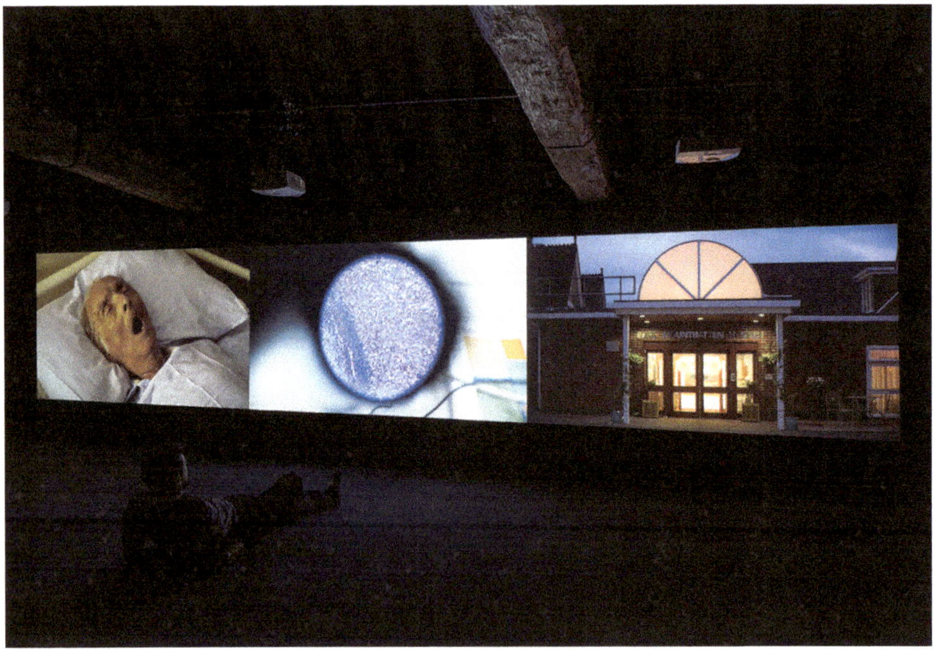

FIGURE 31.3: Steven Eastwood, *The Interval and the Instant*, 2018, Installation, multiscreen video, HD colour, sound, loop 50 min. Exhibition 'Take Care', 2019, Ferme du Buisson Centre for contemporary art. Courtesy of the artist. Photo © Émile Ouroumov.

While terminal illness, as presented in Eastwood's work, brings the temporality of death to the fore, I am struck by the parallels that exist between his role as a filmmaker and that of the clinical ethicist. In communicating disease that is incompatible with life to an extended community that includes family members, physicians and audiences alike, it seems like an important meeting point reveals itself amidst the possibility of confusion that lurks beneath the surface. Through my own work in both clinical and non-clinical settings, I have witnessed and shared the intimate experience of being 'towards death' and the ensuing existential trauma that loops endlessly for the individual. Patients with life-limiting conditions are often invited to take part in surveys that attempt to measure their wellbeing as a process of data collection that has nothing to do with the clinical outcome. Overlooked in such studies is the experience of waiting in illness and the conditions under which the waiting takes place (Cohen et al. 1996: 580). If we scrutinize waiting as experienced it varies considerably in response to who waits and with what illness, disease or sickness. For my film *Waiting* (Figure 31.4), which shows patients and those they share their lives with as they wait underwater on a single breath, my research revealed somewhat expected commonalities that align with the mind/body problem whereby the patients described themselves as data, as incessantly measured, encountered as their condition and not for who they are. The simple act of holding one's breath proved a powerful metaphor for the participants who co-produced *Waiting*. Despite what might be considered a move towards a pervasive narrative, the inclusion of water as both a life-sustaining but threatening material attempted to intensify the involuntary nature of breathing for the audience.

The accompanying sound, composed from recordings of magnetic resonance imaging sequences, echoes the process towards the search for a verdict, whilst scenes of empty, impersonal clinical spaces (as in Figure 31.1) focus and tell again the repetitive narrative bleep of medical settings and clinical diagnosis. In *Waiting*, the

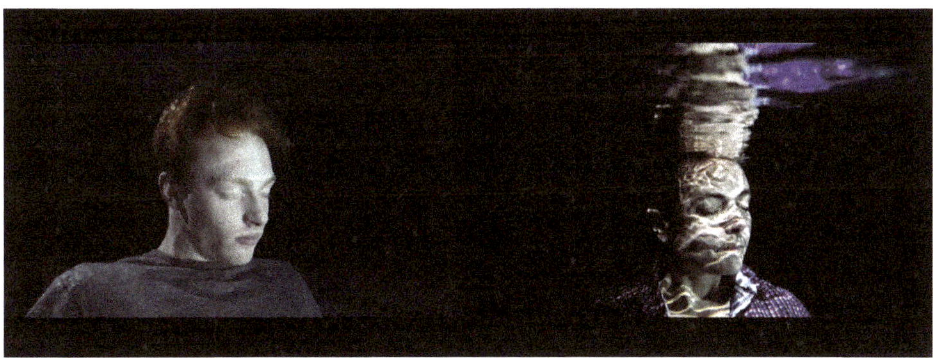

FIGURE 31.4: Catherine Baker, *Waiting*, 2019. Film still, 9:28. © Catherine Baker.

people we see on screen hold onto breath, each other and life as it slips away with surprising speed or is wished away in the hope that certain conditions will reveal their full impact more quickly (Figure 31.5). The unknowns of illness wreak havoc on the temporal experience of being; time can appear to run fast yet pass slowly when waiting for results or the right time to tell others.

When seen through the lens of the artworks herein and in the broader context of contemporary medicine, the mechanistic, orderly separation of mind and body as presented through Descartes's philosophical meditations is clearly inadequate in reflecting the complexity of experience. However, this essay hopes to have articulated that within the problematic inferences of critical illness whereby people, both patient and physician find themselves 'not on the same page', a broader consideration of moral distress is necessary. Maybe repeatedly bearing witness to trauma goes some way towards explaining why the mind/body dilemma has endured in the context of the disrupted body and those who seek to treat and/or comfort. Dr Alyssa Burgart, paediatric anaesthesiologist and bioethicist, in conversation with Professor Ian Sabroe (2021) describes how '[i]t is traumatic to be a clinician, we are sitting in people's pain and suffering'.

What is clear is that adapting to or conceptualizing uncertainty seems significantly more critical than searching for the certainty that Descartes pursued. The patterns that emerge and repeat in the context of disease expose the chaos of temporality and illness, revealing that which was perhaps there all along, the difficult problem of our awareness of time. When data and measurement fall flat and the limitations of language are exacerbated, artistic strategies employing creative metaphors that attempt to enhance our understanding of the intangible and the invisible have the capacity to bring about the creation of a meaningful dialogue attending to the mind/body issue for clinician, individual and future audience.

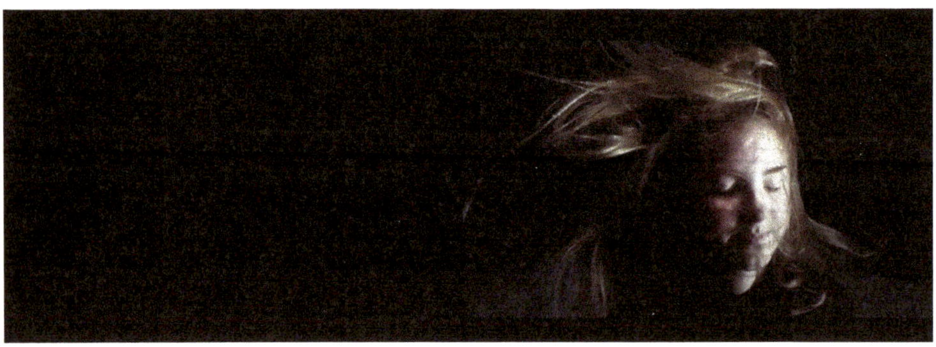

FIGURE 31.5: Catherine Baker, *Waiting*, 2019. Film still, 08:53. © Catherine Baker.

ACKNOWLEDGEMENT

The research undertaken by the author for the artwork *Waiting* (2019) was made possible through a grant from The Arts Council England.

NOTE

1. BackBone: *Interdisciplinary Creative Practices and Body Positive Resilience* (2021–22), research project funded by the British Scoliosis Research Foundation.

32

Unfolding Thinking: Nanotechnology Meets Fine Art Practice

Les Bicknell

The worlds of science and art have specific spaces in which they operate, such as the studio, the gallery and the laboratory. The activities and intentions within these spaces have correlations. The white box of the gallery is a physical and conceptual space that sets out to enable viewers to see art without restrictions or connections. Its function is to neutralize the materiality of the space, to enable the focus to be on the work, placing art at the centre. For the uninitiated entering a gallery, it must be a similar experience to outsiders entering a laboratory. The rituals seem obscure and indecipherable and the space unfamiliar. What do you do in here? How do you perform? Is there a right way to do this? How do you know what to do? How do you know when you have finished?

The images for *unfolding thinking*, 2019 (e.g. Figures 32.1 and 32.2) were made during a residency within the Centre for Interdisciplinary Ph.D. Training and Research in Nanoscience (NanoDTC) at the University of Cambridge (UK). They explore and represent several strands of the research I undertook as an artist during my time there, such as what does science look like, how is science communicated though hand gestures and general laboratory choreography and how might it be possible to map the seemingly disregarded by making the unseen visible?

The project set out to recognize similarities, explore possible confluences and develop connections by researching and deconstructing the physical and theoretical concerns of scientists working in the nanotechnology field within the framework of a fine art practice. The process of working within the laboratories with research students generated new artwork which was informed by engaging with nanotechnology from an artist's perspective. A series of ongoing interventions

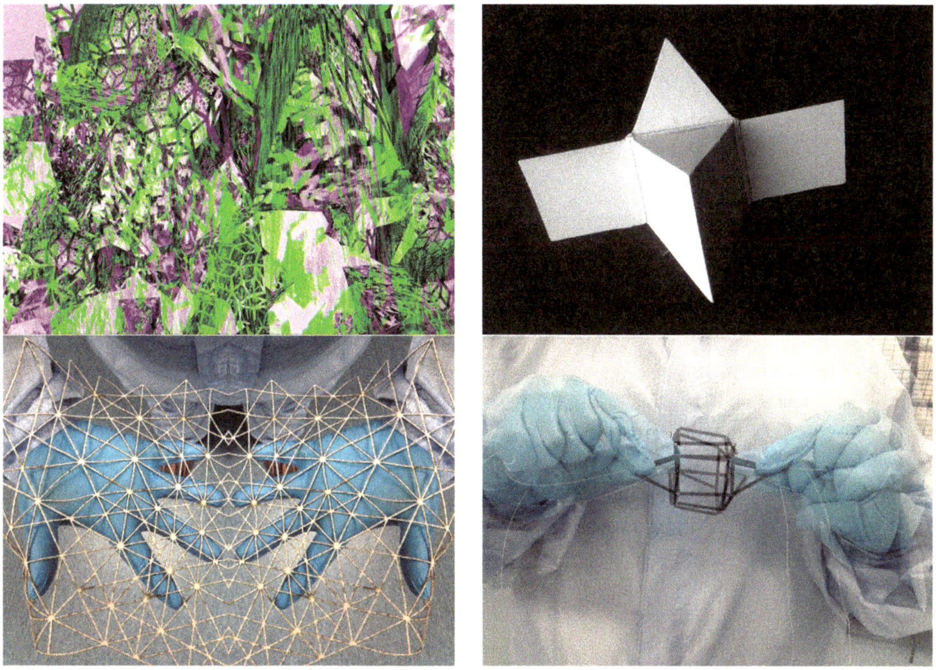

FIGURES 32.1–2: Les Bicknell, *unfolding thinking*, 2019. Images documenting a residency at NanoDTC, University of Cambridge (UK). Photographs, ink on paper.

within the spaces generated discussion and presented new possibilities and ways of thinking, demonstrating possible roles and opportunities for an artist within a scientific situation. I developed the patterns I observed, during lab demonstrations of crystallography, atomic force microscopy, spintronics and the growth of carbon nanotube, within the articulated structures I created.

A laboratory within a scientific environment is scrupulously controlled, every aspect including light levels, moisture content and dust particulates are monitored. Entering through an air shower, wearing head to toe specialist clothes, one becomes conscious of one's body; every action is considered and constricted. Within the laboratory, the glove box, also encountered in the COVID-19-saturated media, conflates fear and science. To operate, arms are forced into rubber gloves within a sealed box that enables the operator to work safely with materials, focusing on their inherent qualities. The space attempts to be a 'non-space' that is neutral and repeatable, in which the environment is controlled to enable the focus of activity to be on the samples and experiments. Just as in an Art Gallery.

ACKNOWLEDGEMENTS
Many thanks to Emma Clarke, Dr Karishma Jain and to the student cohort who were so understanding and supportive: Tommaso Busolo, Bill Stockham, Emily Williams, Michael Jones, Katherine Sanders, Demelza Wright, Ryo Mizuta, Luka Skoric, Tim Van De Goor, Johannes Osterrieth, Taylor Uekert. To those who have been generous with their time and knowledge: Sarwat Baig, Duncan Johnstone, Giorgio Divitini, Carmen Berraquero, Dr Malcolm Longair, James Xiao, Robert Foster, Girish Rughoobur, Diana Vulpe, Heather Goodwin, Richard Howe, David Wharton, Hadi Modarres, Dr Deepak Venkateshvaran. Residency supported by Arts Council and The Maxwell Centre.

33

Instead of the Feeling of Home

Townley and Bradby

We don't know where home is. We know where it used to be but we moved abroad and the feeling of home didn't come with us. We have a place to live, an awkward narrow house with neighbours and a flourishing fig tree in the garden. It's adequate, but it isn't home, and neither are the things that surround it. That's the point of course. It's all different and the difference presses in on us. Even doing something simple, like buying some stamps or making a hair appointment, there's always a step we didn't know about, a point where we trip up. It's exciting and it's wearing and it's not like home.

We talk about what we miss. It's easier to describe an object than a feeling, so we talk about the things we didn't bring: books and favourite t-shirts and do-it-yourself tools. Home is not just the things, it's also the movements between the things. It's the rhythm with which the space is occupied, the intensity of the actions. This creates the space you live in: you carry a mug of tea from the kitchen to the garden gate; you put a book on the stairs to take up to the bedroom later; you lend your drill to a neighbour. When we packed up our home (our former home) we collapsed the relationships between all of our possessions and the way they made our space of living. If writer Henri Lefebvre subjected our life to his technique of 'rhythmanalysis' (Lefebvre 2004: 15), he would diagnose a dramatic shift: from the rich and repeating daily rhythms of our former life to an erratic arrhythmia in our life now.

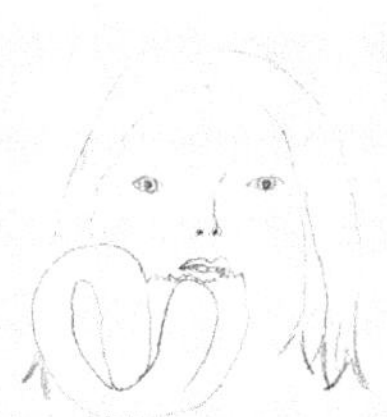

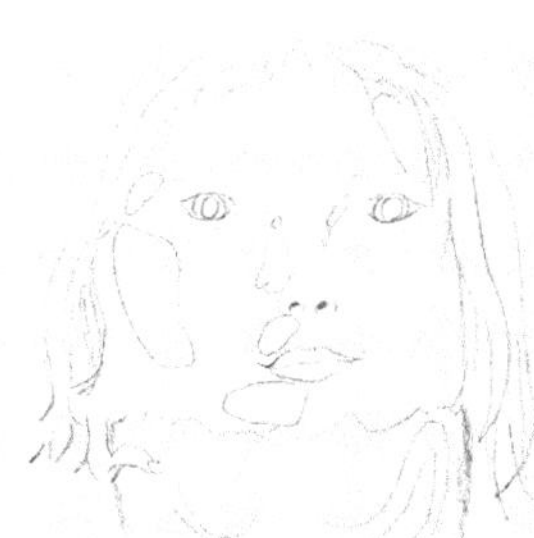

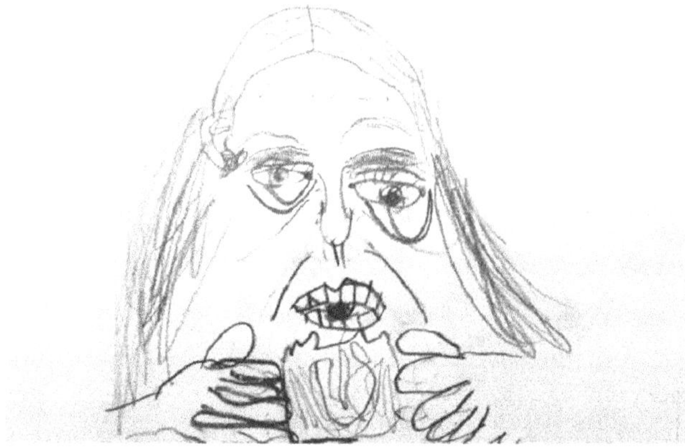

To cope with the absence of home, we watch TV with total absorption. We sit close enough to touch it, close enough for the static from the screen to raise the hairs on our arms. We twitch between channels but never look away. We're like a sleeper mumbling from one dream into the next.

To cope with the uncertainty of the day to come, we make exhaustive lists: what time to get up, what to wear, what hairstyle, what breakfast and what schoolbooks to pack.

To cope with the absence of home, we reward each other with frequent treats from the bakery.

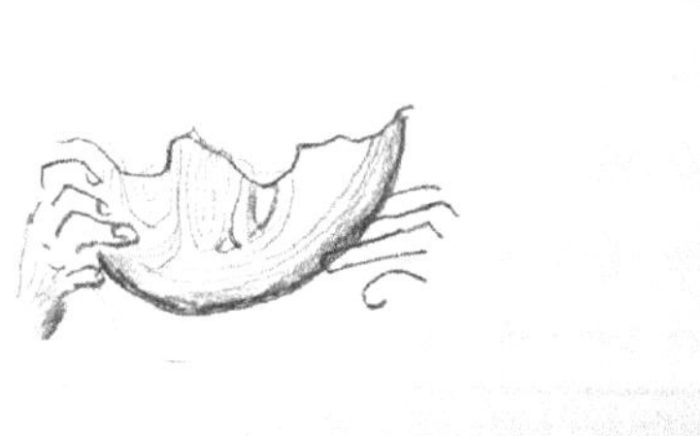

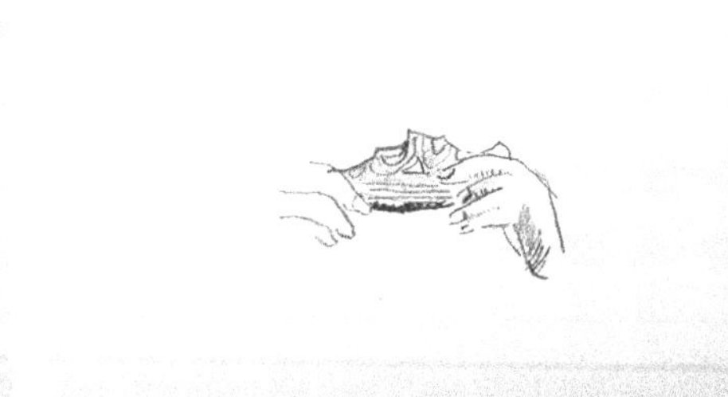

FIGURES 33.1–5: Townley and Bradby, *Instead of the feeling of home* 1–5, 2021. Pencil on paper, each 210 mm × 148 mm. © Townley and Bradby.

34

Designing for the Real World: The Importance of Chaos

Anthony Hudson

There is a side of urban planning and architecture that planners, urban designers, architects and their clients ignore at their peril: the un-categorized, difficult to pigeonhole, contingent *activity* that is the life force of cities and buildings. They see this activity as chaotic and in need of control and moulding. Yet, as we will see later, from this seeming chaos what can emerge is a different, resilient and more humane pattern for living.

In 'Walking in the City' (1984), Michel de Certeau alludes to this poetically, contrasting the desire of authorities to make the city and architecture readable and ordered, as it would have been viewed from the top of the World Trade Centre, whilst below, the city's inhabitants 'make use of spaces that cannot be seen; their knowledge of them [...] as blind as that of lovers in each other's arms' (de Certeau 1984: 93). They elude legibility. His paean, and this essay, is a plea for a way of thinking and designing that understands and encourages this intimacy which is fundamental to healthy life in our cities and buildings; a design process that builds from what is at our feet rather ordered from above, bottom-up rather than top-down.

Structuring our world from the top-down has been the *opus operandi* for design since the inception of the modern world. We have understood order through the prism of classical science and the legacy of Platonic and Christian philosophy. It is a view that sees the world around us and the artefacts we design, such as the city, to be controllable, manipulatable entities to order citizens' lives. It was given a dress rehearsal in the Renaissance by orchestrating the visual world through perspective, then taken on by the scientific community and authorities in the seventeenth century, giving eventual birth to twentieth-century town planning, whose midwife was the Modern Movement. The central tenet is that ordered cities beget ordered citizens. The rallying call of

Le Corbusier summarizes the optimism of ordered life through architecture rather than revolution. Such a desire to impose order from above for purpose of control has, as its current apogee, data use in the 'surveillance society' (Zuboff 2019) where citizens' lives are mapped out, nudged and controlled. It ignores the contingent processes that are central to the vitality and adaptability of cities evolving in time.

This approach has seen city life hollowed out and architectural texture flattened out. Jane Jacobs's seminal book of 1961, *The Death and Life of Great American Cities*, was the first significant critique of twentieth-century town planning that challenged the simplistic thinking we have observed above and suggested that there was an alternative way of looking at cities. She said that

> cities happen to be problems in organized complexity, like the life sciences. They present situations in which half a dozen or several dozen quantities are all varying simultaneously and in subtly interconnected ways [...] the variables are many but they are not helter-skelter: they are interrelated into an organic whole (Jacobs [1961] 1965: 432).

This suggests people making decisions individually and collectively, on the ground, based on the seemingly chaotic and unrelated cues from our environment which result in specific patterns of use. For example, if we have the freedom to do so, we can creatively adapt and use our streets in a way we think fit, whether it is for socializing, doing business or for play.

The life sciences over the past 50 years have been the driving force in understanding how such self-organized, bottom-up processes create fit-for-purpose structures. Complex patterns emerge from the seeming chaos of local interactions both in nature and in artefacts, like cities and buildings. Such complexity is generated not by executive top-down control or 'The Myth of the Ant Queen' as Steven Johnson (2001) calls it in his book *Emergence*, subtitled *The Connected Lives of Ants, Brains, Cities, and Software*, but instead emerges from the bottom-up, from the actions of individual entities (people, ants, neurons) that create highly sophisticated and complex structures. Johnson shows how this process of emergence 'grows smart' by creating organisms (cities, anthills, the brain) capable of resolving complex problems which are inherently fit for purpose.

Ilya Prigogine describes nature's patterns as emerging and giving order to situations which are in flux and which can change into different patterns if the conditions and context change. Moreover, 'the maintenance of organisation in nature is not – and cannot be – achieved by central management; order can only be maintained by self-organisation. Self-organising systems allow adaptation to the prevailing environment', and are 'extraordinarily flexible and robust against perturbations from outside conditions' (Prigogine 1997: 71). These seemingly chaotic interactions resolve themselves in the form of visible patterns. As Philip Ball says, 'pattern making lives on the edge of chaos, it appears when competing forces banish uniformity but cannot quite induce chaos. It sounds like a dangerous place to be, but it is always where we have lived' (Ball 2011: 182).

Such patterns have been computer modelled by Michael Batty at University College, London, based on the crucial observation 'that cities in particular and urban development in general emerge(s) from the bottom up and the spatial order we see in patterns [...] can be explained only in this way' (Batty 2007: 6). He observes that the cities' dynamics and random processes, constrained by context such as history and topography, lead to explicit patterns (Batty 2007). They encompass the multitude of activities within the city at every scale, the one-off and the reoccurring. Not only does this recognize that complexity comes from the bottom-up rather than from the top-down, but adaptation and evolution of the city's form is fundamental to this process.

Against the tendency in architecture to conceptualize and order from above, Robert Venturi's *Complexity & Contradiction in Architecture* (1966) advocated the importance of ambiguity against certainty, complexity rather than simplicity, summed up in his riposte to Mies van der Rohe's maxim, 'less is more' as 'less is a bore'. It was a liberating vision within which 'The Obligation Towards the Difficult Whole' (Chapter 10) is central to making life-affirming architecture. It is about creating order, or pattern, from all the competing elements of a building project. In contrast to the Modernist tenet of 'form follows function', he argued that buildings were considerably more complex than just their function and were equally a product of geography, history, culture, economics and politics.

In this spirit our practice, Hudson Architects takes a 'loose fit' view of design. Here (Figure 34.1), in the town of Castleford, Yorkshire, we designed a public square and furniture that had no singular purpose, to allow for open-ended uses. To our surprise and delight, it is being used in ways and patterns that we never envisaged by these children. For richness, diversity and adaptability, design needs to take account of the complexity of the real world from the observable phenomenon that surrounds us. It is about understanding the context and what is being observed in a particular place and time. But it is also crucial that we recognize unpredictability, the importance of chaos and the competing forces in the world from which emerge new forms, patterns and structures. Joseph Ford, following Charles Darwin, stated that '[e]volution is chaos with feedback' (Gleick 1987: 5). We should encourage a design process that allows forms to emerge from uncertainty and that mirrors nature.

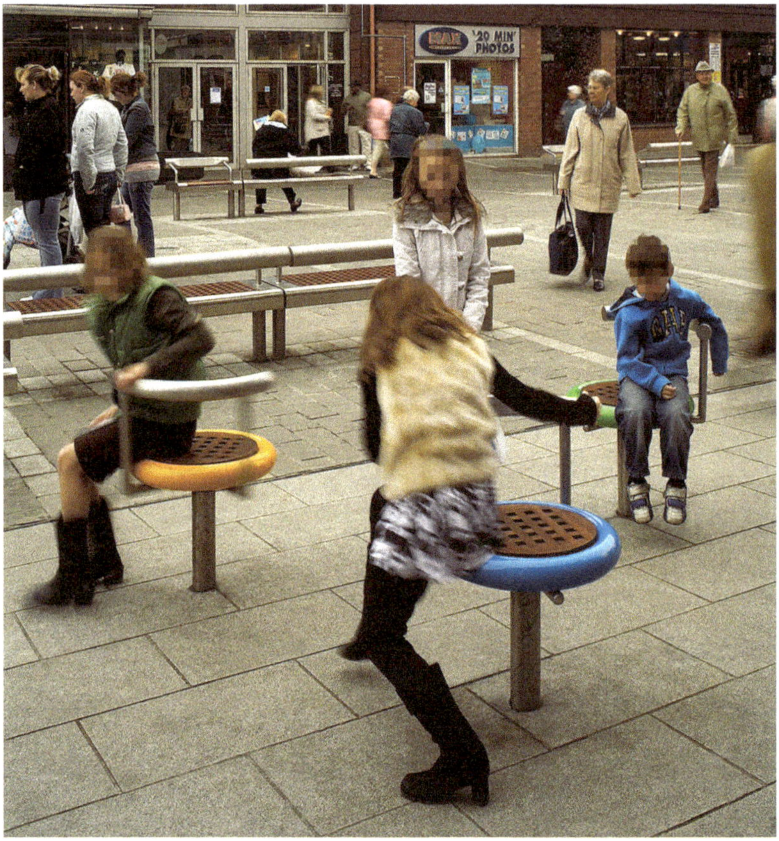

FIGURE 34.1: Hudson Architects. Henry Moore Square, Castleford, Yorkshire, 2006. © Hudson Architects.

The uncertainty of predicting how patterns might develop to adapt to new observations and to our changing environment is crucial to the evolution of our cities and architecture. The snowflake is a metaphor for this process. Each unique flake is itself a result of competing for chaotic forces as it travels through the atmosphere. The chaos surrounding it is unpredictable and unknowable, yet it emerges with the most amazing structure with no blueprint or imposed plan. Let us tease out and work with the forces in our environment in a similar way. Let us allow the invisible actions of de Certeau's citizens to drive the way forward.

35

Order?

Sarah Blair

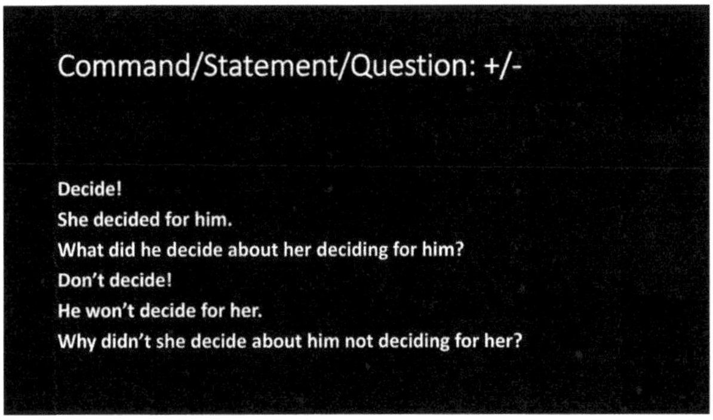

Figures 35.1–7: Sarah Blair, 2021. Digitally generated images. © Sarah Blair, 2021.

There is order in language, and it thrives by means of pattern. Yet it is also very playful and can become happily chaotic in being so. This is my territory.

I originally studied illustration, going on to do a Ph.D. that looked at the language of visual narrative with a strong focus on connections between visual and verbal behaviour. At that stage, I was looking at full-scale narratives – comics, graphic novels, picture books, as well as abstract stories and pictographic sequences of instructions. My level of scrutiny has subsequently become even more microscopic, and I am working now at the level of the sentence, exploring through simple visual translations the narrative patterns conjured by very few words. I am keen on the power of illustration – from Latin *illustrare*, 'to light up' – to illuminate, and my current work is illustration at the level of grammar and syntax, lighting up clause and phrase.

For the last few years, I have been scrutinizing the behaviour of English sentence structures, seeing how these rule-based primaries are idiosyncratically transformed in common usage by disordering decisions. The core structures (simple, compound and complex) and types (statement, command and question) are regularly put through aesthetic permutations. Once pointed out, the sensory character of these is readily appreciable to anyone with a creative-arts education, but this is not generally acknowledged in language-focused contexts – in the academic discipline of linguistics or broader language education, for example, where the terminology for linguistic behaviour has its own academic lineage. Part of the stimulation for my project has come from teaching language-based communication and research skills to art-and-design students, not least in supporting their approach to writing. The objective has been to make much more explicit the potential for the *creative crafting* of language, very much in parallel with a visual practice.

I have written previously (Blair 2019: 143) on how sentences vary in their essential energy according to core aesthetic themes. Regular patterns include effects like balance, contrast, repetition, inconsistency, accumulation, front-framing and sting-in-the-tale. These are categories familiar from the mechanics of picture-making. Sentence design plays with the same logic; it is worth looking at their forms, therefore, with a visual eye.

In what I show here, I have worked with a different set of principles, rooted now in primary metaphors which cognitive linguists locate at the base of our perceptual and conceptual mindset. These are foundational ideas determined seemingly by the intense bodily experience of the first months of life and that go on to dominate the modes by which we feel, think and express. The cognitive psychologist Jean M. Mandler (Mandler 2010: 33) distils them into the following shortlist. I have introduced groupings for ease of reference:

Path
Start path
End of path
Blocked path
Linked paths
Direct path

Motion
Motion transfer
Move out of sight

Contact
Attachment
Link

Container
Into container
Out of container

Surface
Onto surface
Off surface

Thing

Location
Up
Down
Behind

According to Mandler (2010), a baby's attention progressively makes sense of the chaos of the world by observing these repeating patterns in the things she encounters.

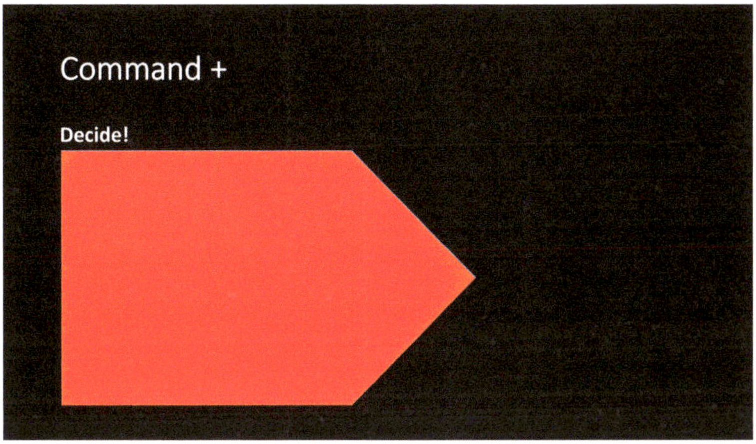
Command +
Decide!

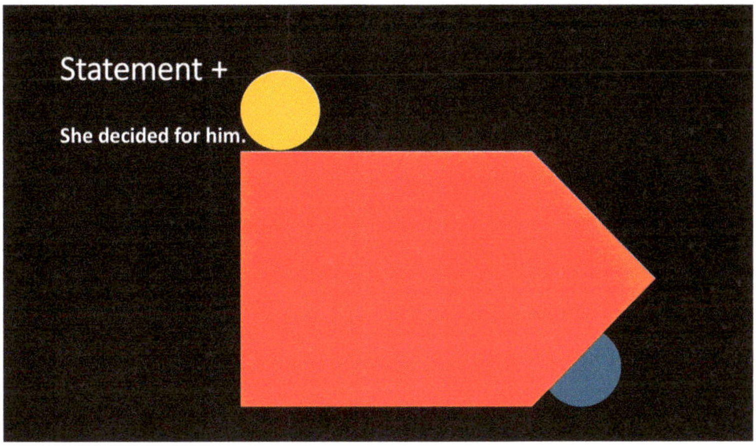
Statement +
She decided for him.

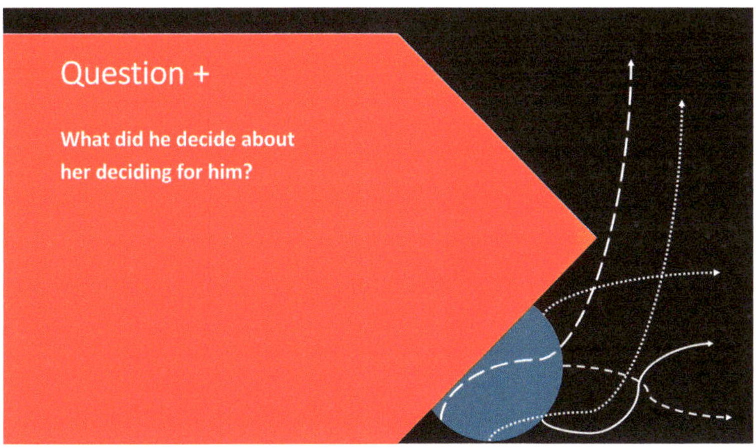
Question +
What did he decide about
her deciding for him?

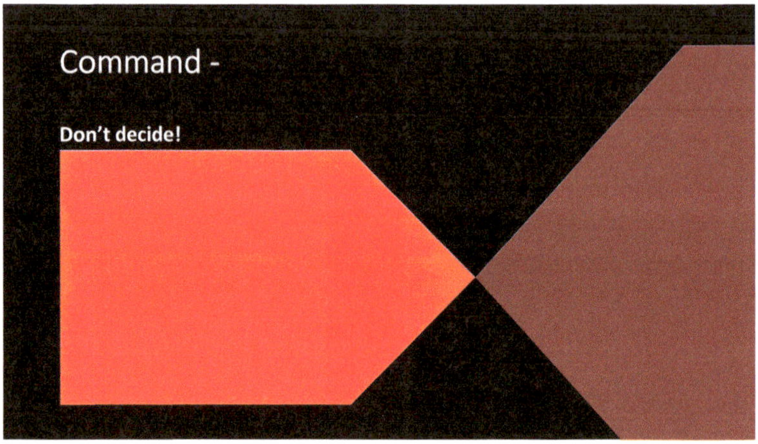
Command -
Don't decide!

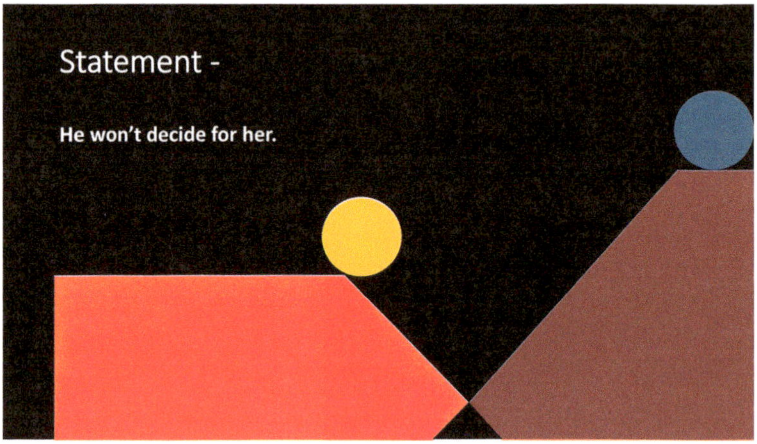
Statement -
He won't decide for her.

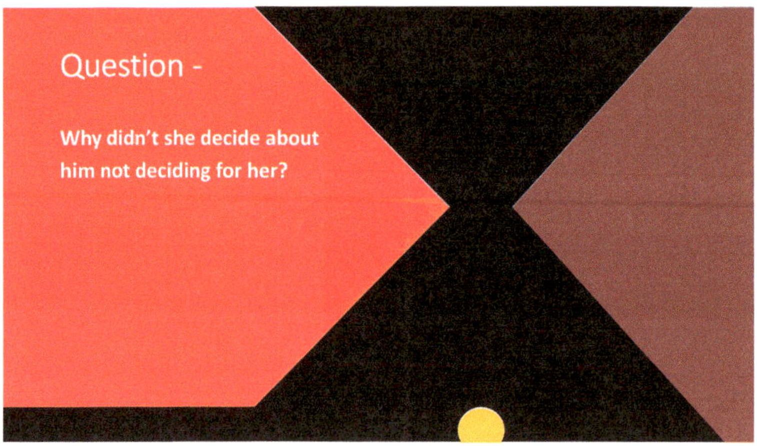
Question -
Why didn't she decide about
him not deciding for her?

Recent discussions in the field of cognitive linguistics (e.g. Górska 2008; Mandler 2010) have centred around how these ideas translate into metaphors, feeding our thought and channelling our communication across life. They inhabit not just the structures of our thinking and language, but are recognizably multi-modal: surfacing in everyday gestures, aspects of visual communication, strategies for performance and musical modes too. Mandler's list quickly suggests, for example, the generic language of maps (most obviously, the **up** and **down** of north and south), and there is an obvious affinity with Paul Klee's 'point' (**thing**), moving itself forwards to create a 'line' (**path**), and then further linking up with another line to engage in **contact** or **attachment** (Klee [1956] 1973: 16–17). In terms of ordinary verbal language, it is easy to characterize the grammatical 'noun' (from Latin *nomen*, 'name') as the discrete **thing** which we wish to name in our mental landscape as we speak – the subject of our sentence. And then the 'verb' becomes the action or state which is a type of **path** undertaken by this noun-**thing**, giving us the sentence's predicate. And so, in this small landscape of words, we have the feeling of a narrative journey being enacted, working like a child's view of a simple **thing** in motion, but taking in no end of complex material according to our fancy and experience.

Elżbieta Górska (2008: 19) has helpfully summarized how these ideas are embodied in common life concepts. On the right, I add some unashamedly hackneyed sentences as examples to demonstrate their familiarity:

STATES of being are LOCATIONS.	He was always the centre of attention.
CHANGE is MOTION.	In later years, she left all this behind.
CAUSES are FORCES.	He was driven both by curiosity and greed.
PURPOSES are DESTINATIONS.	She never got to where she wanted.
ACTION is SELF-PROPELLED MOTION.	He took himself in hand and made a start.
LONG-TERM PURPOSEFUL ACTIVITIES are JOURNEYS.	It was going to be a long haul but everyone was on board.

These are absolute clichés of course, though nonetheless resonant of life being lived. In my present work, I use a repertoire of visual coding that bounces off the back of these ideas: a visual translation of conceptual patterns underpinning sentence shapes. And so this adds another layer of visual signalling with which to approach a sentence analytically.

On this basis, I present in this chapter six images (Figures 35.2–35.7) devised in response to six ordinary sentences (Figure 35.1). To introduce them, some background on the thinking that has guided and been prompted by their construction:

1. Six-sentence structures are presented in a tight symmetrical framework: two commands, two statements and two questions; the first group is positive and the second negative.
2. The choice of words in the initial sentences is deliberately restricted, encouraging repetition and making clear any permutation or re-ordering.
3. The visual vocabulary is likewise limited. Two characters – (pro)nouns/circles – engage in interconnected acts or thoughts.
4. The sentence types are given specific personalities: a command is strongly directional; a statement presents a scenario; questions are more permissive and open.
5. It is worth noting that questions are interesting more generally, by their nature covering speculative, risky territory. They deliver a range of functions in social life, from grovelling petitions to outright commands (Connor n.d.), and they are often powerful plot changers (*How much do I want this job? Will you marry me?*). They may be likened to Anthony Howell's potent notion of 'inconsistency' in his trio of primary colours for performance art (Howell 1999: 71–86). In my images, questions are the grit that determines deviation.
6. With the three negative sentences, a significant shift takes place. A negative brings an obstruction – a **blocked path** – highlighting the force of negation in language and in life. Negation torpedoes intentions and desires; thus while 'not' is a little word, it is a force to be reckoned with. Its pattern is to threaten chaos.
7. In these six sentences, not all the primary metaphors are implemented. Those that seemed most apt are: **thing** ('she' and 'he' as self-contained entities), **motion transfer** (the verb 'decide' in negative and positive forms), **blocked path** (the negation), **paths** real and imagined ('he' wondering and choosing), **off surface** ('she' teetering on the brink), **up** ('she' being superior) and **down** ('he' and 'she' in morale collapse).

I should add finally that my approach here was prompted by one of Anthony Howell's performance exercises (1999: 156), where he makes the beautiful suggestion that 'sentences may be considered as objects'. And in a way, that is exactly what the visualizations I create try to do: to pick a sentence up and turn it to the light so that we might see facets that are otherwise too implicit to notice, or else so far buried that they seem invisible.

36

You Guys Are So Stochastic

Lucy Ward and Karoline Wiesner

Complexity theory describes ways that systems made up of many parts behave. A complex system is one that has many interacting components; examples are diverse, as in bee colonies, democracies, the human brain and the climate. Although they are complicated, complex systems may sometimes be reduced to a few simple rules, from which global order or patterns can emerge.

The example discussed here, concerning the emergence of complex pattern through drawing, is the outcome of a collaboration I undertook with mathematician Karoline Wiesner. The mathematical and philosophical foundations of the formation and evolution of complex systems are at the core of Wiesner's work, and a recent interest in her research is the application of information and network theory to the stability of democracy (Ladyman and Wiesner 2020). My artistic practice takes an expanded view of drawing, and I make intricate drawings mapping the occurrence of anticipated and repeated events in ordered 2D space, intimating a friction between ideal pattern and repetitive behaviour in our everyday experience. I use the practice of drawing and the systems of pattern design to explore our everyday habits and activity.

As drawing was the means for this collaboration, conversations began with notes, sketches, plans and diagrams, and moved on to patterns, mappings and arrays (Figure 36.1). Instead of making observational drawings, drawing was used to indicate and record action, movements and developments, laying down traces of activity or signifying lines of communication. Drawing can be a speculative activity, posing questions and communicating possible answers. Drawing was used in a number of ways to examine different aspects of the theory. One of these was by embodying the behaviours of complex systems. As an autographic practice, drawing can convey actions in the marks left on the page, in the trace of the movement of the hand. As Avis Newman says,

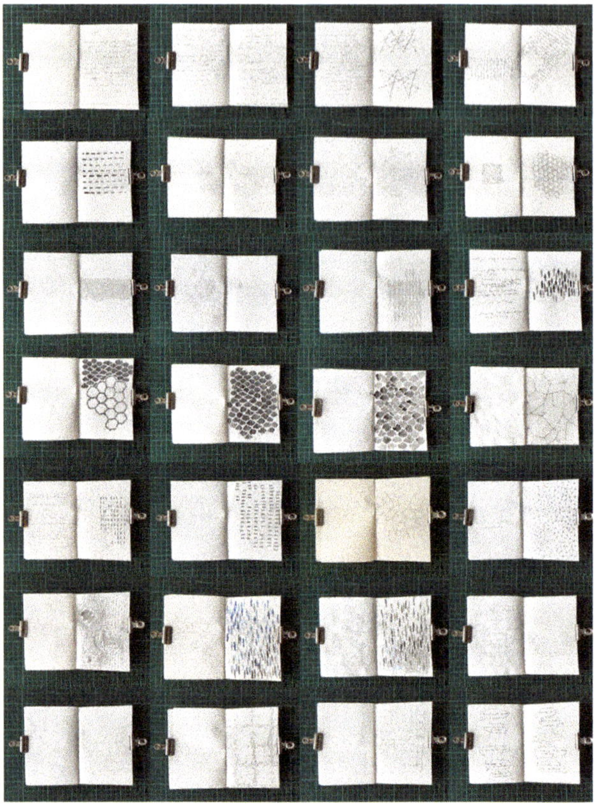

FIGURE 36.1: Lucy Ward, sketchbook pages, 2019. Pencil, ink and watercolour on paper, pages 10 cm × 15 cm. All images courtesy of the artist.

> [g]estures as they are experienced when looking at drawing are actions that are registered and suspended in time. And as such they encapsulate for the viewer an intimate moment of recognition that informs the self of the self through the actions of another (Newman and de Zegher 2003: 76).

An example of this comes in the exploration of concepts like 'disruption and recovery'. In nature, this might be witnessed as a large group of birds negotiating a church steeple. The birds are aware of the obstruction by communicating with their neighbours, and without any bird taking charge or directing the others. They are able to fly around the steeple, and the system recovers without any of the birds hitting it, and the group carry on their flight. An analogue to this can be found in drawing: when drawing out a pattern of marks, then encountering, negotiating and passing around a drawn 'obstacle' on the page, the artist can perform the actions of the birds when faced with the steeple. The pattern of marks and their disruption also indicates the pattern of behaviour by the artist. The negotiation of the drawn 'obstacle', by the squashed and manipulated pattern, relates to that of the compromising actions of the artist's hand while drawing. In this way, the artist can embody a complex system, using the traces, constraints and rhythms of drawing.

The rules and patterns that are characteristic of complex systems can also be performed when drawing by following repeated movements or systems of practice and translating these to the page. By working within a set of constraints, responding to the emerging drawing and applying rules to the process of making marks, the drawing becomes a complex system, playing out as the image develops. As noted in *Performance Drawing* (Foa et al. 2020: 65), '[d]rawing can acknowledge movement not only in the visual traces of an action carried out in time, but also by utilising philosophical and theoretical concepts to conceive a form of mark making'. Exemplifying this method, and following necessary rules while making the drawing, the need to work repeatedly from left to right, laying down layers of marks, so as not to smudge the ink, became essential.

To conclude the collaboration the aim was to make a drawing (Figure 36.2) by working in the manner of a complex system to build a coherent grid pattern on the page, made up of many small marks. There was no guide drawn on the paper, but the eventual emergence of the grid would be determined by the rituals of the drawing process and the placement of the marks as the drawing progressed. The small marks representing

FIGURE 36.2: Lucy Ward, *You Guys Are So Stochastic*, 2009. Ink and pencil on watercolour paper, 77 cm × 56 cm.

people were laid quickly so there wasn't time to consider exactly where they were placed, in a move to make their placement as random (or stochastic) as possible. The hexagonal lattice emerged through instinct, by reflecting on where connections between marks were forming, rather than with a pre-determined plan. This reflexive nature of drawing is another tool that can be used to explore and describe the nature of complex systems. 'Drawing plays with appearance; it oscillates between seeing, thinking, remembering and imagining, controlling and being controlled as the image emerges. It is continuously and simultaneously shifting itself in the course of its making' (Downs et al. 2007: xii). As the process of drawing is reflexive, the artist responds to the drawing on the page as it emerges, reviewing the build-up of marks, the emergence of the composition and the form of the work. Similarly, the complex system is in constant review, with the system's parts in communication, adapting their actions.

The aspects of complexity theory explored during the collaboration and in this drawing are expressed in the conversations held by the people in it and reflect the conversations that took place between the collaborators (Figure 36.3). Examples include: 'complexity comes from simplicity' ... 'simplicity comes from complexity', 'disruption' ... 'and recovery', 'we are so predictable!' ... 'yet so unpredictable!', 'macroscopic order' ... 'but microscopic disorder' and, tellingly, 'you guys are so stochastic!'.

Thanks to Jake Langham and CREATE-REACT for facilitating this collaboration, which was funded by the University of Bristol's EPSRC Impact Acceleration Account.

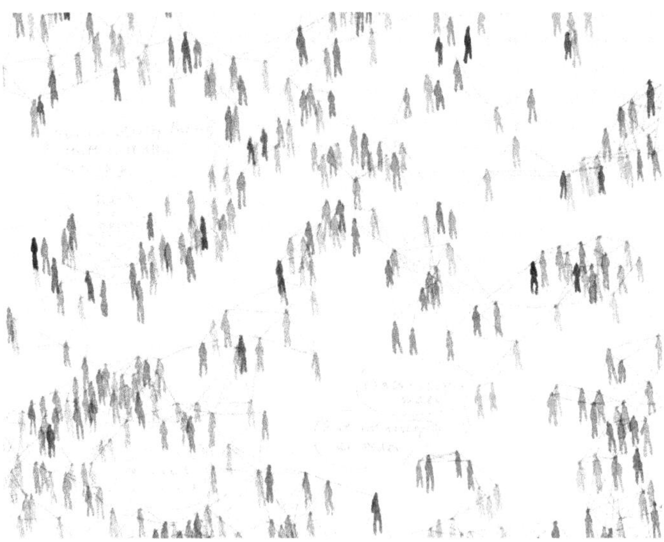

FIGURE 36.3: Lucy Ward, *You Guys Are So Stochastic* (detail), 2019. Ink and pencil on watercolour paper.

37

Clouds in the Machine

Sarah Horton

A polysemous motif

A cloud is a form that shifts in shape, morphs, dissolves and reassembles. Its mutability lends itself perfectly to pareidolic visions – the tendency to see images in random stimuli. It is a ubiquitous symbol, easily understood and assimilated and its shapeshifting propensities lend it endless appeal. Its position above the earth separates us from the lesser-known mysteries beyond which has led it to be characterized as a portal to other religious or mythical worlds.

Here I describe and probe the cloud as it figures in a body of artwork made for an industrial-scale laundry in Norfolk, UK. The research was designed to test a dual function of pattern: as something that could disrupt the appearance of the laundry by appearing out-of-place, and second to provide a visual metaphor for the repetitive processes of the laundry. By repeating the cloud motif across the workplace and negotiating my role as an artist in an unfamiliar environment, the project also problematizes assumptions around repetitive labour and the duality of habit as both accomplishment and constraint. The specific motif used for the pattern was the *cumulus humilis* cloud, a 'fair weather' cloud and probably the simplest and most recognizable of cloud formations (Figure 37.1). John Ruskin referred to the cumulus as the 'Cloud-Chariot ... the most majestic of all clouds' (Ruskin ([1860] 1905a: 144). Coincidentally cumulus means 'heap' or 'pile' (*Collins Dictionary* 2021) and this meaning is visually and semantically echoed in the bundles of clothes, bed linen and tablecloths that are stacked high in crates awaiting processing at the laundry.

The modern-day nomenclature for clouds dates back to Luke Howard's 1802 typology (Hamblyn 2017: 49), based on his observation that although clouds change shape endlessly they can be captured within a few basic forms: the *cirrus*, Latin

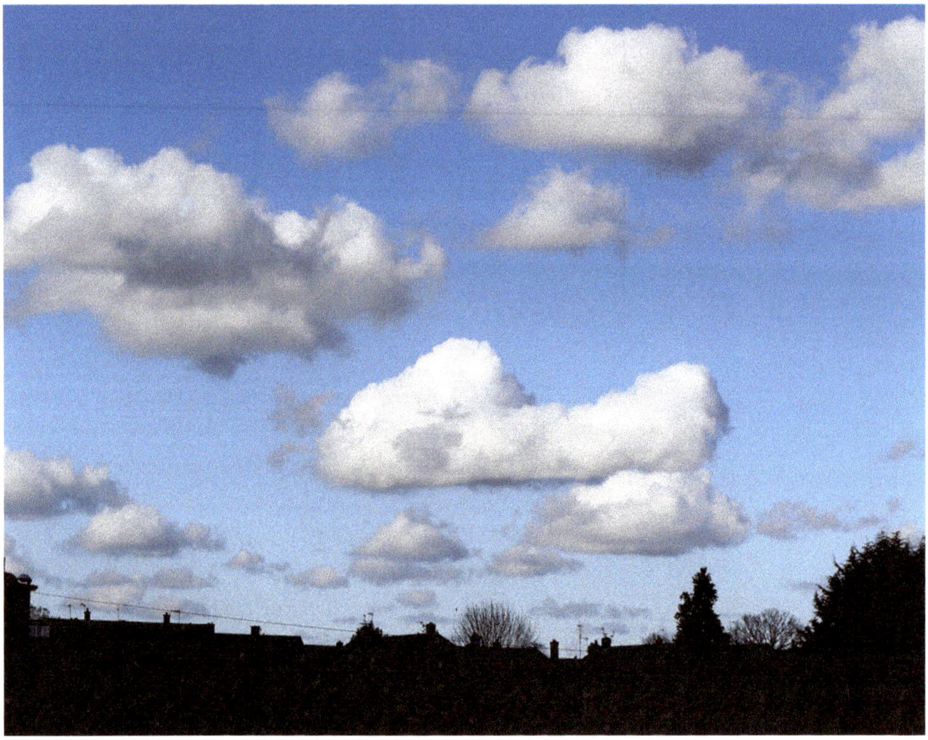

FIGURE 37.1: *Cumulus humilis* clouds. Photograph © Sarah Horton.

for 'fibre' or 'hair'; *cumulus* meaning 'heap' or 'pile' and *stratus* – 'layer' or 'sheet'. Thus, the chaos of the clouds is defined and brought into a system of categorizable pattern. This typology captures the changeability of cloud shapes within a number of basic patterns that, far from being random, are related to the cycle of the rise and fall of vapour and can change within minutes, like a continuously changing animation. What is especially notable in the context of the laundry is that they can all be related to textiles in some way.

Providing a foil for other-worldly scenarios as one shape morphs into another, the cloud has the ability to be whatever form the beholder imagines. The clouds carry the suggestion of alteration, affording a sense of hope, dreaming and escapism. Thus the complexity of the cloud as both repeating, but shape-shifting, as ethereal and as unearthly, and as a vehicle for day-dreaming, make it the ideal symbol for use as a patterned motif in the context of the laundry. When I mis-type the word 'cloud' its anagram 'could' conjures and confirms my sense that here is a motif that has the potential to indicate endless possibilities.

Cloudworks

Roland Barthes argues that 'all images are polysemous: they imply, underlying their signifiers, a "floating chain" of signifieds, the reader able to choose some and ignore others' (Barthes 1977: 38). In this context it might be said that Barthes's chain of signifieds 'floats' with the clouds. The denotative meaning of a cloud image is of a mass of vapours, but it also has the potential to connote associated symbolic meanings – images that are replete with potential signifiers. When discussing the 'rhetoric of the image', Barthes makes clear that connotative signifiers are generated through different kinds of knowledge: 'practical, natural, cultural, aesthetic, and that these can be classified or brought into a "typology"' (Barthes 1977: 46). Gavin Pretor-Pinney (2006), Richard Hamblyn (2017) and Hubert Damisch (2002) gather many of these cloud 'types' or associations together: the cloud as an intermediary between heaven and earth; the cloud as furniture of the gods and the cloud as a vehicle for simulacra-seeking are just a few of these. The artworks in this project, although denotative by design, are intended to carry additional meanings, partially dependent on their form, materials and locations.

Within this industrial laundry, the products that are washed are frequently hard-wearing and durable items of workwear which are plain and often white, along with table and bed linen from the hospitality industry. In contrast, *A Break in the Clouds* (Figure 37.2) was made of a 3-m piece of silk printed with a repeated cloud pattern generated from my own painting. Whereas the fabric in the laundry is put to the service of utilitarian functions, this artwork has no practical use value. The title of the piece hints at its potential to break with the utilitarian milieu of the workplace. It is a singular piece of fabric unlike the garments and linen which are processed in their hundreds, the silk contrasting strongly with the other fabrics seen in the laundry, billowing ever so slightly when walked past. The blue dye is a riposte to the magnolia-painted walls as well as providing a view that is never present in this part of the environment as it is without windows. Its height breaks up the wall space. These clouds, in their own way, let the light in.

As a single outline, *Dream Cloud* (Figure 37.3) might appear like a thought or speech bubble, an association emphasized in its reflection in the office door where it appears to float above the desk of an office administrator (Figure 37.4). Here fluorescent lighting was used to produce an instantly recognizable cloud, a linear outline depicted in glass and containing gases. The neon is both pleasurable and industrial, both art material (think Dan Flavin and Tracey Emin, for example) and non-art material.

The repetitious nature of work, both in the laundry and in the office, is unmistakable. In the office space of the laundry, for example, when administrative staff answer telephone

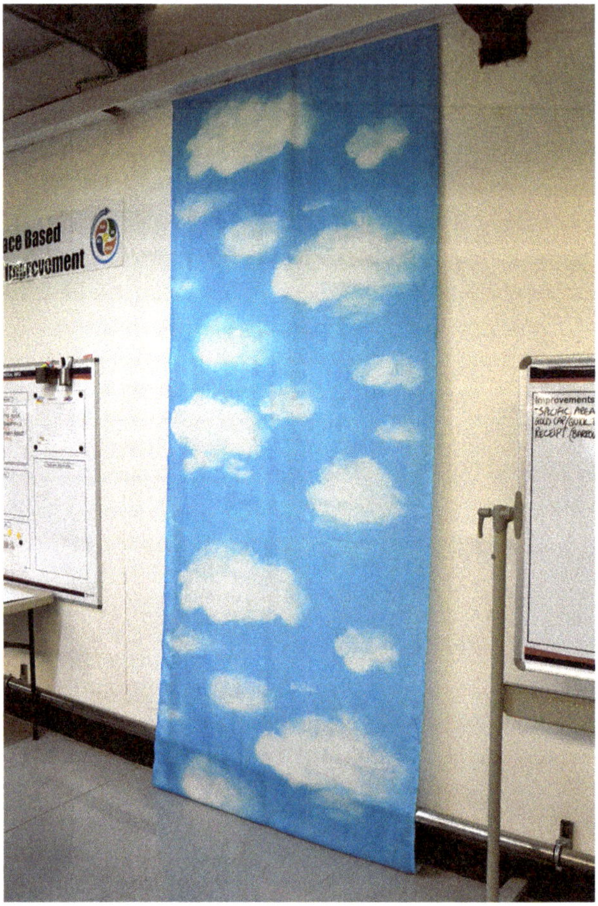

FIGURE 37.2: *A Break in the Clouds*, 2014, installed in the main laundry. Digital print on silk. 350 cm × 160 cm. © Sarah Horton.

calls their responses are almost identical. In the laundry, the gestures and movements are efficient and precise: hanging, folding, pressing, sorting and patting. Used over and over, they are familiar and automatic, not in a dissimilar way to the machines they work alongside. These repetitious behaviours in the workplace are reproduced metaphorically in the artworks, in the repeated patterns of clouds and their recurrence around the laundry in various locations. The manual repetition of painted clouds reiterates and mirrors the manual repetition of folding fabric, even as they also break the rhythm of work.

Through their different locations the cloud could be repeatedly seen across the whole site emphasizing the morphing, mutating nature of the cloud and connecting one

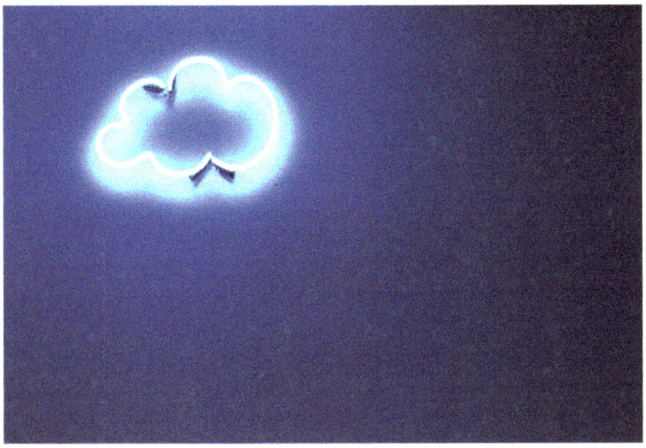

FIGURE 37.3: *Dream Cloud*, 2014, installed on the main staircase. Neon. 40 cm × 60 cm. © Sarah Horton.

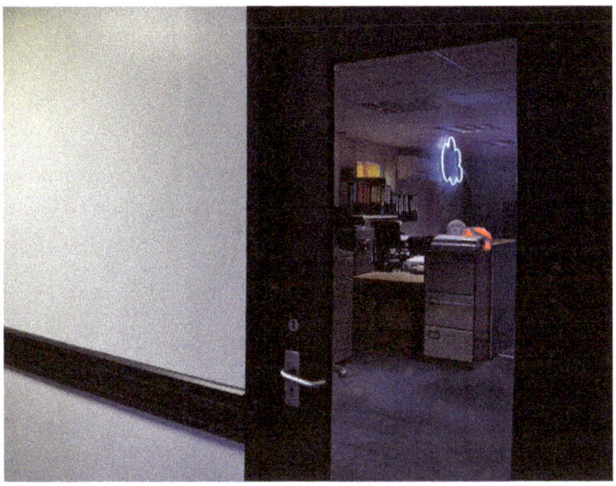

FIGURE 37.4: *Dream Cloud* reflected in the first-floor office window. © Sarah Horton.

space or function to another. *Dream Cloud* was positioned at the highest point of the laundry on a staircase used by all staff to access the washrooms and canteen; two further cloudworks were positioned on the main factory/laundry floor; another in the reception area; another in a showroom of uniforms and one in the main sales office. The artworks act as a portal or aperture for the imagination of the workers, both a broadening of vision away from the artificially-lit space of the factory to the arguably unlimited interiority of the mind and the exterior of the sky, the clouds functioning as a metaphor for both the outside and the inside – a literal sign of the outside or exterior and a symbol for interior thought.

The transitional aspect of the cloud, and the contrast between the shape-shifter and repetition, is also mirrored throughout the laundry in the transition between the chaotic mounds of material to finished, folded pieces, where the cloth is changed from dirty to clean, crumpled to pressed, disordered to ordered. The cloud and its repetition throughout the laundry is a reminder of these repetitive but shape-shifting aspects of the labour. Interestingly, the cloud resembles soapsuds in both its appearance and its make-up. Thus, in true pareidolic fashion, the clouds appear to be soapsuds fomenting and agitating, as they do in the oversized industrial washing machines. Interestingly, although the word pareidolia only goes back to 1962 (*Merriam-Webster* 2021), the sensation is itself described in many earlier sources including Shakespeare's *Hamlet* (1609) and da Vinci's *Treatise of Painting* ([c.1540] 1877).

Head in the Clouds (Figure 37.5) alludes to the cloud in its more recent iteration as a virtual space for storing and accessing data. Although the cloud as a mass of data

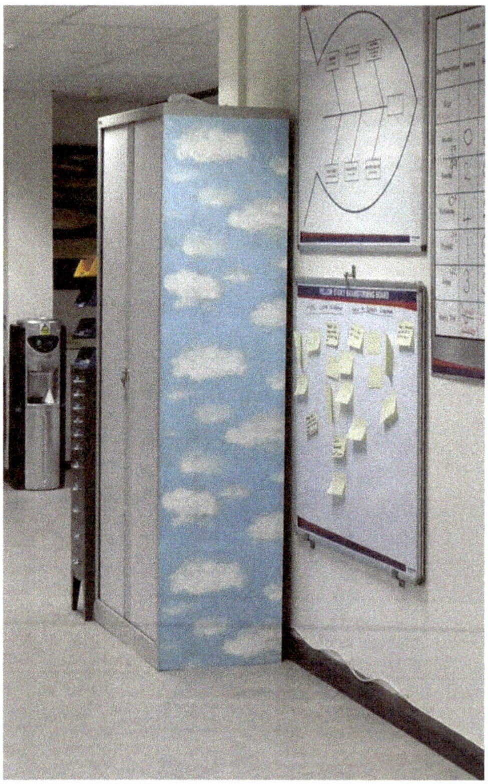

FIGURE 37.5: *Head in the Clouds*, 2014, installed on the side of a cabinet in the sales office. Digital print on vinyl. 160 cm × 40 cm. © Sarah Horton.

storage is a relatively modern concept, Thomas Hearne, as far back as 1705, referred to 'a cloud of informations' (Hearne 1885).

In the context of the laundry sales office, the cloud pattern is situated alongside whiteboards that are used to record production targets as well as for brainstorming ideas. In an environment where sequence and order are vital to the laundry process the office is one of the few places where 'blue sky thinking' is actively encouraged, presumably to visualize new strategies for maximizing efficiency and profitability.

Straitjackets and corsets

Although many of the processes at the laundry are automated, human labour is still required to load and unload the machines. This is repetitive and fast work in a warm and fairly airless environment. Whilst large industrial machines do the washing, and giant rollers steam and press, tea towels and large tablecloths are meticulously folded and stacked in large quantities by hand. The processes used are both hands-on and mechanized, the body of the person working alongside the body of the machine.

The polysemous richness of the cloud, its shape-shifting qualities and potential for pareidolic interpretations have made it the subject of many literary and artistic references. In 1820, Percy Bysshe Shelley wrote of the cloud 'I change but I cannot die' (Hutchinson 1907: 596), referring to the circularity and infinity of the water cycle. In the laundry, the sorting, washing, drying and pressing processes continue throughout the day and night. The same items return week after week, the same workers arriving day after day. The painted repetition of the cloud mirrors these never-ending processes and the cyclical nature of the industry that takes place in the laundry. Henri Lefebvre describes the way the rhythms of the body are superseded by the rhythms of capital (1992). In the laundry, the activities of washing, rinsing, pressing and folding need to continue day and night to keep pace with demand. So the laundry operatives suppress and alter their own circadian rhythms to comply with the machines they work alongside and the demands of the business.

The distribution of bodies in the earliest factories is described in detail by Michel Foucault (1977) who argued that spatial arrangements were made according to function and located in relation to the machines operated so that all could be supervised through central spaces for observing workers or for 'the individualising fragmentation of labour power' (Foucault 1977: 145). The control of activity was also governed by the use of the timetable: 'the temporal elaboration of the act ... [and] correlation of the body and the gesture' used to maximize efficiency (Foucault 1977: 150–51).

In the laundry, the workers are incredibly dextrous, being completely in control of the unitized and fragmented tasks they undertake. At the same time, however, their boundedness to time and strict production quotas removes from them any real autonomy.

Repetition and compliance to routine are often cited as positive: orderly, logical and constructive. Predictable patterns in behavioural terms are deemed polite and accommodating. Chaos, on the other hand, is disorderly and disruptive, out-of-control. However, repeated patterns of behaviour also signal passivity and a lack of criticality. Chaotic tactics, as employed in Dada, Surrealism and Situationism, for example, conjure imaginative, critical and questioning characteristics that reverse the pattern/chaos pecking order. In the context of industry, in this laundry, the cloud patterns are rogue elements that don't belong. In this sense, the pattern reverses its usual function and introduces a chaotic, disorderly and irreverent element because decorative pattern here is out-of-place.

Routine and repetition are not always indices of mindless, unthinking behaviour, where the individual has little or no agency. According to psychologist William James (1890), we learn to complete repeated complex behaviours, e.g. driving, by delegating to muscle groups. As these actions become automatic, we can use our brains to daydream, problem-solve, lust or be creative. The brain's plasticity means that actions that were initially complicated and difficult eventually lead to accomplishment.

Billy Ehn and Orvar Löfgren similarly point out that routines can be perceived as either of two polarities: 'constraining straitjackets or supportive corsets', such that routine can actually provide a form of reassuring familiarity and also allow for multi-tasking (2009: 101). They argue that the mastering of routine behaviours can work as a liberating force, providing energy and focus for other things. Maybe in the context of the laundry, the routine nature of many of these repetitive tasks allows the workers to daydream, and these artworks may have become an active conduit for this.

'I would build a cloudy house / For my thoughts to live in / When for earth too fancy-loose / And too low for Heaven' (Barrett Browning 1870: 94). If the cloudscapes began to inhabit the minds of the staff within the laundry, I found that the laundry consumed my own thoughts through the duration of this project, even at times becoming the home for my own labour as I photographed the site, spoke to staff and installed the artworks. Here I found that my own presence was also out-of-place in this context. The experience of working at the laundry altered my understanding of my role as an artist. Many of the tasks I undertook for the making and installation of these artworks were repetitive: pinning, tacking, stitching and so on. They were administrative: photocopying, sending emails, making telephone calls and meeting with personnel. These activities are not especially appealing or creative but are fundamental for the realization of the work. They are laborious and protracted and challenge any romantic idea of what it is to be an artist.

Where the labour of the artist and other professions can be seen to overlap, therefore, is in the number of repetitive tasks that have to be completed to work effectively; also the hidden administration and health and safety concessions that were common to the tasks of many at the laundry, as well as to myself. Thus, although initially feeling discombobulated by this displacement I began to see many ways in which my own labour echoed that of staff within the laundry (Figure 37.6).

If, as James posited, habit can be positioned as both problem and possibility, Ben Highmore (2014) adds that developing new habits can foster enriching experiences (such as my placement within industry), leading to fresh understanding and nurturing a sense of resilience. As we have seen in recent history, in response to crisis new patterns and habits can emerge – new forms of collectivity as we prepare for and deal with the unexpected.

Conclusion

In Aristophanes's play *The Clouds* (423 BC), the chorus of cloud characters have two roles: first, they are the patron goddesses of thinkers and, second, when they remove their cloud costumes, they are conspirators with the audience. Acting in this dual capacity, the clouds negotiate between the audience and the make-believe characters on the stage. The clouds at the laundry also possess a dual character. They are, on the one hand, imposters, placed to unsettle the visual appearance of the laundry and offer new ways of thinking about it. On the other hand, their decorative appearance enables them to be assimilated and accepted through their pleasurable qualities, to be able to 'speak' or negotiate. If Aristophanes's clouds are interlocutors, so too are these artworks. Designed to insinuate themselves as alien but accepted, even appreciated, elements in the workplace, they occupy a unique space, conceptually and physically, enabling a visual dialogue to take place that challenges the décor, the furnishings and fixings of the workplace, able to oscillate between the world of work and leisure in their suggestion of 'otherness'. Disturbing the efficient decorum of the workplace, the somatic, affective, sensual and bodily aspects signified or captured within these artworks are key components. Thus, the 'dirty' business of these artistic imposters/interlocutors can be seen as ameliorative, playful and unfamiliar interventions that insert and assert themselves, presenting a resistive act (de Certeau 1984) in the workplace.

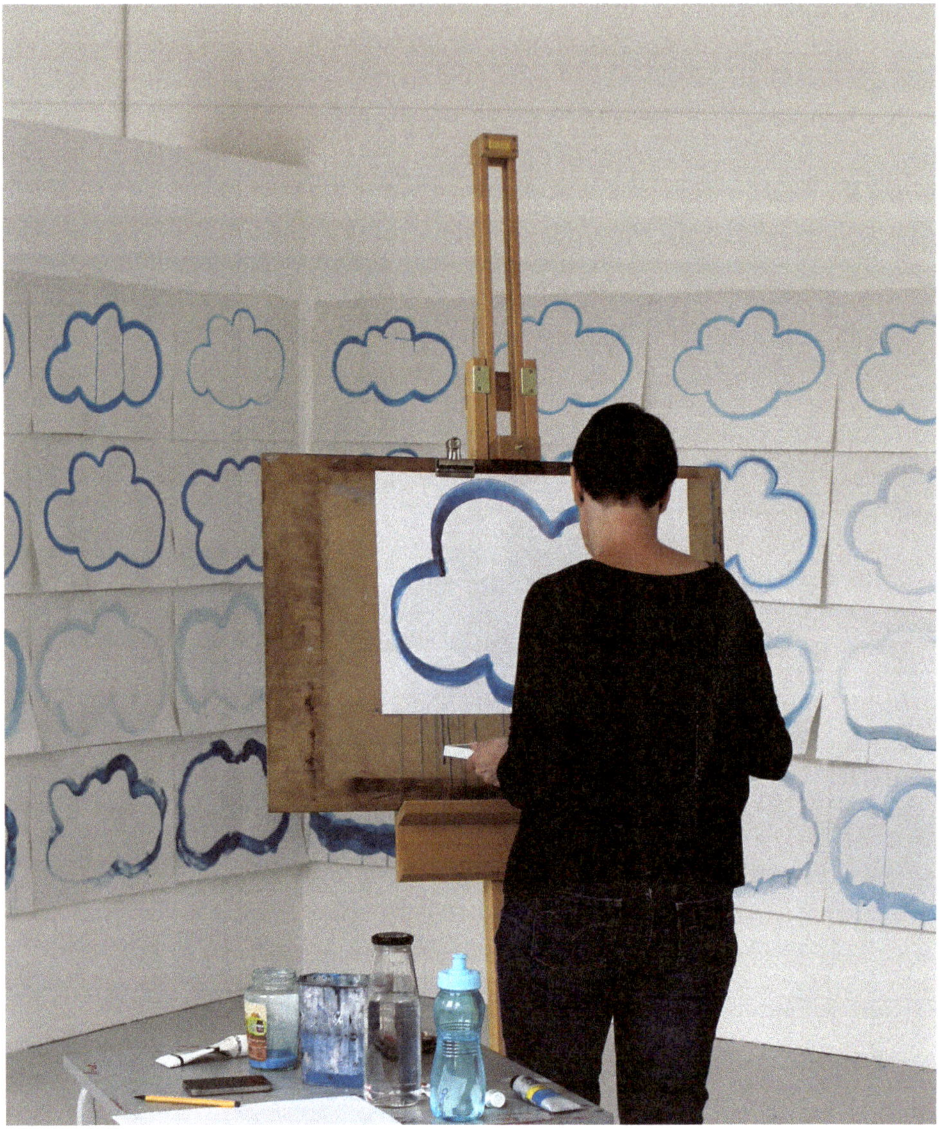

Figure 37.6: Sarah Horton, *How to Paint the Perfect Cloud*, 2014. Performance to camera. © Sarah Horton.

PART 6

NOTHINGS IN PARTICULAR

Introduction

Whether at the edges of the finite world or lurking in the interstices of matter, the inconsequential here assumes focus. The random, ghostly, transitory or experimental become starting points from which patterns of reason or stability might be conceived or disrupted. A speck of dust or an incidental action might disturb order and the ordinary but give shape to patterns beyond, whether mystical, psychological, mathematical or dangerous to life. As materiality and immateriality intermingle and pattern and chaos intertwine, nuances of subjectivity and perception are also shaped, for better or worse.

'Nothings in Particular' is a phrase that Bill Prosser incorporates into his contribution here and which draws on notions of 'nothing in particular' as contemplated by Samuel Beckett and Martin Heidegger (see also Prosser 2012). Within the remit of this book, the notion of nothings in particular signals a density and complexity of detail, an aporia, that begs more questions of the intersection between pattern and chaos than it answers. For Beckett, 'to think of myself as somewhere on a road, moving, between a beginning and an end, gaining ground, losing ground, getting lost, but somehow in the long run making headway' gives way to almost complete self-annihilation, thus 'I have nothing to do, that is to say nothing in particular' (Beckett [1958] 2010: 25). His words form a kind of existential refrain or continuously circulating pattern of thought at the edges of reason.

For Doris Rohr, the 'lowly matter' of dust is both formative and destructive – symptomatic of fraying emotions and a metaphor for 'dark matter' in the contexts of the environment and health, yet also that from which crystalline diamonds are formed, or gold, or the carbon in the pencil used in her *Melencolia* drawings. The subtleties and varieties of graphite are subject to a different kind of reflection for Prosser. Set against the measured like-for-like of mimesis, his forensic attention not only focuses on insignificant things first caught, like a speck, out of the corner of the eye but also on the 'cocktail' of possibilities and the chaos of raw data that the concentrated act of drawing unleashes.

In the pattern-rich carpets which Victoria Mitchell considers as hypnotic in their ambiguity and as opening the mind to dreaming and the unconscious, worn or speckled fibres have agency, activating perception of pattern-chaos and drawing the subject into hidden worlds. Medium and message similarly go hand in hand in evanescent images and allusions flickering 'at the edges of perception' in the carefully processed artwork and poetic-narrative writing of Andrea Stokes. In a subtle combination of pencil drawing, net curtains, Vaseline and photography, the magical, historical associations of Cottingley are both caught in the geometry of the netting but also float free, not unlike the gossamer web of unknowing that language makes possible.

In the typestracts and 'cosmic dust laminate poems' of Dom Sylvester Houédard, curated and described by Nicola Simpson, spiritual transmission and infinite wisdom are grounded in the seeming chaos of all objects (also including Vaseline), a disrespect for syntax and an acceptance of all manner of transgressive actions. Houédard's Tantric collages 'wove together the chaos of forces, events, consciousness and objects that appeared in the daily and impermanent flux of life' but were nevertheless sustained by and rooted in a pattern of daily rituals.

In ancient Egypt, the mythological serpent Ouroboros was in a state of constant rebirth from the act of continuously devouring itself. In the meeting between pattern and chaos explored in this volume, this state of constant becoming and equally constant unravelling is not dissimilar. It is therefore prescient to note that Houédard's dissembling of reason as explored in the final essay is not without reference to the motif of the spiral which was the subject of the first essay in this volume. His typestracts were occasionally composed in the form of a spiralling mantra and in one of his laminate poems a spiral form swirls in fragments as if in an endlessly mobile vortex. As the poet T. S. Eliot reminds us in 'Little Gidding' (1942), 'the end of all our exploring Will be to arrive where we started And know the place for the first time'.

38

The Shape of Dust

Doris Rohr

> Home! She looked round the room, reviewing all its familiar objects which she had dusted once a week for so many years, wondering where on earth all the dust came from. Perhaps she would never see again those familiar objects from which she had never dreamed of being divided (Joyce [1914] 1968: 34–35).

Dust sticks to the hand. Film of dust on the old photographs, at the bottom of the crystal glasses, in the boxes with the unpacked library. For Eveline dust evokes shrouded memory and the comfort of habit. Dust's persistence in settling and remaining indicates Eveline's lack of courage to leave her father's home. She dreams of a new destiny in America but cannot leave.

Dust implies absence, neglect and stagnation. Dust, as a non-thing or lowly matter, is a spurious concept. In the hierarchy of all things scientific, it ranks low, and lowlier still in its metaphorical or moral evaluation. Paralysis is a commonplace association with dust: it connotes slovenliness or an unhealthy anchoring of a person in the past (Charles Dickens's character Miss Havisham, in *Great Expectations* [1861], for different reasons to James Joyce's Eveline, is caught in emotional paralysis). This is a state of mental stasis, where dust acts as a literary metaphor. Dust also causes physical damage: miners suffer from coal dust and builders' lungs are infected by asbestos dust. Whole industrial communities have become affected by the legacy of CO_2 emissions. Health matters are inextricably linked to a causal relationship between pollution with respiratory illnesses. Dust's danger denotes ill health in more than physical ways. Regardless of whether dust is a product of natural disaster or man-made, emotional associations tie dust to memory, loss and forgetting.

Scientific evaluations 'objectify' dust as a medium for analysis. These in turn feed eschatological fantasies and predictions of an environmentally depleted earth, the horror of Gomorrah, or Last Judgement, the Endgame. In this – *our* – age patterns of normality in everyday life have become frayed and disjunct. Visual metaphors such as mountain debris, and transformative agents of weather, chance events and destiny become conflated, mixed, hybrid. We seem condemned to watch impotently familiar structures unravel. Up in flames do not only go our remaining primaeval forests with their precious wealth of diversity of species but also our understanding of everyday life, personal security and trust in political representation. This is the vortex: dust being swept up by the whirlwind of the twenty-first century.

Cleaning up the world, or our cultural legacy, takes on a different order from dusting, wiping, removing. The polluting agent informs new ways of seeing. Does this offer solutions to problems and afford a paradigm shift? One example is conceptual artist Marcel Duchamp's *Large Glass* (1915–23) as seen through the eyes of Man Ray in *Dust Breeding* (Man Ray [1920] 1967: n.pag.), another is contemporary artist Jorge Ortero-Pailos's architectural preoccupation with dust on buildings. Both these examples speak of a desire to reincarnate the original artwork, giving it new life and adding surplus meanings. Man Ray presents an altered vision of the *Large Glass* as *Dust Breeding* as a horizontal planar landscape, as a mythological map, even. Ortero-Pailos makes new 'surplus' art out of restoring the old. Man Ray does not appear to wish to restore Duchamp's object, taking it in its abandoned presence as a quasi-archaeological find, a readymade dust drawing. Ortero-Pailos's efforts, although restorative, make us think about dust in a new way, as both attractive and repulsive. In *The Ethics of Dust: Westminster Hall*, Ortero-Pailos (2016: n.pag.), one might say, creates Bruno Latour's *quasi-objects*[1] (2004), dust as dark matter that needs removing and becomes in turn the exhibit. An ultimate form of recycling or an alchemical transformation? Dust captured, locked into yellow honey-coloured latex – dust, like sticky film on flycatchers. Unlike those, Ortero-Pailos's latexed dust is mesmerizing, attractive and seductive. Otero-Pailos's exhibits offer a different way of appreciating the grime of time, making dust desirable. His objectives are twofold – through applying the latex, architecture becomes cleaned and preserved, thereby he makes the viewer reconsider a stale relationship, turning the unwanted and abject into artwork. Robert Hewison (2016) relates Otero-Pailos's title to John Ruskin's *The Ethics of the Dust*, subtitled 'Ten Lectures to Little Housewives on the Elements of Crystallisation' ([1866] 1905).

The Ethics of the Dust

Ruskin's text makes for strange reading. A mature male figure, 'the lecturer', a thinly disguised John Ruskin, is introduced as the voice of reason, speaking to a small group of young girls. The format reinforces patriarchal structures of pedagogy. Despite these misgivings, perseverance with the text offers fascinating glimpses of Ruskin's thinking. The lecturer speaks to one of the young girls called Florrie about 'mixed dust of snow and gold' settling in the 'serpent forests' that contain 'great cliffs of dead gold, which form a labyrinth winding always higher and higher, til the gold is all split asunder by wedges of ice; and glaciers' (Ruskin [1866] 1905: 214). This fairy tale landscape is rich in the use of oxymorons ('golden ashes', 1905: 215) that evoke Ruskin's moral universe. Gold and diamonds, pure, desirable, speak of a heavenly kingdom, the language of precious and semi-precious stones of the Old Testament's description of the temple, and of Revelation's mystical images of a heavenly kingdom. Yet a diamond's double identity is coal, and gold becomes dust, lacking stability and form. Ruskin's language makes clear that this is a corrupted universe, and the pilgrim must navigate the path on grass lanes, resisting the temptation to become distracted by the lure of gold and riches. This is a repeated motif in Ruskin's writings: soiled Eden, the corruption of the industrial revolution and of capitalist gain (a motif that Raymond Fitch explores in depth in *The Poison Sky: Myth and Apocalypse in Ruskin* 1982). In Ruskin's own *Notes* to *The Ethics of the Dust* he directly references 'the souls of all who had perished in misery through the pursuit of riches' and the 'serpents [...] the souls who had lived carelessly and wantonly in their riches' (1905: 367).

In his essay 'A Dry Black Veil' (2009), Brian Dillon investigates the metaphorical function of dust in Victorian Britain. Dust becomes a larger category, 'visionary and prophetic, attuned to the utopian and apocalyptic possibilities that inhere to earth and air' (Dillon 2009: 8). Ella Mershon (2016) speaks about Ruskin's interest in the transformation of debris or dust into renewable form. In her essay 'Ruskin's Dust' she uses *The Ethics of the Dust* as a prime text for her proposal that Ruskin's interest in dust goes further than geology, requiring a broader scientific and cultural perspective. Her argument runs that dust 'signifies decay's release of chemical potentiality, which thus encodes inorganic matter's formative power' (2016: 464). Mershon comments on Ruskin's wavering views on whether erosion leads to entropy or permits the possibility of reforming matter into new mountain ranges (Rohr 2016; Mershon 2016). *The Ethics of the Dust* signifies a moment of hopefulness bracketed by periods of despair on either side of its time of writing (2016: 466). Ruskin's subsequent lectures on the *Stormcloud* ([1884] 1908) are a late example of utter despondency and pessimism about humanity's incapacity to look after natural resources and lack of care for nature.

Mershon situates *The Ethics of the Dust* as a didactic and hopeful text within Ruskin's overall more prophetic and pessimistic perspective on the power of nature to restore itself. In 'Lecture 1' Ruskin teaches his cohort of young women that coal and diamond are two forms of the same substance, they are allomorphic (based on the same chemical compound) (Mershon 2016: 475). This causes incredulity. Ruskin aka 'the lecturer' performs a magician's task of alchemical transformation through rhetoric. He conjures a rich tapestry of minerals and of such witnesses of geological time as amber and fossils, introducing scientific concepts that fascinate through their transformative poetic potential. Dull and sticky become clear, mesmerizing; dark and dust become splendid, refracting light into the spectrum of the rainbow. Moreover, there is the humble pencil, a 'formless diamond' ('Your pencils in fact are all pointed with formless diamond, though they would be HHH pencils to purpose, if it crystallised') (Ruskin 1905: 219).

Formless

For the author of the *Elements of Drawing* the pencil is a tool to examine and understand the laws of nature (Rohr 2016). With this it becomes apparent the potential of the pencil's carbon for the resurrection of what is fragile, expecting transformation, destruction or erosion, and moreover, the capacity to remember, recall or imagine, drawing as a means of giving form to loss. In *The Ethics of the Dust*, the lecturer's tutee Sibyl finds it hard to believe that diamond and carbon are chemically the same and this acts as a lever for Ruskin to expand on the hybridity of coal as a metaphor – its obvious association with the industrial world and its chemical other in the shape of diamond belonging to the world of earthly (corrupt) or heavenly (unattainable) riches, as formless potential – an inscribing tool, opening the world of observation of natural form and design, and of cultural imagination.

Is it possible to see continued relevance in Ruskin's *formless*? Recent philosophical discussions on new materialism embrace a reordering of animate and inanimate matter (Bennett 2010). Dust, actual and metaphorical, may help us to understand the restitution or reordering of our world.

Mershon introduces dust as negative space for potential: 'Dust's morphic ability to take on 'other forms' conjures a negative space, the empty outline of what is not – but might have been – present' (2016: 482). Loss of form, negative space – this terminology sounds familiar to anyone who has studied late modernist art and viewed Yves-Alain Bois's and Rosalind Krauss's seminal exhibition catalogue *Formless: A User's Guide* (1997). In the introduction, 'The Use Value of "Formless"' (1997: 13–40), Bois applies avantgarde writer Georges Bataille's concept of formlessness as de-definition of matter, to radically reposition art processes and products, breaking with conventions such as 'style, theme, chronology, and [...] oeuvre as the total body of an artist's work' (1997: 21). The radical new categories are divided into 'operations', 'horizontality', 'base materialism', 'pulse' and 'entropy' (1997: 21). The curators, Bois and Krauss, selected artworks they considered effective in deconstructing art historical norms, avoiding the presentation of works based on an individual artist's oeuvre, thereby deconstructing a sense of an artist's identity or authorship (1997: 24). This principle of reorganization creates freedom to re-evaluate irreverence in art, and repositions artworks, dismantling established art historical and theoretical modernist hierarchies: 'The "tactile" that art history addresses is only the visual representation of tactility: matter does not exist for it except as in-formed, made over into form' (1997: 25).

'[B]eing purely visual', art is the privilege of the eye, tapping into the ocularcentrism of modernity. This situates art experience in the verticality of the 'erect' human being. 'Horizontality' in contrast permits to re-establish the 'horizontal axis', akin to the perspective of animals (1997: 25). This provides the anchoring of horizontal experience with dust; Bois reminds us that 'the modernist picture is still conceived as a vertical section that presupposes the viewer's having forgotten that his or her feet are in the dirt' (1997: 25).

The curators, Bois and Krauss, juxtapose Robert Rauschenberg's *Dirt Painting (for John Cage)* (c.1953) with *Untitled (Gold Painting)* (c.1953) (Rauschenberg 1953: n.pag.), questioning inherent material values: gold dust equals dirt or vice versa, dust, mould, is equal to silver and gold leaf: diamonds are coals, metaphorically speaking. Here the curator's irreverence follows the artist's subversion in the first place: artworks are no stable entities, and as much as pollution (dust) hastens the demise of an artwork's originally intended shape, the transformative power of uninvited 'guests' such as mould, become accepted by the artist, wilfully embraced, welcomed as living comment on the fugitive 'nature' of materiality (Jennifer 2016).

> One day or another, given its persistence […] dust will probably begin to gain the upper hand over the servants, pouring immense amounts of rubbish into abandoned buildings and deserted stockyards: and, at that distant epoch, nothing will remain to ward off night terrors, in the absence of which we have become such great bookkeepers (Bataille 1929a: 278; Bois 1997: 38).

Since dust settled on Duchamp's *Large Glass* and the residue of air raids pulverized the old order, if not completely by 1916 then certainly by 1945, not much remains, neither of servants, nor of the comfort of material structures, housing, home and such securities. The pouring of rubbish into places abandoned takes on a contemporary and prophetic dimension in the twenty-first century, in our world drowning in waste. Artists offer paradigms to work with the agents of transformation, so Ortero-Pailos uses a fragile method of conserving the façades of buildings we cherish, as they offer us materiality and memory, history, style and comfort of cultural survival. Yet the proposed artworks, the latex sheets, are subject to transformation and lack curatorial stability. Architectural writer and researcher Teresa Stoppani (2007: n.pag.) proclaims that 'Bataille "sees" dust and activates it as an agent of change, while modernity endeavours to remove it from the bourgeois interior and from the streets of the city'. Does this mean that an art practice such as Ortero-Pailos 'satisfies both bourgeois' capitalist demands (restoration) *and* subversive art aesthetics? Or is dematerialization the magic password that absolves his practices from the accusation of commodification? In the essay 'Ray Guns', Bois talks about collecting trash and recycling as the 'business of public sanitation'; recycling, the very height of capitalist alchemy, turns everything into grist for commodification's mill; at the same time, it continues to be a subversive 'strategy of aesthetic sublimation', a 'rehabilitation' project (1997: 172).

Rehabilitating dust

> *Drawing dust. Objects shaken by the sea, ground by pebbles in tidal streams, washed and etched by sea salt. The organisation of objects is circular. The fine line pencil captures their disintegration of form into broken strokes, dashes and dots. Albrecht Dürer's* Melencolia 1 *(1514) – the contemplating winged, angelic personification of broodiness about the (lacking) shape of things, the multi-faceted other of a jewel, the polyhedron, a leaden object irregular, a dark presence. Only now I realise that I draw dust through dust:*

FIGURES 38.1–3: Doris Rohr, *Melencolia, composite drawing in layers*, 2014. Pencil on Modelspan paper, each 70 cm × 50 cm. © Doris Rohr. Collection of the artist. Photo credit: © Bryan Rutledge.

The drawings were meant to layer one on another, but now I prefer them separate, three telling the same story as repetition, like a fragment or movie stills. Repetition evokes time passing and a sense of eternal return.

Artist Johanna Love (2020) speaks about fragments of her grandparent's bodies congealing with the dusts of matter, of rubble, of the fallout from the air raids in Hamburg in 1943 (*Operation Gomorrah*). Drawing permits an embodied understanding of historic 'facts' and scientific 'objectivity'. This embodiment of what appears formless becomes Love's drawing approach. Drawing the remains of history, touching the past. Dust 'signals the disappearance of something else that we can no longer see' (2020: 198).

Not letting the dust settle. It fluffs up in the air, the granddaughter unearthing the dust in the cellar, to awaken memories and make connections with the past. Contained in the entropy of memory, the archaeology of unpacking an unknown past. Where former lives have hybridized with coal, ashes and carbon.

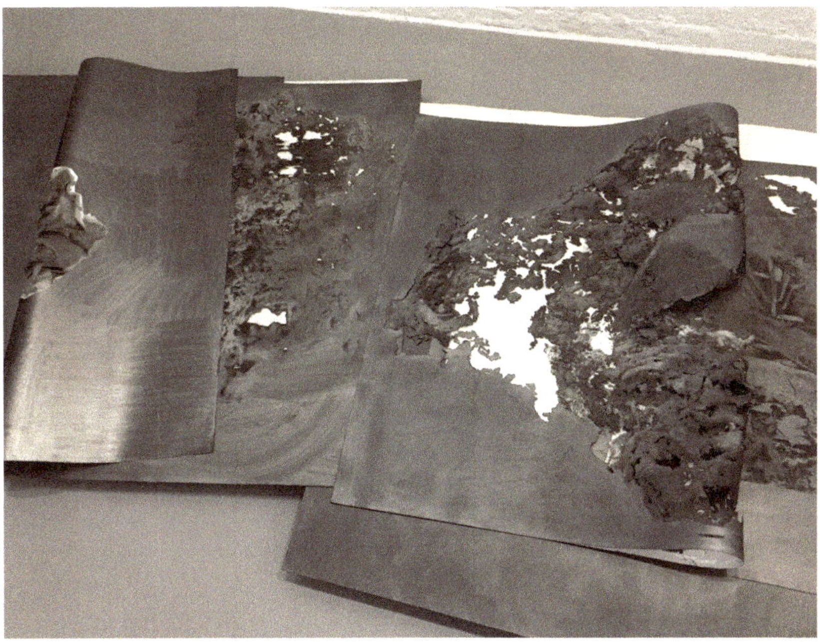

FIGURE 38.4: Johanna Love, *Der Engel Schwieg* (2019). Installation, laser etching with graphite and ink on paper, individual paper sheets 120 cm × 80 cm, installation sizes variable. Copyright with permission of Johanna Love.

We keep alive memories associated with the objects, we dust, we dust and reawaken the ghosts, and from this reconstitute the shapes of things, houses, objects and lives – a parallel universe of dreams, hopes and resurrection.

NOTE

1. According to Bruno Latour, quasi-objects are ambiguous and are neither cultural nor natural or both; frequently they are simultaneously positively and negatively loaded. So for example asbestos or coal or diamonds may become the quasi-object of cultural appropriation of a substance or object, agents of sorts for action or referents.

39

Mimesis: Nothings in Particular

Bill Prosser

What I have not drawn, I have not seen (Goethe cited in Cartier-Bresson 2017: 137).

'Mimesis' first appeared in Books 3 and 10 of Plato's *Republic* ([*c*.375 BC] 1974: 149–57, 421–31), where it refers to representation across the arts. Visually, mimesis is to make 'an appearance [...] that resembles something of a certain kind but is not something of that kind itself' (Hyman 2006: 61). Shadows and reflections qualify, and anyone can become an artist simply by holding up a mirror.[1] Famously, Plato held a dim view of art, thinking it was more likely to tell us lies than Ideas of the Good. More positively, his pupil Aristotle believed that imitative art could instead be aspirational, providing ideals of truth and beauty: 'Art [...] completes what nature cannot bring to a finish' (Aristotle cited in Clarke 1956: 9).

From its classical birth until the nineteenth century, imitation matured into art's standard definition. Reduced nowadays to only one style amongst many, its topics range from the trite to the disquieting. That said, its roots are deeper – more interesting and influential – than simply historical. Reflecting on its origins, Aristotle notes that 'the habit of imitating is congenital to human beings from childhood' (Aristotle 1970: 20); a natural desire to represent the world is there from the beginning, and is, to all intents and purposes, instinctual.

This is borne out by the universal patterns of children's drawings across cultures, as demonstrated, for example, by Rhoda Kellogg (1969).[2] Within these make-believe worlds, imitation is so embedded that the thought a drawing might not represent anything seems odd to children, so 'even when they have drawn something to which they cannot give an exact interpretation they claim that it represents "something"' (Luquet 2001: 77).[3]

Children draw for fun, and it's regrettable that, as adults, spending lengthy periods doing anything just for the fun of it is often looked down upon, even sneered at. These jibes are best ignored. Any attempts to skate over the downplaying of playfulness by glossing it with veneers of grown-up respectability (e.g. art equals 'serious' play, as suggested by Grayson Perry [2014: 110] and Julia Cameron [1995: 196]) seem superfluous, over-egging an activity that is really quite ordinary.

More pertinently, the child psychoanalyst Donald Winnicott introduced the concept of 'transitional objects' – everyday items (scraps of cloth, soft toys) of special value to very young children. He believed that these help in bridging the developmental gap between 'me' and 'not-me', essential in recognizing a world beyond the infant's misplaced belief in its own omnipotence. This step is crucial because it kick-starts 'living creatively' (1991). Jonathan Fineberg neatly focuses Winnicott's broad idea onto making images: 'In art, children find ways to embody their experience [...] in order to explore the relations between themselves and the things they find in the world – just as adults do' (Fineberg 2006: 88). Yet regarding mimesis, children have advantages. The old saw 'familiarity breeds contempt', impels adults, lacking the open wonder of childhood, to stare harder when drawing things around them.[4]

Playing hard is second nature to children, who mix immersive concentration with physical dexterity in losing themselves. We can rekindle this early passion by applying our energies to drawing, reliving some of the explorations and discoveries we once relished – '[t]hrough mimesis [the] pleasure of knowing is illuminated' (Nancy 2013: 61). Having begun, however, recognizing that the delights fashioned through hard graft in childhood must be even harder won as adults can produce a paradox. This is not due to rigorous scrutiny, perhaps satisfying in itself, but rather that any pleasure is all too often leavened by laborious monotony. The root of 'to draw' is related to *trahere*, to drag, as time does when drawing becomes necessarily repetitious and boring. But we should not despair; with a little mental gymnastics, we may turn even this deadweight drawback on its head, because 'Boredom is not far from bliss seen from the shores of pleasure' (Barthes 1975: 25–26). It has long been appreciated that meditation, contemplation, and boredom are siblings; their values indivisible (Simmer 1981). Seen this way, boredom is revealed as an unexpected opportunity to draw reflection closer.

Compounding things, not only can drawing's drudgery be boring, but its subject matter too. That said, banishing all humdrum topics and plumping for alternatives might be equally dispiriting: the easy gratification sought in novel or sensational themes is heavily exaggerated. Besides, once examined, the insipid possesses qualities that by definition are excluded in the immediately alluring; 'Blandness characterizes the real in a way that is complete, positive, and natural' (Jullien 2004: 45). Banality, like boredom, has its rewards.

Bringing unnoticed items to the forefront of appreciation has long been one of art's roles. 'The feeling of pleasure was already there in you, unconsciously, at the sight of a humble existence [...] otherwise it would not have risen within you when Chardin [...] happened to summon it' (Proust 2017: 13). More attuned to contemporary life, Boris Groys attests firmly that 'the role of art in the materialistic age is to make things visible' (Groys 2016: 116), while seen from Terry Eagleton's perspective, paying rapt attention can be psychologically redemptive: 'art seeks to disclose the divinity in things, re-enchanting a world gone sour' (Eagleton 2014: 102).

For my part, a predilection for the trivial has grown in tandem with savouring all manner of pictures showing day-to-day paraphernalia – a cavalcade of Nothings in Particular.

As to my own drawings (e.g. Figures 39.1–39.4), they represent mundane items first caught out of the corner of my eye by some chance snag or speckle. In the same way that authors search for the *mot juste* (or perhaps just the wrong word in the right place), to move seeing into drawing requires finding a means to shift the jumble of shape, tone and texture seen *over there* onto the piece of paper *down here*. It's like a type of scrying, searching out the graphite forms hidden inside the things in front of me. Despite its shortcomings, to help peel my eyes I use what Philip Rawson's crisp alliteration calls 'post-Renaissance peep-show perspective' (so-called because linear perspective's accuracy depends upon the viewer looking through a small hole with one eye, as in early drawing machines) (Rawson 1969: 204). I also frequently obey common-or-garden proportions (perhaps equal divisions or the golden section) to compose each cluster. Organizing them in this prosaic fashion mildly orders the disorderly, and, by focusing my attention, picks up the fecund interplay between banal neighbours. These exchanges fall under obvious headings, including spatial relations (proximity and proportion), contours (angular or sinuous), shades (bright to murky) and surfaces (scuffed, shined, fluffed). To draw these jags and tangs I explore on an ever-decreasing scale. Starting with overall differences (the floor is wider than the bin [Figure 39.1]), before moving to objects themselves (the bin is lighter inside than out), I finally become finicky (the glints on the bin's liner vary).

Staring hard enough to distinguish these major and minor fluctuations is one thing, but the trick of translating such looking into drawing is still quite another – not least because it requires noticing something else. Although I use the most ordinary materials

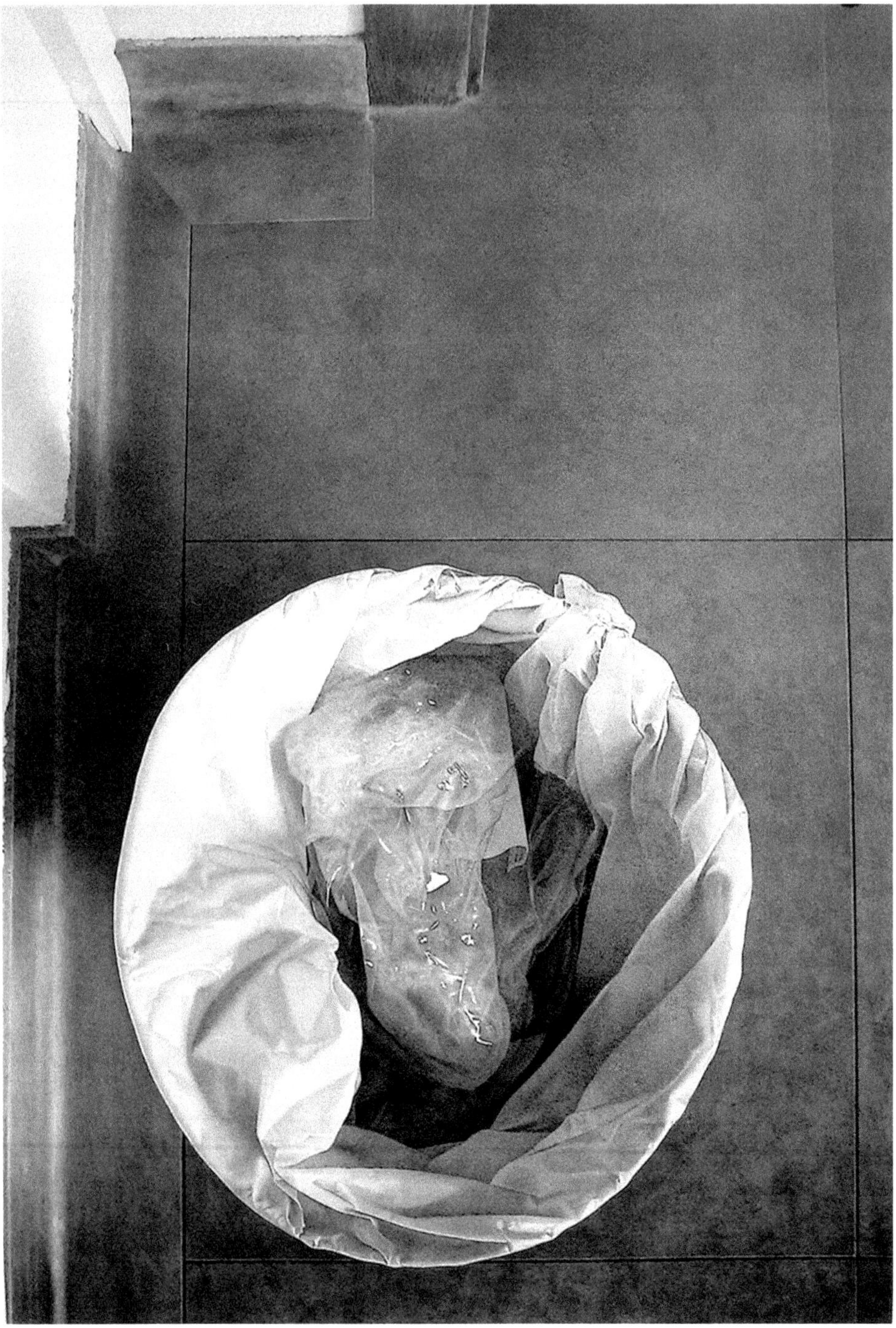

FIGURE 39.1: *Landing*, 2011. Pencil on paper, 60 cm × 42 cm. All drawings by and © the author.

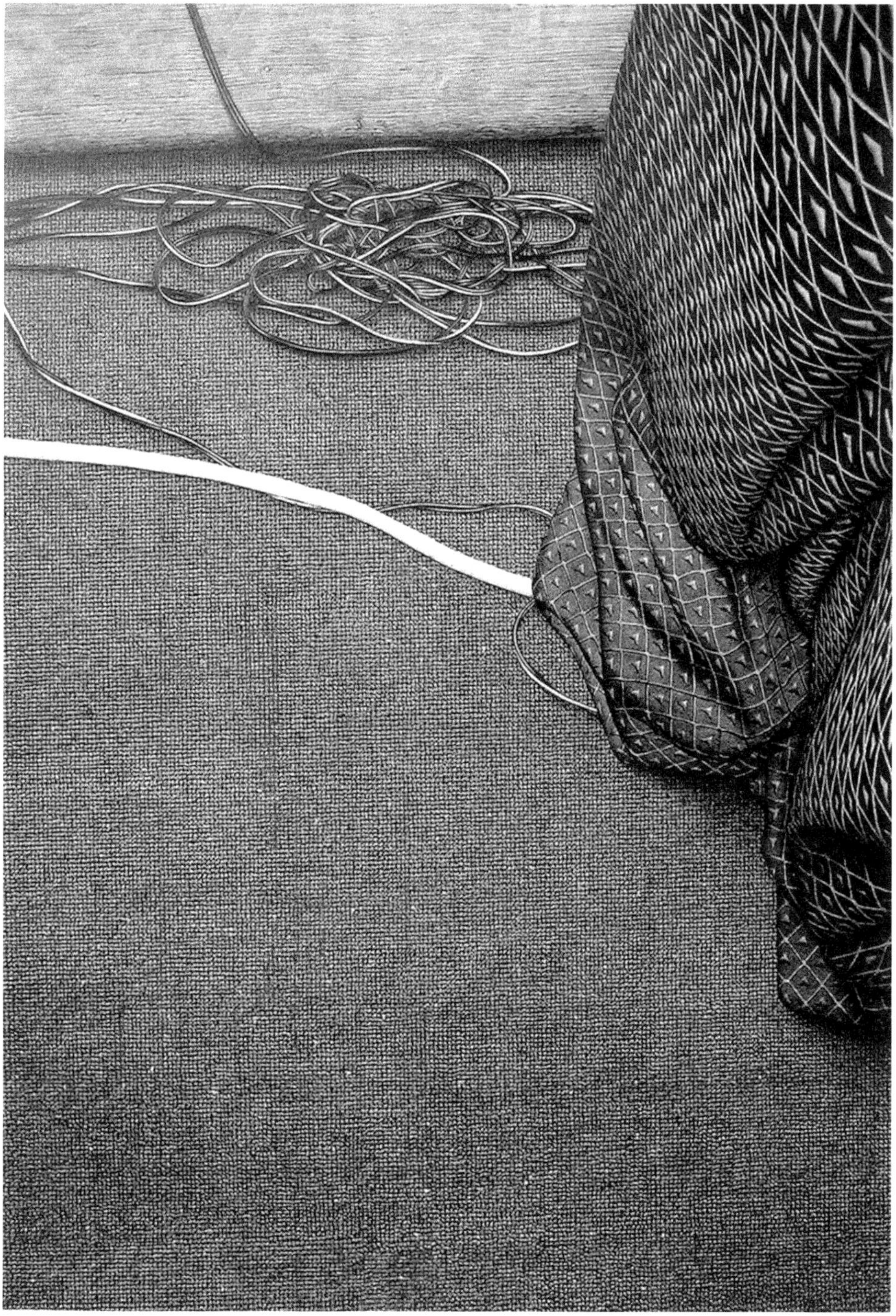

FIGURE 39.2: *JCR*, 2010. Pencil on paper, 60 cm × 42 cm.

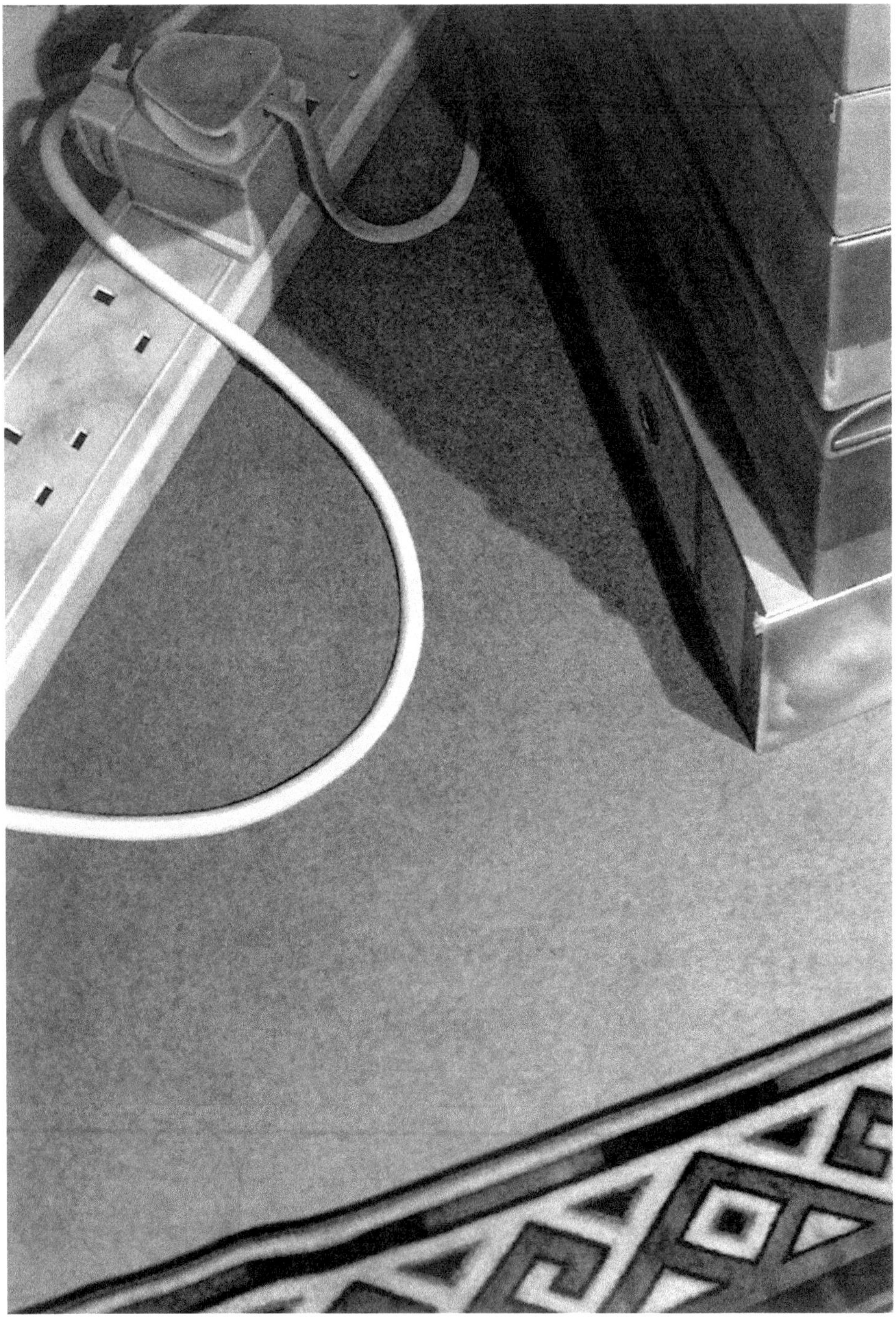

FIGURE 39.3: *Room 48*, 2019. Pencil on paper, 60 cm × 42 cm.

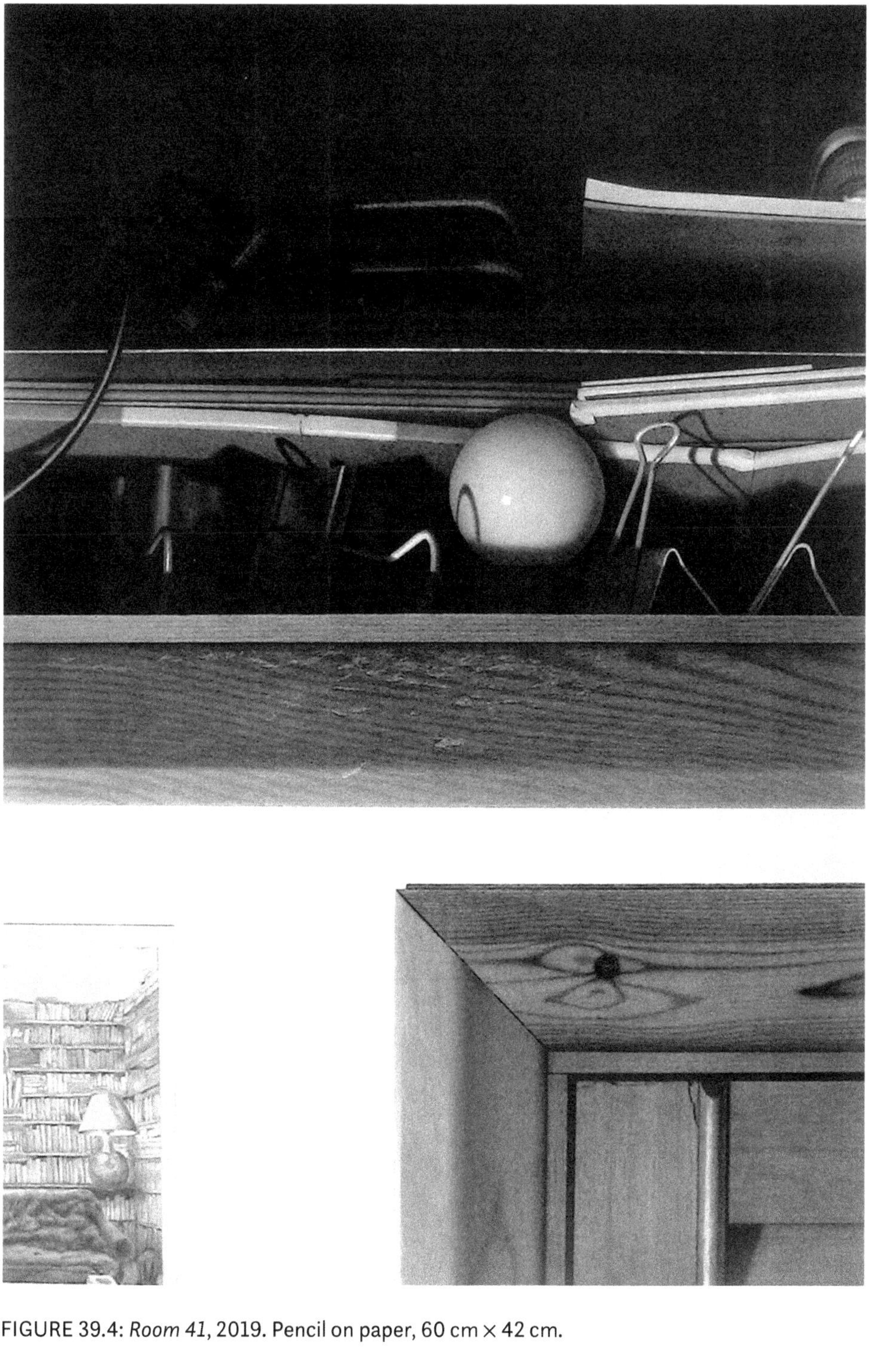

FIGURE 39.4: *Room 41*, 2019. Pencil on paper, 60 cm × 42 cm.

(pencil and paper), mimicry requires watching how their various idiosyncrasies interact. More common fusions and frictions soon become second nature. Otherwise, I fumble along following hunches, struggling with trial and error to solve awkward aesthetic puzzles, nearly always feeling I don't know what I'm doing. Often bored.

Relative tones are habitually built by laying one pencil's marks over another, but every sequence has its subtleties. For example, hard pencils flatten the paper's weave, so that when, say, a 2B pencil covers a 2H it produces a lighter tone than a 2H over a 2B. In the first instance, 2H pushes down the paper's ridges, making it less abrasive, reducing the deliverable amount of 2B's softer graphite; whereas in the second, dark graphite already on the surface is simply pressed in further, augmented by 2H's minor contribution. Overall, the medium has its advantages (broad malleability – erasable errors, usually) and disadvantages (limited tonal range – it's impossible to get a really dense black).

Applying a variety of graphite cocktails (a shot of 5B here, a squeeze of rubber there) is invariably inadequate. The image and objects don't match at all. Thankfully, it is now that the drawing itself really helps towards a mature mimesis; only with its individual wrongness can nitty-gritty noticing begin.[5] Moving beyond error requires looking harder still, then removing or adjusting what's already done through mixing some different cocktails.

Once tediously elaborated, shapes and tones snap back to mimic the bits and pieces which provoked them. Like a mime, the drawings require no linguistic explanation or theoretical underpinning. I agree with Iris Murdoch that 'we like art when it seems to be pointless' (Murdoch 1997: 207), and share William Kentridge's aspiration 'to make something impervious to cross-examination and assert the primacy and necessity of stupidity' (2014: 15) – as in the old French saying 'stupid like a painter' (Alberro and Stimson 1999: 170).

Drawing subdues the constant badgering of my ego. Yet intuition has propelled me to shadow a personal canon, however tentatively – writers (Samuel Beckett's 'to restore silence is the role of objects' [1959: 13–14] and Georges Perec's precise enumeration of the inconsequential [2011]), as well as artists (from Chardin to the Precisionists, Pop Art and the Spanish realists), photographers (Walker Evans, Wright Morris) and even theologians – Simone Weil's 'attention, taken to the highest degree, is a form of prayer' (1952: 170).

More prosaically, despite working alone I feel in the swim; as endless snapping on mobiles proves, fascination with trivia is timeless, and, like death and taxes, mimesis will always be with us.

Stare. It is the way to educate your eye, and more. Stare, pry, listen, eavesdrop. Die knowing something. You are not here long (Walker Evans cited in Hill and Liesbrock 2015: n.pag.).

NOTES

1. In Greek mythology a shadow was integral at drawing's inception, when the daughter of Corinthian potter Butades drew around the thrown silhouette of her lover as a memento.
2. Sadly, the proposal that a 'neutral' standard of mimesis can measure children's intellectual development can be found, for example, in Dale B. Harris (1963).
3. Wittgenstein gives one such, rather challenging, account: '[a] grown-up had drawn pictures for the child several times and said: "this is a man", "this is a house", etc. And then the child makes some marks too and asks: "what's *this* then?"' (Lubbock 2012: 165, original emphasis).
4. Phenomenologists say that we are often entirely unaware of items under our noses and have to focus quite deliberately to appreciate them, e.g. Martin Heidegger (1962).
5. John Berger emphasizes the projective impetus of drawn marks: 'A line, an area of tone, is not really important because of what you have seen, but because of what it will lead you on to see' (Berger 1979: 23).

40

Mottled Geometries: The Lure and Allure of the Pattern in the Carpet

Victoria Mitchell

As the mind wanders, it also wonders. You find yourself engaging with the patterns in a carpet, in an encounter that might shift from close range to distant view, perhaps from an oblique angle, or activated through touch as well as sight. They are rich and complicated. Is that a bird or a plant? Why is that curious shape on one side of a motif different from the one opposite and why do the colours seem to dance? You might marvel, too, at the construction of the carpet – how it is made and who it was who laboured to arrange the multitudinous knots of coloured, textured yarn into those patterns. From what distant context and through what complex technical, social, cultural and psychological passages did we arrive here?

Michel Foucault considered the carpet's Persian incarnation as a garden which evokes a microcosm, 'the smallest parcel of the world', 'onto which the whole world comes to enact its symbolic perfection' and furthermore 'can move across space' (Foucault [1967] 1984: 6). Albeit shaped by an unfathomable cross-pollination of references, the agency of the carpet also conjures up the time-rich essence of the everyday, perhaps even registering 'the soul of the apartment' (Poe 1840: 244). The intensely and brilliantly coloured hand-woven Qashqai rug covering the analytical couch in Sigmund Freud's domestic-feeling consultation room may be considered exemplary in these respects (Figure 40.1). Made in a harsh dry landscape by people (women) who had no gardens, it is 'festooned with flowers and birds' in such a way as to depict 'a garden in paradise' in which 'geometric patterns "enclose" the flowers in the same way a garden is organised with borders, paths and fences' (Burke 2006: 135). Referring to the complex flower-bird patterning of this rug, the French artist Anne Deguelle considers how 'the shape of the motif' not only constitutes a familiar

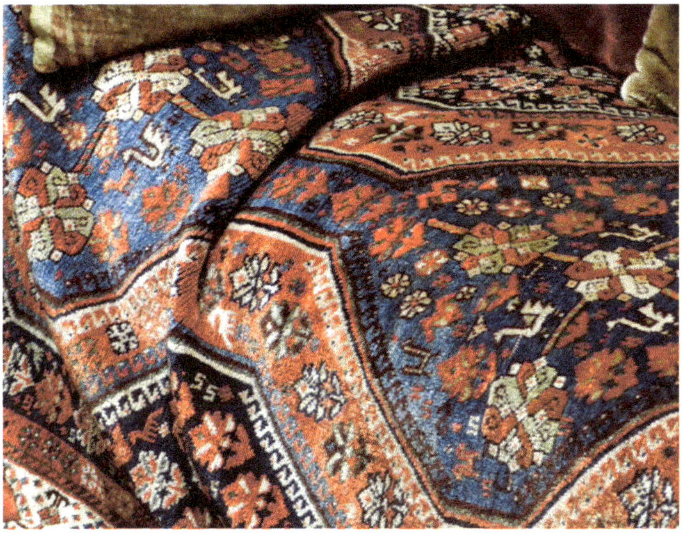

FIGURE 40.1: Qashqai rug on Sigmund Freud's couch (detail). Photo: Victoria Mitchell, 2020.

language for the weavers but also establishes the decorative as a 'configuration with variable geometry' akin to the movement of constellations (Deguelle 2011: 10).

For the many recent and contemporary fine artists who have appropriated such carpets, if only as quotation, their cultural and material potential figures as a means to address critically resonant issues such as war, empire, identity or modes of production. In this, they articulate otherwise veiled and often uncomfortable truths while also alluding to the colour-saturated associations of princely workshops or the imagined pleasures of nomadic lives. Perhaps less well acknowledged are the ambiguities afforded by the textured matter of their patterning in the context of perception.

In the focus outlined here, the carpet is subjected to the intricacies and intimacies of close, textural looking in which encounters with patterned carpets are conceived as affecting patterns of perception and consciousness as they dissemble, dissolve and disappear. Simultaneously they attract and even derange the viewer through complex combinations of knots, fibres, textures, colours, patterns and figurative or symbolic forms. A faded, worn or threadbare carpet might further embed the wandering and wondering. As the dense pile of a hand-knotted carpet gradually wears down or its colours fade away, the structure of the threads that support the knots can become exposed and the patterns more indiscernible, beckoning a state of 'becoming-imperceptible' wherein 'all becomings are already molecular' (Deleuze and Guattari [1988] 2020: 318). The effect on the mind of this maze-gaze might induce an 'obscure realm of unconscious determinations' (Caillois [1935] 1984: 31–32), hovering between a destabilized sense of self and a state of healing or reverie.

To return to the carpet before us, the 'scrutinizing gaze' associated with Islamic geometries (Necipoğlu 2015) leads imperceptibly to the mind glazing over in a haze of complex associations or lured into a sense of enchantment. Sometimes the patterns themselves are puzzling to behold, as when the figure-ground relationship is indeterminate in what it indicates, hovering between bird, flower or sign of heaven, for example, or perhaps obscured by wear and tear or false memory. Such patterning plays games with perception, causing a state of incomprehension or instability that refuses resolution, often taking the eye on a seemingly endless journey from one association to another. In the introduction to the catalogue of an exhibition of 400 carpets in the Handels-Museum, Vienna, 1891, Arthur von Scala, in period mode, noted the difficulty of perceiving 'the clear separation of decorative motifs into dominant and merely subsidiary types and the subordination of the latter to the former'. He adds that it is 'a matter of debate, even among experts, as to whether they might represent a leaf or a bird, a vegetable or an animal model from nature' (von Scala cited in Vasold 2015: 175). Similarly, in the biographical short story 'The Persian Carpet' (1983: 109), Hanan Shaykh is brought face-to-face, in her estranged mother's house, with a carpet remembered from childhood, recalling how

> I used to lie on it as I did my lessons; I'd be so close to it that I'd gaze at its pattern and find it looking like slices of red water-melon repeated over and over again. But when I sat down on the couch, I would see that each slice of melon had changed into a comb with thin teeth.

Questions of viewpoint and distance, raised in Shaykh's account, are especially significant in affecting ambiguities of perception. Ernst Gombrich, in *The Sense of Order*, draws attention to Walter Crane's 'rapid sketches of a Persian Rug' of 1902 (Figure 40.2) as examples of (what Leonardo describes as) 'the perspective of disappearances' ('*perspectiva de' perdimenti*'); the further we stand back from a rug, the more indistinct become the motifs, and similarly, if we magnify too much, Gombrich adds, 'it will cease to be an image and become a piece of textured matter' (Gombrich 1979: 115). It is precisely these 'disappearances' of outward geometry that concern us here.

The complexity arising from the relationship between structure, texture and pattern represents reciprocal activity between the maker and viewer, with the carpet active as a mediating agency, affecting the mind of the recipient and causing psychological change. Endorsing Shayk's account we recognize that complex

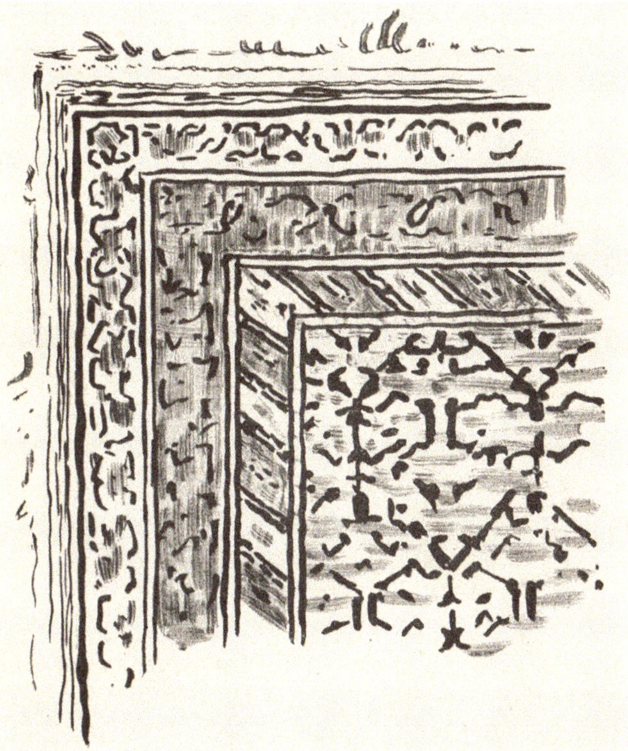

SKETCH TO ILLUSTRATE VALUE OF DIFFERENT QUANTITIES IN PERSIAN RUGS.

FIGURE 40.2: Walter Crane, 'Sketch to illustrate value of different quantities in Persian rugs', 1902.

patterns and their construction can also give rise to a*maze*ment, a puzzle which is also a form of entrapment or a lure, thus '[t]he process can continue interminably; the pattern is inexhaustible, the relationship between carpet and owner, for life' (Gell 1998: 80). Alfred Gell asks '[w]ho, possessed of an intricate oriental carpet, can say they have entirely come to grips with its pattern; yet how often the eye rests on it and singles out now this relation, this symmetry, now that' [*sic*] (1998: 80). For Gell this amazement is also inscribed in the manner in which they (objects, carpets) are made, the enchantment of technology having the capacity to bring 'the technology of enchantment' to the fore (Gell 1992). They 'are difficult to make, difficult to "think", difficult to transact. They fascinate, compel, and entrap as well as delight the spectator' (1998: 23).

In teasing out the 'unconscious cerebration' that is 'the figure in the carpet' (James [1916] 2016: n.pag.), the intricate relationship between the technology of weaving and pattern is fundamental, especially bearing in mind the marked difference between the pattern manifest on the surface of the pile and the fibrous threads of wool as they are worked (spun, dyed, knotted, etc.) in a time-consuming manner to produce. Considering late-nineteenth century carpets made by the nomadic Qashqai sheepherders of southern Persia (under which category the carpet that covers Freud's analytic couch would fall), P. R. J. Ford notes that 'it is difficult to believe they were made by the human hand' ([1981] 1997: 122). Intricacy and complexity can bewilder and overwhelm the ability to comprehend any logic in the arrangement before the eyes but becoming lost in technical detail may also be productive and affective, a form of agency not to be discounted. Uncertainties or irregularities that inform the relationship between the *technē* and texture of textile and the textility of perception might find accord with the nerve fibres that imperceptibly shape our becoming. Metaphor, forming a fine web between textile technique and perception, may aid comprehension even as we lose our way.

Abrash

A descriptive term concerning an aspect of carpet-making, abrash refers to the dappled, mottled or speckled colouring caused by dye variations (of hue and saturation) in yarn, traditionally associated with hand-spun, naturally dyed, hand-knotted carpets produced in nomadic or village contexts. It can indicate a change from one hank of wool to another, where these have been separately dyed but used within the same colour field, or the effect of ageing and exposure on certain dyes (Galafassi 2009: n.pag.). The word is borrowed from Turkish *abraş*, and in turn from the Arabic أَبْرَش (*ʾabraš*, piebald, mottled), and where once it signalled a defect, lack of refinement or variability in quality it has come to be sought out as aesthetically desirable, associated with subtlety, warmth, richness, vitality and shimmer (Denny 2014: 32; Thompson 1988: 62). The inconsistencies of abrash are at odds with the outwardly distinctive patterns and regular knotting characteristic of most professionally produced carpets, though it can now be incorporated artificially into machine-made rugs.

Dyeing is absorbing. Dye uptake on the yarn can be affected by the heterogeneous structure of wool and the intrinsic characteristics of natural dyes (the shades of which are unpredictable) or by the circulation of the dye in and around the yarn, as well as by challenges of ensuring that different batches are consistent (if that is desired). Pierre Galafassi (2009: n.pag.) begins his account of hand dyeing by considering wool's molecular composition and its relationship with dye from the perspective of chemistry, thus '[c]hemically speaking, wool is keratin, a protein copolymer containing about seventeen different amino-acid monomers' and the 'covalent cross-linking of adjacent cysteine residues by disulphide bonds is a major factor for the mechanical properties of keratin fibres'. These bonds, which are critical in the dyeing process, are very fragile at high temperatures. In contrast, a maker's knowledge of the wool and its subsequent receptivity to dye is addressed through knowing from within, thus

> [t]he quality of wool, its textile and dyeing properties are determined by fibre fineness, length, scale structure, natural shade, brightness, cleanliness and freedom from damage. It depends on many factors, mainly sheep breed, age, health and stress of the sheep, season of shearing, body part it comes from (finest comes from shoulders, coarsest from breech), the climate it was exposed to, food, the thoroughness of fibre scouring (full or part removal of lanolin) etc (Galafassi 2009: n.pag.).

Not surprisingly, he adds that Central Asian nomads distinguish 'dozens' of wool qualities by name. The intricacies of the process are open to subtle meander and variation at every stage. Subtle, from the Latin *sub tela*, is literally beneath the web or warp of the cloth, but subtlety is also a property of the fibre itself.

When a dye is evenly dispersed it is known as 'level' dyeing – the opposite of abrash. Levelling, related to the even balancing of scales, implies an undifferentiated, smooth surface, whereas abrash, in contrast, indicates unevenness of effect and inconsistency. These characteristics affect the mind of the beholder: an otherwise stable ('level') state of mind becomes displaced or altered through the kind of haptic interference produced by such effects. As a fibrous undertow of order, abrash here figures as a notional means to destabilize norms of perception, a corollary to Deleuze and Guattari's concept of haptic perception as a composed chaos, distinguished from perception which is 'striated', optic, repetitively patterned and above all culturally bounded (Deleuze and Guattari [1988] 2020: 551–83).

Looking closely at abrash, the eyes dance and regularity disintegrates. Perspective recedes from view, and imagery is reconfigured. The mottling, as earlier remarked, brings the unevenness of a carpet's texture into visual contact with the formal elements of its pattern, softening borders and shifting associations. Like Roland Barthes's 'grain' (1977: 179–89), it forms a semiotic code, the text of the tissue (or the tissue of a text) wherein 'each thread, each code, is a voice' (Barthes 1974: 160). This is particularly significant in the light of the decoration of carpets which have evolved through generations of makers for whom the oral and tactile are fundamental (Deguelle 2011: 10). In short, we are absorbed by a mobile, intimate, textured geometry, where mottled markings meet distant constellations, on a path to dissolving into dreams.

Fading into reverie

Between texture and motif, distance and detailed scrutiny or subject and object, a camouflaging of consciousness emerges. Laura U. Marks has proposed that the perception of pattern in Islamic decorative art enables a more active, performative engagement in the viewer than iconographic subject matter and that it more directly embeds the contemplative subject in the process of viewing (Marks 2010). Following Deleuze and Guattari, among others, she considers this perception operating as or in a haptic space, 'that invites not distant contemplation but intimate involvement, the eyes moving over the surface as though touching it' (Marks 2010: 54).[1] It is not so much what a pattern represents as how it is encountered that shapes perception, refiguring consciousness by bringing to the surface ambiguities between pattern motifs and embodied perception. In looking closely, in absorbed looking, we are close to touch if not in touch, intimate in a form of granular contemplation that unleashes the texture of vision as textured matter, becoming imperceptible and enabling a concomitant blur or disruption of consciousness as our hypnotic susceptibility succumbs to reverie and the awakening of dreams. As patterns dissolve before our eyes the mind unravels.

To return now to Freud's consulting room, Hilda Doolittle (H. D.) described the carpet-covered couch as 'the homely historical instrument of the [...] science of the unravelling of the tangled skeins of the unconscious mind' (H. D. [1956] 2012: 16). Andreas Mayer notes that although 'the patient's gaze was attracted, if not provoked, by the multitude of objects surrounding him [*sic*] in the consulting room' (2013: 216), 'the most promising path to the unconscious was found by lying on the couch' (221). Here, depending on the hypnotic susceptibility of the subject, resistances might be dispelled and by blanking out the surrounding objects the patient might enter into transference through free association (218–19). Unravelling skeins of the unconscious enabled this transference; as Freud himself noted: 'I have laid it down as the task of dream-interpretation to replace the dream by the latent dream-thoughts, that is, to unravel what the dream-work has woven' (Freud [1901] 2001: 686).

In the context of abrash and related instabilities of perception, it is fitting to consider the view not only of the couch but *from* the couch and the extent to which this view might contribute to the unravelling of the dream-weave. In the late nineteenth century, the couch, or the *diwan* (divan) as it was termed in reference to its 'oriental' origin, was associated with instability, psychic unrest and moral laxity, a site on which one might be positioned half seated or half lying down, in an 'unsettling' position, while also sinking into its amorphic softness, with intimations of domesticity and seduction (Marinelli 2006). Lydia Marinelli notes that late nineteenth-century debates relating to 'lying down treatments' (*liegekur*) and the position of the body as pertaining to certain psychic states and nervous illnesses were circulating in medical fields (Marinelli 2006: 18–21). Psychic calming was to be encouraged. Freud's couch, gifted by a patient in 1891, was covered in a now-familiar manner with the Qashqai rug he already possessed (perhaps a gift of or purchase through his brother-in-law who was a carpet dealer). It was positioned in such a way that the therapist would be unseen, behind the head of the patient, perhaps reflecting contemporary debates concerning the dynamics of people in therapy situations (Marinelli 2006). Correct positioning of the patient was crucial if therapy was to be effective.

In 1908, when he moved his practice to an upstairs room in Berggasse 19, Freud hung another Persian rug on the wall adjacent to the couch (Burke 2006: 134) (Figure 40.3), less technically bedazzling or bejewelled in its colouring than the Qashqai rug, but none the less significant, not only for adding to the semi-enclosed,

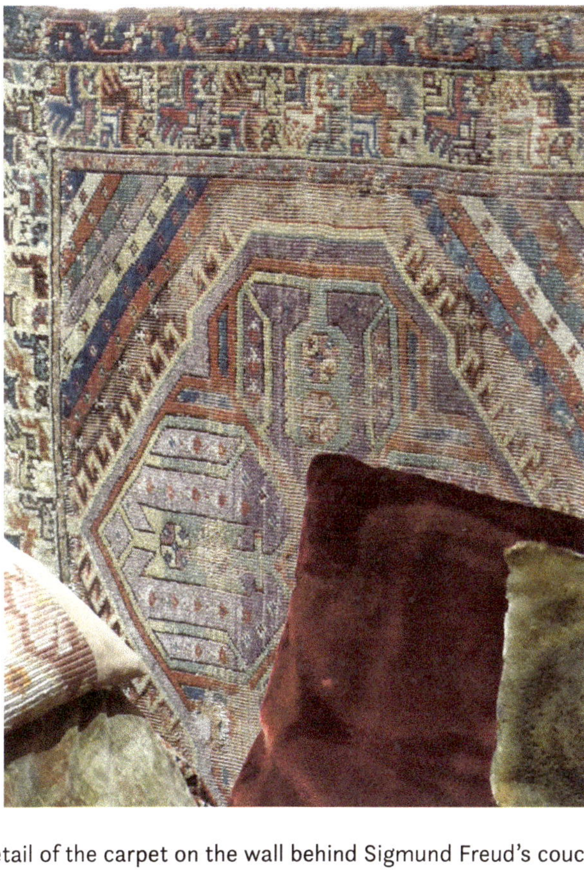

FIGURE 40.3: Detail of the carpet on the wall behind Sigmund Freud's couch. Photo: Victoria Mitchell, 2020.

cocoon effect of 'a protected enclave, a room within a room, a private interior space' for the recumbent or semi-recumbent patient (Fuss 2004: 89) but also because it would be seen by them at very close quarters, obliquely, as they reclined on the couch, hovering between waking and dreaming, in 'a twilight state of drowsiness' (Marinelli cited in Warner 2011: 148). The carpet pile is worn away and is markedly threadbare and faded above the area closest to the head end of the couch, as if, as Marina Warner has noted, 'one patient after another must have stroked or rubbed or patted it with their left hand as they lay there and talked' (Warner 2011: 152–3). The knots of its making are laid bare (Figure 40.4). For the supine patient, the soft Qashqai carpet on the couch might be comforting, but it would not be in view to the same extent as the carpet hanging on the wall. Here, the patterns of the central areas are (now) less distinct and the colours blurred with age and abrash, encouraging a loss of focus and an anaesthetized drifting from consciousness as well as from the surrounding space and its clutter of objects. In the borders of the rug, the patterns are more distinct, but also more beguiling – a maze of ambiguous motifs. A site of free association and a sight of free association blend together, albeit unevenly, as with (and as if succumbing to) the effect of the textured matter.

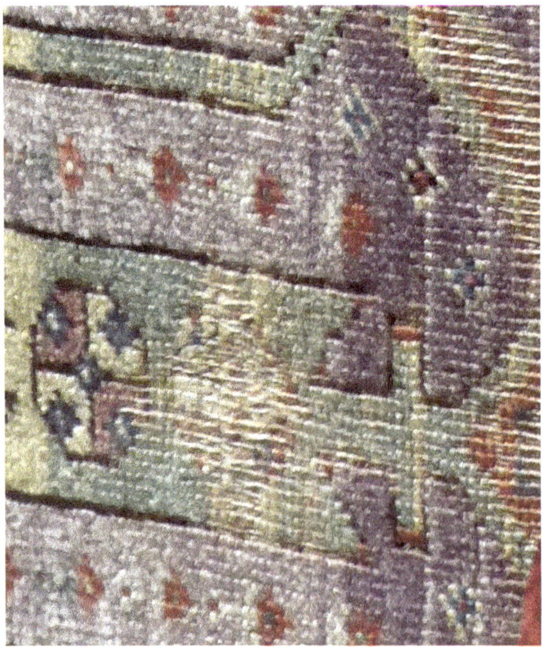

FIGURE 40.4: Detail of the carpet on the wall behind Sigmund Freud's couch. Photo: Victoria Mitchell, 2020.

Perhaps Freud himself was led to hidden recollections and ruminations in the maze-gaze, carpet-pattern-saturated world that he had created – for example that his father Jacob had been a wool and clothing merchant who had a particular interest in the dyeing of wool (O'Donoghue 2019: 285) or that, as he believed, the relationship between women and weaving was a fount of knowledge (Freud [1933] 1968: 132). Bringing together Freud's subjects and the patterned objects which they encountered at close range in the consulting room thereby offers illumination into the relationship between pattern and chaos in shaping the social and the psychological, over time and across space. The 'unstable subject' of Freudian psychoanalysis figures as a carpet-in-the-making on the loom, with the uneven patterns of the surface revealing the knots and fibres whereby both subject and carpet are formed.

NOTE

1. Marks also notes that '[c]arpet designers and weavers are like genetic software artists in the ways they quasi-genetically modify received motifs' (Marks 2010: 311).

41

Ghost Flower 3

Andrea Stokes

FIGURE 41.1: *Ghost Flower 3 (draw)*. Pencil on paper. 25 cm × 33 cm. © Andrea Stokes, 2020.

I have known about you for as long as I can remember. You left a mark. You were present in the names of the streets, and the collective memory of our village. We said we did not believe your story, but I went out searching anyway. I didn't tell anyone. I stood in the cold beck, aware of the outline of my body in the fading light. That was nearly 50 years ago. I came back, to print Vaseline flowers on hotel windows for you.

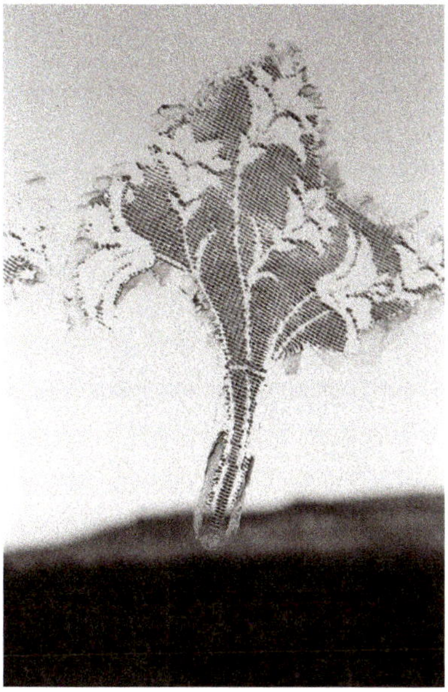

FIGURE 41.2: *Ghost Flower 3 (grain)*. Black and white analogue photograph of print on hotel window overlooking Cottingley. 36 cm × 54 cm. © Andrea Stokes, 2020.

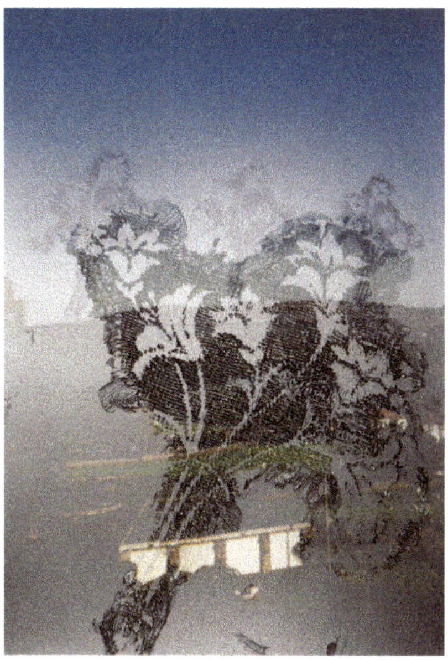

FIGURE 41.3: *Ghost Flower 3 (vaseline)*. Vaseline print on top of colour photograph. 18 cm × 27 cm. © Andrea Stokes, 2020.

I am the Frances of the two photographs you say you like. The window had no curtain and the view in the daytime was good, with a range of hills in the skyline. But it frightened me to death at night. A blank shiny piece of black glass, with nothing beyond but the roaring of water. (I never tried to make friends with them). A face in a bird's nest and other unfocused faces in odd places. I was staring with a penetrating eye toward the camera. Sticking the hatpin into the soft soil, she made sure that I had the camera flat on the ground, then she was ready. We destroyed the evidence, tearing up the cut outs, floating them down the beck and poking the hat pins into the Earth (Griffiths and Lynch 2009).

We are about my age when it happens. We notice it first at quiet times, in flickers at the edges of perception. And then at other times in sharp bursts like a strobe in a disco. Over the years it comes more and more into view. It dawns on us. We sense how it extends in all directions. We feel sure that it is a shape, as big as a continent and as thin as the indentation of a pressed flower. Not long afterwards we wonder if it is an edge. But come to think of it, it is not so much an edge, but a gap. We see that this gap is deep. As deep as the opening between tectonic plates on a seabed, over which we float, vertiginous and disorientated. As we radiate towards the gap in barely perceptible increments, we start to make out the other side. This has been a long exposure. We stare with a penetrating gaze towards each other; shooting across a century, hot with velocity and cooled by time. One day we go to the foot of our stairs. We lay on the carpet and the thin, deep jagged gap fills up with liquid metal. When it is set, we pull it out with our bare hands and it looks like an asymmetrical crown for a giantess.

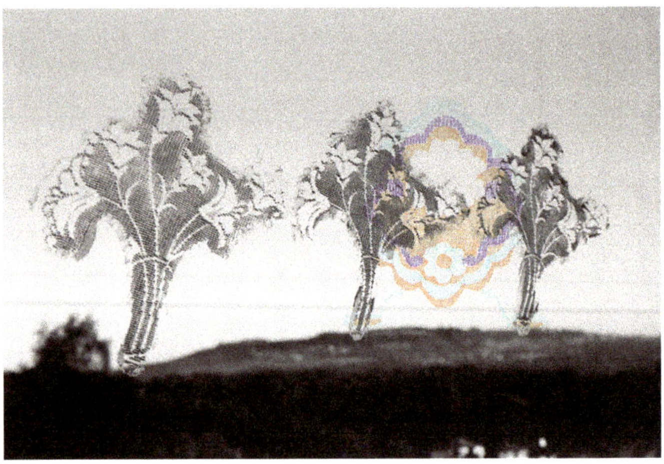

FIGURE 41.4: *Ghost Flower 3 (felt).* Felt tip drawing on black and white analogue photograph. 81 cm × 54 cm. © Andrea Stokes, 2020.

42

Dom Sylvester Houédard: Exhibiting Spiritual 'architypestractures' and Cosmic Dust

Nicola Simpson

Dom Sylvester Houédard (1924–92) was a Benedictine monk, artist and poet from Prinknash Abbey, best known for the typestracts and concrete poems he made on his Olivetti Lettera 22 typewriter in the 1960s and 1970s. In March 2020, I was invited to curate an exhibition, *Dom Sylvester Houédard: tantric poetries*, from the extensive private collection of his work held by the Lisson Gallery.

Houédard's practice of making typestracts began in 1963 with grids built up from the punctuation keys on his typewriter. In his notebooks and papers from this period there are grids of colons, apostrophes and quotation marks. As his practice developed experimental circles are made from the back slash and forward slash keys, rudimentary forms gradually develop and change initially into spatialist landscapes and constellations and gradually towards the deliberately repetitive geometry of the three-dimensional circles, ensōs, pyramids, triangles, ladders and stupas of his fully developed work. Forms are endlessly revisited yet each typestract has its own identity and its own *pattern* (Figure 42.4). In the words of Houédard, each typestract is a yantra, 'tantric perhaps ...' and together they form an extensive, almost daily, record of a monk at prayer, typewriting his inner contemplative experiences.

In 1966, alongside this routine of composing typestracts, Houédard began making a series of poemobjects he called 'laminate poems' and 'cosmic dust poems' and that incorporated many small found objects alongside cut-up prose and cut-out pictures, pasted onto bright techni-colour sheets of transparent plastic (Figures 42.1 and 42.6). They are visually very different from the precision and permutational variations of the typestracts and seem far removed from the yantric patterns made on his typewriter;

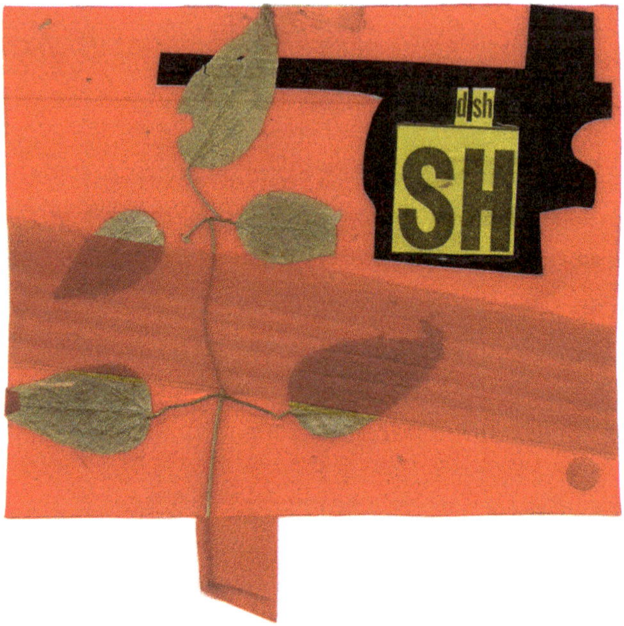

FIGURE 42.1: Dom Sylvester Houédard, *ON THE ENCLOSED LEAF OUT PETITION TO TIME BEFORE THE NO ACT OF UN DOING*, 1967. Vinyl plastic laminate (newspaper cuttings, miscellaneous material and PVC plastic). 5.6 cm × 15.2 cm. © Dom Sylvester Houédard; Courtesy Lisson Gallery.

however, the laminate poems also engage with tantric practices of transforming the mind and the objects which appear to mind. Through a seemingly stochastic and kaleidoscopic juxtaposition of forms, often found in his immediate environment, Houédard developed a methodology of tantric collage, that wove together the *chaos* of forces, events, consciousness and objects that appeared in the daily and impermanent flux of life.

There is an inter-relationship between these two distinct methods of making concrete poems, understood in the context of Buddhist, and in particular Tantric Buddhist, epistemology. Houédard's intentions to exhibit both together is evidenced in an exhibition leaflet from his show with Antonio Sena and David Medalla at the then fledgling Lisson Gallery in June 1967 which states that 'dom sylvester houédard is exhibiting a selection of his typestracts, some extracts from the mantra "jrim, hum, ho, ho, ho phat", some cosmic patches (attempts at repairing the universe), and some particles of antimatter from gloucestershire' (Lisson 1967: n.pag.).

The meditative precision involved in engaging and disengaging the typewriter roller, moving paper carefully a fraction at a time before pressing a key, evaluating and leaving blank the spaces of the foolscap clean and un-inked in the composition of a yantric typestract is a methodological contrast to the making of a laminate poem: cutting out a shape of coloured cellophane; spreading the surface with glue, spit or other sticky and receptive bodily fluids; pasting layers of cut-up text, sand, petals, leaves, jam, pills, combs, Christmas cracker toys and other found and visible objects and waiting patiently to capture layers of the invisible such as spoken prayers, the trace of a bird call from the tree outside his cell window, particles of dust and anti-matter.

Taking the 1967 Lisson show as curatorial prompt, and in the words of one of the laminate poems, to 'clarify the intentions of the original' in the making of *Dom Sylvester Houédard: tantric poetries* (Figure 42.6), I foregrounded the knowledge and practice of tantric spiritual methods and the way in which Houédard utilized these to produce a body of work that can be seen as pivotal in the emerging narrative of a twentieth-century, post-war, transhistorical avantgarde and its engagement with Tantric Hindu and Buddhist practice. Houédard, with his early contacts with the Tibetan Lamas, Chögyam Trungpa Rinpoche and Akong Tulku Rinpoche, his autodidactic study of the first Western commentaries to the practices of Tantric Buddhism, and his initiation into the path by the current Dalai Lama, was instrumental in acting

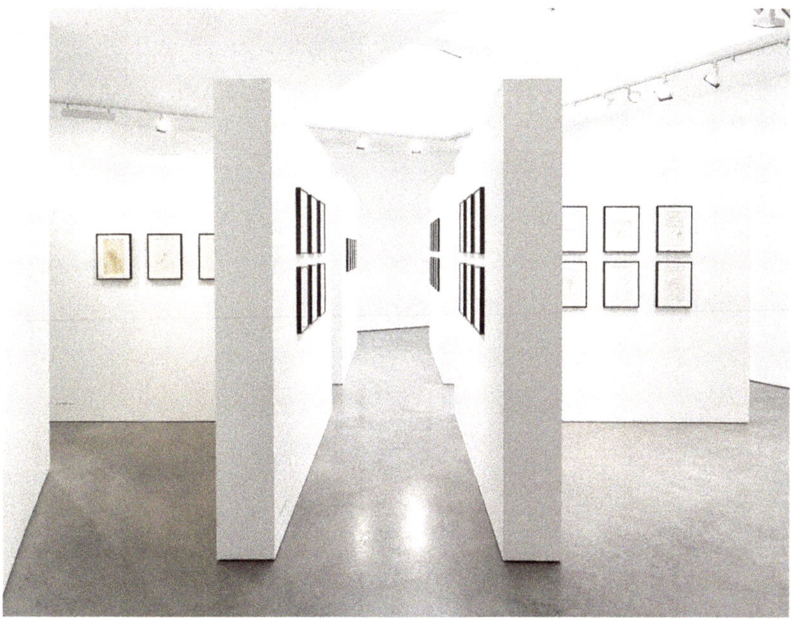

FIGURE 42.2: Dom Sylvester Houédard, *tantric poetries*, 12 March–31 July 2020. Installation view. © Dom Sylvester Houédard; Courtesy Lisson Gallery. Photography by Jack Hems.

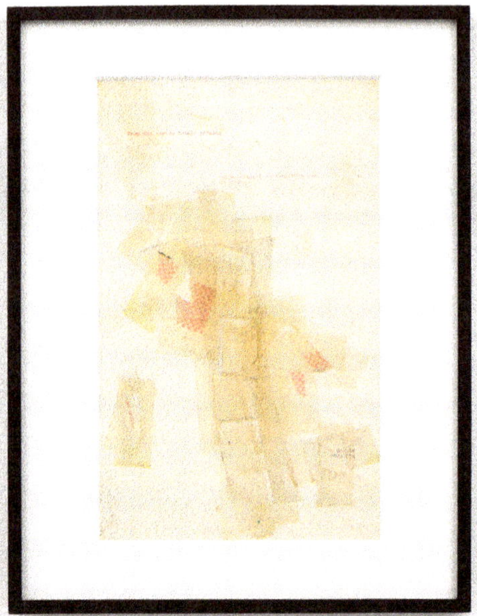

FIGURE 42.3: Dom Sylvester Houédard, from the series *cosmic patches. some attempts at repairing the universe. For david medalla*, 1967. Typed page. 41.5 cm × 32.5 cm × 2.5 cm (framed). © Dom Sylvester Houédard; Courtesy Lisson Gallery.

as a conduit for the transplantation and transmission of this doctrine to the transnational avantgarde artists and poets of the 1960s.

Spiritual architypestractures

The focus of this gallery space was on the movement of the viewer through the space and how the path of each 'co-creator' brought into being the spiritual patterns of energy and flow that underpin Houédard's typestracts. The four sections directed the viewer to walk from the edges to the centre and out again thereby mapping out in physical footsteps on the gallery floor the imaginative choreography within the mind of a Buddhist or Hindu Tantric practitioner as they engage with a mandala diagram (Figure 42.2). The four sections and sixteen walls also enabled the typestracts to be very carefully curated into groups that reflected the mathematical and geometrical ideas that structure most of Houédard's permutational and combinational poetry.

Houédard had been interested in the eastern contemplation traditions and Tibetan music since a small child and his studies of Zen and Tibetan Buddhisms ran parallel with his Catholic theological journey. He saw his Christianity and any practice of Tantric Buddhism as dovetailing in the mystic's journey to knowing his own heart (Figure 42.4). In one of his first published essays, he outlined what he understood by the Tibetan Tantric Buddhist practices:

> Tibetan mysticism [...] aims at liberation from all that is unreal [...] it seeks attainment of a blissful knowledge of the Ultimate Reality [...]. The aids used in Tibet are based on the tantra (net, web, woveness) between the inner and outer worlds. Forces and their events,

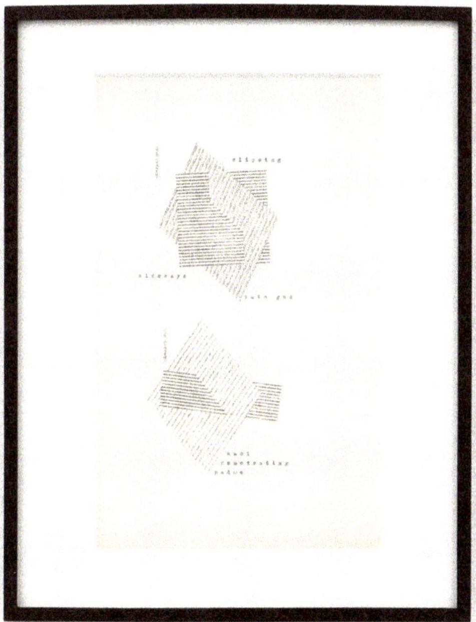

FIGURE 42.4: Dom Sylvester Houédard, *slipping sideways into god*, 1967. Typed page, carbon copy. 41.5 cm × 32.5 cm × 2.5 cm (framed). © Dom Sylvester Houédard; courtesy Lisson Gallery.

consciousness and its objects, all form a single weave; and tantrism is the discovery or establishment of inner relationships between the matter and spirit worlds, between ritual and reality, between mind and the universe, between the microcosm and the macrocosm (Houédard 1960: 95).

A mind that is liberated from all that is unreal, that has attained a blissful knowledge of the Ultimate Reality, has attained this realization. Tantra is not unique in asserting this position of 'ultimate truth' as it is understood in Buddhist traditions. What characterizes Tantra is the methods used to experience this truth. Houédard continues: 'Mudra, mantra, and yantra are translations of the inner experience into movement, speech, and *pattern*; they link inner and outer reality, they perfect the correspondence between micro- and macro-cosm' (Houédard 1960: 96, emphasis added).

In this explanation Houédard establishes his understanding of the performative reality of Tantra. Buddhist contemplative method practices were incorporated into an increasingly performative artistic practice that engaged directly with Tibetan Tantric ritual and, as such, locate Houédard as one of the very first Western Tantric artists, if not the very first British artist to acknowledge that his work may be understood as yantra, mantra and mudra. Therefore, the typestracts in this exhibition can be considered an 'intentional' Tantric language that asks questions about reality (Eliade 1969: 251).

Art historians such as Ajit Mookerjee have established that a yantra is essentially a geometrical composition. But to understand the 'true nature' of these geometrical drawings, Mookerjee clarifies that 'one has to go beyond the notions of geometry into those of dynamics'. He continues, 'a *Yantra* then represents a particular force whose power or energy increases in proportion to the abstraction and precision of the diagram' (Mookerjee 1971: 20). The Tantric practitioner, through his ritual use of yantra or power diagrams can create and control mental and physical forces.

The visual 'intentional' concrete language of Houédard's typestracts knowingly engages with a transhistorical Tantric discourse of coded language, thus 'a tantric text can be read with a number of keys: liturgical, yogic and tantric [...]. To read a text with the "yogic key" is to decipher the various stages of meditation to which it refers' (Mookerjee 1975: 252).

Houédard provided 'keys' to his texts in his prose. This exhibition presented a selection of those keys in relation to groupings of his work: moire, thunderbolt vajra, yantra, chakrometers, chakra wheels, mantra, bija, tantric staircases. Houédard's development of this powerful and creative contemplative language, which the critic Guy Brett identified as 'geometric figures', 'architectural spaces', 'orientation figures' and 'signs of energy' (Hunt et al. 2017: 47), took place prior to even the concept of 'Tantric art' in the West.[1]

The Tantric practitioner, through his ritual use of yantra or power diagrams can create and control mental and physical forces (Mookerjee 1971: 20). Therefore, if these works are to be understood as being 'similar in nature to mandalas and other cosmic diagrams' (Hunt et al. 2017: 47), then an understanding as to *how* they function as yantra is helpful. Again, Mookerjee provides a succinct definition of the yantra:

> Such basic geometrical figures as the point, straight line, circle, triangle, and square, have a symbolic value in representing the basic energies of the universe. They can be combined in increasingly complex figures to represent particular forces or qualities embodied in some aspect of creation, evolution, dissolution (Mookerjee 1975: 33).

Houédard's typestracts become diagrams of forces, dynamic graphs of the process of conceptualization itself, understandings of how the mind creates and destroys form, moment by moment. Writing about the typestracts, Houédard states 'in the moments of making [...] they step by step control me & pose ultimate questions of their own identity dependence destiny & independence'. As forms 'they should probably be viewed like cloud-tracks & tide-ripples – bracken-patterns & gull-flights – or simply as horizons & spirit levels' (Simpson 2012: 172). Yet to do so within the framework of Tantric practice that informed them, these 'cloud-tracks' and 'tide-ripples' are as equally un-selved as a tide ripple, which cannot be separated from the sea or a gull-flight, which cannot be separated from the sky. Realizing his un-selved self the Tantric practitioner can impute his 'I' on any form. The typestracts are yantras when Houédard, or another Tantric practitioner, imputes their 'I' upon their forms. However, this 'I' is as inseparable from the emptiness out of which it arises as the typed spiritual geometrical forms that are continually undone by a compositional grammar that engages the space on the page as referent to its own emptiness.

environmentpoem

The focus in this gallery space was on the interpenetration of all forms: the interdependence of, and dance between, all objects inside the gallery space and all objects outside the gallery – initially those changing forms seen through the gallery window, the calls of the primary-aged children tumbling through the local school gates, the spring trees emerging into leaf on the pavements outside, the intermittent traffic on Bell Street and the grey March London skies beyond and then beyond to all phenomena in the universe. It was intended to be a space where Houédard's playfulness would come to the fore: his profound knowledge of the Tantric practice

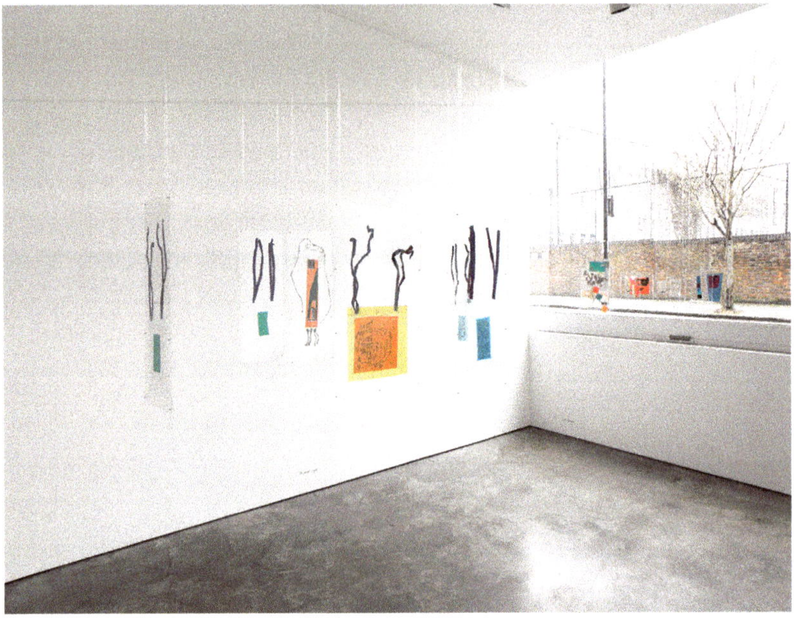

FIGURE 42.5: Dom Sylvester Houédard, *tantric poetries*. 12 March–31 July 2020. Installation view. © Dom Sylvester Houédard. Courtesy Lisson Gallery. Photography by Jack Hems.

of transforming all objects into the spiritual path and his connection to the Tibetan spiritual lineage of 'Crazy Wisdom' as embodied in the 1960s and 1970s by the Tibetan Lama Chögyam Trungpa Rinpoche. The laminate poems exemplify a Tantric practice where all objects (leaves, sand, chairs, tin foil, Vaseline, jam) and all actions, however seemingly transgressive for an ordained Catholic monk (sex, eating, dancing, taking drugs) are methods for spiritual transmission.

The laminate poems serially and collectively enact this transformation. The relationship between cut-up text and the objects and images collaged together in each separate poem at first may seem random, even chaotic, representative of the impermanent and unexpected flux of daily life. However, in this opportunity to exhibit from such an extensive collection I was able to present the viewer with some of the same relational concepts of permutation and combination in these laminate poems as in the typestracts. There are reoccurring motifs of leaves, sand, antimatter, alongside reoccurring themes of death, sex, god and the Buddha in the poems.

These laminate poems also function as observations of the microcosmic, in that they tantrically weave together tiny particles of substances such as talcum powder and dust but also an invisible world that remains very deliberately invisible. Forces and energies are collaged and sandwiched between layers of plastic, as is the invisible sub-atomic world of matter and quantum physics to become, alongside visible forms of buttons, Vaseline, leaves, sand, bodily fluids and detritus, carriers of a spiritual meaning within Houédard's eastern counterspace.

In this space, there was no prescription as to how the viewer should move. The laminate poems were grouped together in small collections and hung from the ceiling like collaged prayer flags. In Houédard's 1967 Lisson Gallery exhibition, these laminate poems were described in the accompanying exhibition leaflet as 'cosmic patches (attempts at repairing the universe)' and a typestract, with the same yogic key was included too (Figure 42.3). At one level, they are counter-culturally 'cosmic' and

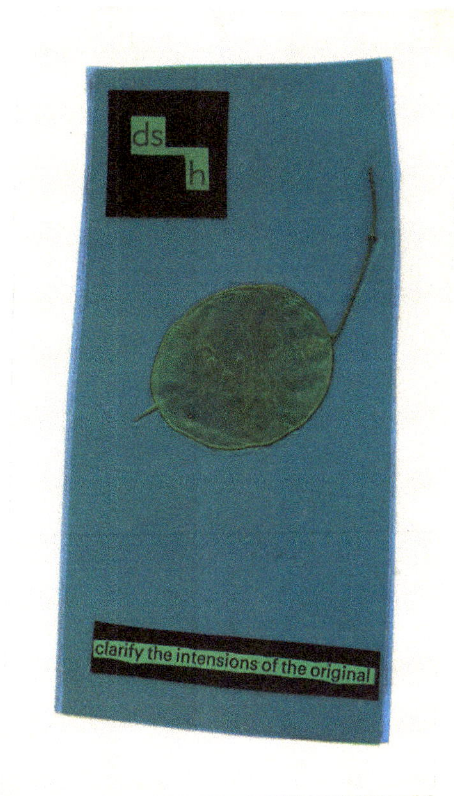

FIGURE 42.6: Dom Sylvester Houédard, *clarify the intensions of the original*, 1968. Vinyl plastic laminate, newspaper cuttings, miscellaneous material and PVC plastic. 14.6 cm × 8.3 cm. © Dom Sylvester Houédard. Courtesy Lisson Gallery.

many of the poems were made for prominent figures of the 1960s psychedelic scene. Works featured in the *Dom Sylvester Houédard: tantric poetries* exhibition include those made for John Hopkins ('Hoppy'), his girlfriend 'Suzy Creamcheese' or members of the hippie Galactic commune and artistic collective such as David Medalla, Edward Pope and John Dugger.

In hanging together the laminate poems, like an assemblage of hallucinogenic prayer flags, in the impermanent environment of the exhibition, this gallery space is a transient mandala, where the viewer temporarily encounters an installation that activates the macrocosmic and the microcosmic, beyond the world of ordinary appearance, such as 'some particles of antimatter from gloucestershire'. To activate this mandala, I felt it was important that these laminate poems were exhibited alongside 'some extracts from the mantra "Jrim, hum, ho, ho, ho, phat"', just as Houédard intended. In this contemporary exhibition, the mantra was printed in the exhibition guide enabling the viewer to recite it out loud, or silently in their head, as they participated in this environmentpoem, and maybe use it to generate the wish to attain the mind of bliss and emptiness, the Buddhist mind of enlightenment.

This was an instruction to bring the visual, the aural and the kinetic together in one environmentpoem that again would place the viewer consciously at the centre of the experience, the mandala and Houédard's spiritual transmission. Here, at this centre, all phenomena are born and return, the macrocosm of the seen gallery space, the attendees, invigilators, artworks, floorboards, walls, windows and the microcosm of the unseen vibrations of spoken word and breath, the impermanent thoughts and emotions of the persons present and the further hidden collisions, appearances and disappearances of the sub-atomic quantum world.

NOTE

1. Rebecca Heald discusses how the art historian and collector Ajit Mookerjee may have devised the term 'Tantric art' 'as way of neatly packaging artefacts for a Western market' in the 1970s. 1970s (Heald 2016, published on the occasion of the exhibition *Thinking Tantra*, Drawing Room, 24 November 2016–19 February 2017).

Contributors

GEMMA ANDERSON-TEMPINI is an artist, academic and mother. She is currently a co-investigator on the art/science/philosophy AHRC-funded project 'Representing Biology as Process' (2017–21) at the University of Exeter and has collaborated on a number of innovative art/science projects including '4D Eye' with the Mathematics Department at Imperial College London; 'Isomorphology' and the 'Cornwall Morphology and Drawing Centre' at CAST with the Natural History Museum, London; and 'Portraits: Patients and Psychiatrists' (Wellcome Trust Arts Award 2009) in collaboration with psychiatrists and patients at Bethlem Royal Hospital. Her work has been exhibited at the Victoria and Albert Museum, the Freud Museum and the Wellcome Collection, London. Most recently, her work was part of the exhibition 'Critical Zones; Observatories for Earthly Politics' at the Centre for Art and Media (ZKM), Karlsruhe, Germany and 'The Botanical Mind: Art, Mysticism and the Cosmic Tree' exhibition at Camden Arts Centre, London. Peer reviewed books include *Drawing as a Way of Knowing in Art and Science* (Anderson-Tempini, Intellect, 2017) and *Drawing Processes of Life* (Anderson-Tempini and Dupré, Intellect, 2023).

*****

KATARINA ANDJELKOVIC is a theorist, practising architect, researcher and painter. She served as a visiting professor, Chair of Creative Architecture, at the University of Oklahoma, Institute of Form Theory and History (Oslo), Institute of Urbanism and Landscape (Oslo), University of Belgrade, guest-lectured and mentored at the Delft University of Technology, Faculty of Architecture and the Built Environment, Oslo School of Architecture and Design, Faculty of Architecture, University of Porto, Istanbul Technical University, Anhalt Dessau and School of Making Thinking, New York. Andjelkovic has published her research widely in international journals (*Web of Science*) and won numerous awards for her architecture design and urban design competitions. She has published two monographs, an upcoming book chapter and several journal articles with Intellect United Kingdom. Andjelkovic has exhibited her artwork at five solo and many international architectural, fine arts and photography group exhibitions. https://independent.academia.edu/KatarinaAndjelkovic.

*****

CATHERINE BAKER is Reader in Art and Health at Birmingham City University in the United Kingdom. Based at Birmingham School of Art, she founded and co-leads the Material Encounters Research Cluster (https://www.bcu.ac.uk/art/research/material-encounters). As an artist/researcher, her early Ph.D. work involved working with neuroscientists with whom she shared an interest in the neurobiology of the eye and the perceptual encounter. Her interest in this encounter as a phenomenological consideration remains although since 2010 her research has become more focused on the impact of art practice in clinical contexts. Baker is currently the principal investigator leading an ambitious multi-partnered interdisciplinary project investigating environment and embodiment, and health/wellbeing in relation to adolescent scoliosis funded by the British Scoliosis Research Foundation. She is currently leading a special section for the world-leading journal *Leonardo* with Neuropsychologist Professor Iain Gilchrist on the process of interdisciplinary collaboration across the arts, sciences and humanities.

*****

LES BICKNELL has a history of making that encompasses an extensive range of working practices. His hybrid, interdisciplinary activity includes large-scale public art commissions and a sustained history of working within community engagement initiatives. The work has been exhibited widely and is held in many national and international collections including The Rijksmuseum, The V&A Museum, M.O.M.A. New York, Yale Centre for British Art, The Tate Gallery.

*****

SARAH BLAIR teaches writing and communication skills across a range of educational contexts, including the Royal College of Art in London and Hertford College's International Programme at Oxford University. She was formerly programme leader for MA Drawing for Fine Art Practice at Swindon College and has a Ph.D. in the language of visual narrative, applying concepts from linguistics to the realm and detail of drawing. With a continuing interest in narrative delivery, she has recently pursued a research focus on how graphic modes connect with systematic tendencies in verbal language. She is working on a book that explores the visual potential of English grammar.

*****

ANDREW BRACEY's work explores the relationship between curator and artist, art and reproductions and the reworking of historic artworks in contemporary art. His current Ph.D. explores how contemporary artists use historical paintings to initiate a dialogue between the past and the contemporary, through the metaphor of the parasite and mutualism, and his term 'parasitical painter'. One-person exhibitions include: Project Space Plus, Lincoln; (2019); Isherwood Gallery, Wigan (2017); Nottingham Castle (2014); Manchester Art Gallery (2009); Wolverhampton Art Gallery (2007); Transition Gallery, London (2007) and Firstsite, Colchester (2006). Bracey is co-leading a research project, *Bummock: Artists and Archives*, with Danica Maier; and working on a (potentially lifelong) curatorial project, *Midpointness*, with Steve Dutton. He has a studio at General Practice, Lincoln and is a senior lecturer and the programme leader of MA Fine Art at The University of Lincoln.

*****

GLYN BREWERTON is a senior lecturer in illustration at Norwich University of the Arts. His current research ideas aim to use oral history and documentation of living history to create drawings, maps and recordings of real-life events and rural landscapes. His aspiration in the future is to explore how he may apply gathered oral histories and reportage drawings in imaginative and authentic ways to digital oral history archive and mapping projects, to be used in museums and public spaces. Between January 2017 and January 2020, Brewerton developed and exhibited Heritage Lottery-funded work for the project *Water, Mills and Marshes* in association with the Broads Authority and the Landscape partnership. The work was exhibited at the Norwich University of the Arts, East Gallery in 2020 and consisted of drawings that mapped sightings and written commentary from the visitor's book at Wheatfen Broad in Norfolk.

*****

CHRIS BROWN is a senior lecturer in graphic design at Nottingham Trent University and is currently the course leader for MA Graphic Design and MA Branding and Identity. He has a background in art and design history and has had work published by Phaidon and The Conversation. His doctoral research was in design semantics and he has given conference papers on this in the United Kingdom and abroad. He has had an interest in photography from an early age and has had creative work published by UK-based interior design and greeting card companies and stock images published in national newspapers.

*****

GILL BROWN is a former geophysicist, now freelance graphic designer and illustrator, specializing in the field of scientific visual communication. In 2019, she was awarded a Ph.D. in graphic design from the University of the Arts London, during which she collaborated with the Centre for Neuroimaging Sciences (CNS) at King's College London to devise a new methodology for the creation of scientific conceptual figures (www.neuroscience-graphicdesign.com). Brown continues to work with the CNS and runs workshops on good graphic design practices for research scientists at King's and other institutions. She also works with individual scientists to help them create effective and adaptable figures for their work. Brown applies her research experience in all her work and gives presentations to design students about the importance of graphic design practice as research, www.graphical-science.com.

*****

PAULINE CLANCY is a lecturer in design and course director for Foundation Studies Art and Design at Belfast School of Art, Ulster University. Born in Co. Leitrim, Clancy studied B.Des. (Hons) Visual Communication at National College of Art and Design, Dublin, 2006, before completing her MFA in Multidisciplinary Design (Graphic Design) at Ulster University, 2013. Her current Ph.D. research explores the materiality of language through typographic form. Her work is held in collections including Hamilton Wood Type and Printing Museum, United States and Manchester Metropolitan University Special Collections of Artists' Books. She is a member of the International Society of Typographic Designers (MISTD) and was awarded a Society of Typographic Arts (STA)100 Award as part of the 100 best examples of Typographic Excellence in 2019 (United States America). In 2021, she also received a Communication Arts Typography Award of Excellence.

*****

CRIMP are DEWI BRUNET and GWENAËL PROST. Brunet is a Belgian artist based in Brussels. Prost is a French designer based in Rennes. Both are fascinated by the folds surrounding us; they explore folding techniques through different mediums and materials. Between education, philosophy, traditional crafts and new technology, they conceive folding as a creative and universal tool to understand the world. They are members of the CRIMP (International Research Centre for Folded Creation/Centre de Recherche International en Modélisation par le Pli), an art collective renowned for its research on crumpling techniques. Through scientific research, large-scale art installations and workshops, the CRIMP has, for twenty years, raised the awareness of the crumpling signification and potential. www.le-crimp.org / www.dewiorigami.com / www.gwenaelprost.com.

*****

Anne Eggebert is an artist and curator focusing on the imaging and imagining of place, both local and distant. She co-curated *Beyond Orientation: Eight Views* with Almut Rink at the Korea Kulturhaus, Vienna, and with Polly Gould ACE National Touring Exhibitions *Topophobia: Fear of Place in Contemporary Art* at Danielle Arnaud, London, Spacex, Exeter, and Bluecoat, Liverpool, and *Nature and Nation*. Eggebert has exhibited at Bizkaia Aretoa, Bilbao, Aine Art Gallery, Tornio, Haus am Lutzowplatz, Berlin, the London School of Hygiene and Tropical Medicine, Pitt Rivers Museum, Oxford, Tate St Ives, and on village greens across Essex. She collaborated with Polly Gould at the Cambridge University Botanic Gardens, the Royal Institute of Great Britain and the British Library, with Sarah Cole at Valentines Mansion, Ilford, and with Julian Walker at Kettle's Yard, Cambridge. She lives and works in London leading the BA Fine Art XD Pathway at Central Saint Martins, University of the Arts London.

*****

Kate Farley, associate professor of design at the Norwich University of the Arts, combines a design practice with the academic role of course leader of BA Textile Design. Farley's academic practice-based research explores the notion of pattern communication as narrative through printed design structures. This combines Farley's interest in traditional pattern making, cultural heritage and contemporary design opportunities. Through drawing and design development Farley investigates rhythm and compositional variations for contemporary surface design solutions. With clients such as Transport for London, David Mellor Design, Barbican Centre London, Birmingham International Airport and Queen Elizabeth University Hospital Birmingham, Farley has fulfilled commissions for bespoke pattern, designing for retail products, large-scale exterior surface design projects and public spaces. Farley has launched three collections under her own brand since 2012 featuring giftware and interior products with a focus on British manufacturing. Farley's work is held in collections including Tate, British Library, Yale Centre for British Art and London Transport Museum.

*****

Krzysztof Fijalkowski is professor of visual culture and teaches in the BA Fine Art programme at Norwich University of the Arts, United Kingdom, where he is also a member of the Pattern and Chaos research group. His research concentrates above all on the practice, history, theory and legacies of international surrealism, and current areas of focus consider their relationships to fields such as art, design, photography, film, sound and the everyday. Recent publications include contributions to *The International Encyclopedia of Surrealism* (Bloomsbury 2019), *Objects of Desire: Surrealism and Design 1924–Today* (Vitra Design Museum 2019), *The Situationist International: A Critical Handbook* (Pluto Press 2020), *Surrealism and Film After 1945* (Manchester University Press 2021), *Surrealism Beyond Borders* (Metropolitan Museum of Art/Tate 2021) and *The Routledge Companion to Surrealism* (Routledge 2022).

*****

MARK GRAVER, RE, is an award-winning artist printmaker based in Kerikeri, New Zealand. Born in the United Kingdom he studied at Leeds Polytechnic (BA Hons Fine Art) and Camberwell College of Arts, London (MA Printmaking). Moving to New Zealand in 2003 he established the Wharepuke Print Studio, New Zealand's first and only dedicated non-toxic acrylic resist etching workshop in 2005, an Art Gallery (2009) and Sculpture Park (2015). He is the author of *Non-Toxic Printmaking* (2011, London, A&C Black), a member of the UK-based Prism Print International, an editorial board member for the Intellect journal *Drawing: Research, Theory, Practice,* published by the University of Ulster and an elected Fellow of The Royal Society of Painter-Printmakers (RE) London. His work is held in many public collections including the Victoria and Albert Museum, London, The Ashmoleum Museum, Oxford and in Australia, Asia, New Zealand, the United States, Mexico, South Africa and Europe.

*****

DAVID GRIFFIN (born in Kingston, Jamaica) works with drawing, colour and writing. He has presented and exhibited his work across North America and the United Kingdom and has published articles on notational drawings in the journals *Drawing: Research, Theory, Practice* and *Tracey: Drawing and Visualisation Research* in the United Kingdom, and in the *Leonardo* journal in the United States. In addition to drawing fire, current projects include using laser to inscribe a colossal network diagram between planetary bodies in our solar system, as well as developing a music notation system using data gathered from human pulmonary systems. David is currently assistant professor in the Faculty of Art at OCAD University in Toronto, Canada, and is chair of the university's Research Ethics Board. He holds a Ph.D. from the Glasgow School of Art, an MFA from the Pratt Institute and a BFA from Parsons School of Design. david@davidgriffinart.com.

*****

LESLEY HALLIWELL is an artist based in the North West, United Kingdom. She trained at Dartington College of Arts, Nottingham Trent University (BA [Hons] Fine Art), Goldsmith's College, University of London (MA Art History) and Manchester

Metropolitan University (MA Fine Art). She completed a practice-based doctorate at Manchester Metropolitan University in 2020. Exhibitions include: *Artists' Camps*, The Sidney Nolan Trust; *Tilted Plane,* Alexandria Library, Manchester; *Beauty is the First Test*, Pumphouse Gallery, London; *The Drawing Show*, Castlefield Gallery, Manchester; *Jerwood Drawing Prize; Superabundant*, Turner Contemporary, Margate; *Pattern Recognition*, Leicester City Art Gallery; *Bloomberg New Contemporaries*. Halliwell has presented her research at numerous conferences and contributed to publications including 'Pattern: Entry and Reflection' in *Drawing: Research, Theory, Practice* (2019) 4:1, pp. 97–102 and 'The Depth of Surface' in *Surface and Apparition: The Immateriality of Modern Surface* (2021), Lee, Y. (ed.) London: Bloomsbury Visual Arts, pp. 62–76.

*****

KATY HAMMOND is an artist and lecturer in photography at Norwich University of the Arts. Her practice is focused on alternative processes and the materiality of photography, with a current interest in sustainable working practices. Hammond was awarded the Photofusion Select/17 Bursary Award and her imagery has been selected for numerous exhibitions including the Association of Photographers (AOP) Awards, Shutter Hub *Home*, *Hidden in Plain Sight* at The Gallery Norwich University of the Arts, Photofusion Salon and *Cosmos* as part of the Lumen Collective. Her paper 'The Anthropocentric Gaze and the Animal as Spectacle' was presented at the Animal Gaze Constructed Symposium with images from the corresponding series 'Denizens' exhibited at the London Metropolitan University.

*****

ROBERT HILLIER was the MA Communication Design Subject Leader at Norwich University of the Arts until 2021. He designed the Sylexiad range of fonts for adult dyslexic readers as part of his doctoral research. Sylexiad is used throughout the world and has been featured in many design journals, including Ultrabold, Novum, Étapes, Communication Arts and Baseline. The font is currently used by a diverse range of organizations and individuals including Bloomsbury Publishing for the High Low book series aimed at struggling and reluctant readers, the Australian sculptor Gail Hastings for her artist book *Space Practicing Tools*, through to the Marxist-Leninist Red Fightback website. His most recent research is *Imago Images*, a hybridized visual investigation of butterflies. Hillier is a volunteer at the ZSL London Zoo, which is in and of itself an embodiment of pattern and chaos. See www.sylexiad.com for further details about Sylexiad.

*****

ZOË HILLYARD is a Birmingham-based practitioner and academic seeking to foster resilience through pedagogy, community engagement and material transformation. Hillyard undertakes commissions, creating bespoke collections and working with precious possessions that individuals send her to revive. Her 'Kiln-Cracked Collection' is an ongoing body of work making use of the flawed pieces ceramicists generate that would otherwise end up in a landfill. Her work is inspired by the resourcefulness of nomadic cultures, her dismay at the excesses of consumer culture and a desire to encourage more circular thinking in the way we engage with possessions. Hillyard is a member of Contemporary Applied Arts and the Society of Designer Craftsmen, her work is in public and private collections and she exhibits internationally. She is a senior fellow of the Higher Education Academy and is currently course leader for BA (Hons) Textile Design at Birmingham City University.

*****

CHARLOTTE HODES is a professor in fine art at London College of Fashion, University of the Arts London. She was an associate artist at The Wallace Collection, London 2005–07, supported by the Arts and Humanities Research Council (AHRC) and Arts Council England and winner of the Jerwood Drawing Prize in 2006. She has held solo exhibitions at NCCD, Sleaford 2020; Ruthin Craft Centre, Wales 2019; Wolverhampton Art Gallery 2017; New Hall Art Collection, Cambridge 2014; Clara Scremini Gallery, Paris 2012 and 2019; Marlborough Gallery 2009; The Wallace Collection, London 2007 and University Gallery, Northumbria 2008 (an AHRC collaborative research project *Drawing Skirts* with Professor Cathy Treadaway). She has exhibited at the V&A 2002; Design Museum London 2003; Jerwood Gallery (three-person show) 2010; Venice Biennial 2009 and 2013 and Millesgården Stockholm 2012. Research centre and studio collaborations include Centre For Print Research, University of the West of England, Bristol; Berengo Glass Studio, Murano, Venice; The Clay Studio, Philadelphia; the National Glass Centre, University of Sunderland and Spode, United Kingdom.

*****

SARAH HORTON is an artist-researcher whose practice includes sculpture, drawing and painting often resulting in site-specific artwork. Her doctorate 'Decoration: Disrupting the workplace and challenging the work of art' indicates an ongoing interest in the way pattern, decoration and ornament are used in fine art and in a wider sense to indicate value and identity. She is a reader in fine art and champion for the Pattern and Chaos research group at the Norwich University of the Arts, UK, and an editorial board member for the *Drawing: Research, Theory, Practice* journal. Published papers include 'Hard Graft: The Office Chair as a Site for Decorative Art Interventions' (*Home Cultures*), 'Serial Killers? Investigating Some Modernist Myths About Decoration, Pattern and Ornament Through Workplace Interventions' (*Journal of Illustration*), has edited a special issue of *Drawing: Research, Theory, Practice* and co-edited (with Judith Stewart) a special issue of *Journal of Writing in Creative Practice*. www.sarahhorton.co.uk.

*****

ANTHONY HUDSON is the founding director of Hudson Architects. Over the past 30 years, the practice has designed buildings that have continually received recognition and awards for setting new standards in architectural design. He is Visiting Professor of Architecture at the Norwich University of the Arts and has been an external examiner for Cambridge, Kingston and Newcastle Universities. Hudson is Chair of Design Review panels in the South-East of England and Suffolk, past president of the Norfolk Association of Architects and a fellow of the Royal Society of Arts. He writes extensively on architecture and design and its impact on economic issues and the positive development of society. His book, *An Open Mind: The Work of Hudson Architects*, was published by Artifice Books in 2012.

*****

ALUN KIRBY is an ex-immunologist who makes photographs without using cameras, exploring the possibilities of the cyanotype process. Through collaborations with philosophers and people living with dementia, he explores what we consider memory to be, and how memory influences our sense of individual and cultural identity. These collaborations have led to the publication of his recent work on photography and memory in an academic philosophy journal. Kirby's current work addresses the maintenance of identity over time and the extension of memory through objects. Kirby lives and works in York. AlunKirby.com, Twitter @AlunKirby, Instagram AlunKirby, ORCID Alun Kirby.

*****

GEOFF DIEGO LITHERLAND is a Mexican-born artist based in Wirksworth, Derbyshire. Having spent his early years in Bolivia and Ecuador, he moved to England and graduated from Falmouth College of Arts in 2002 with a first-class honours degree in Fine Art Painting. In 2012, he completed an MFA in Fine Art at Goldsmiths University of

London and is currently a senior lecturer in fine art at Nottingham Trent University. www.geoffdiegolitherland.com.

*****

Sarah Lowndes is a writer, curator and lecturer. Research fellow at the Norwich University of the Arts and Director of Kunsthalle Cromer, her publications include *Contemporary Artists Working Outside the City: Creative Retreat* (2018), *The DIY Movement in Art, Music and Publishing* (2016), *All Art is Political: Writings on Performative Art* (2014), *Social Sculpture: The Rise of the Glasgow Art Scene* (2010) and *Crowds, Community and Contagion in Contemporary Britain* (Routledge 2022). Her curatorial credits include *Votive* (CCA, Glasgow 2009), *Studio 58: Women Artists in Glasgow Since WWII* (Mackintosh Museum 2012) and *The Glasgow Weekend: Art, Design and Music from Glasgow* (Volksbuehne Theatre and BQ, Berlin 2013). Projects under the auspices of Kunsthalle Cromer include *Panoramic Sea Happening* (2017), *Esplanade: A Procession for Women* (2018), the Arts Council England-funded publication projects *Like the Sea I Think* (2019) and *Field Work* (2020) and the Norfolk & Norwich Festival commission *The Sea Cure* (2021). www.sarahlowndes.net.

*****

David Mabb is an artist and occasional writer and designer. He holds a readership in art and is MFA Fine Art programme co-director at Goldsmiths, University of London. He works with appropriated imagery to rethink the political implications of different aesthetic forms in modern art and design history. His work has been exhibited internationally, most recently as part of *Perpetual Uncertainty/Contemporary Art in the Nuclear Anthropocene* at Bildmuseet, Umea (2016–17); Z33, Hasselt (2017); Konstmuseum, Malmö (2018) and as part of *Splitting the Atom*, Centre for Contemporary Art, Vilnius (2020). Recent solo exhibitions include Focal Point Gallery, Southend (2014), the William Morris Gallery, London (2015) and Handel Street Projects, London (2016).

*****

Louise Mackenzie is a British artist, curator and writer. Drawing on experience from a former career in communication and change management, her research explores human relationships with the material and nonhuman world through process-based and participatory art practice. She is the founder and curator of interdisciplinary cinema, Black Box and co-founder of Alive Together, an interdisciplinary community for research in human/animal relationships. Her artwork encompasses film, sound and sculpture and her works, *The Stars Beneath Our Feet*, *Zone of Inhibition* and *Lively Material* have been exhibited in the United Kingdom, United States, Canada, China and Australia. Mackenzie received a first-class honours in Fine Art from Newcastle University in the United Kingdom and a Ph.D. in Fine Art from BALTIC and Northumbria University's Institute of Contemporary Art. Mackenzie lectures and writes and has published in *Leonardo* journal (The MIT Press), *Journal of Technoetic Arts* (Intellect), *PUBLIC* journal and *Artnodes, Journal on Art, Science and Technology*.

*****

DANICA MAIER is an artist and academic currently living and working in Lincolnshire, England. Her practice focuses on the unrepeating-repeat, material processes, transposition, conflating expectations and how an audience looks/listens; as well as the dialogical nature of collaborative projects that foster independent artworks alongside wider group outcomes. Maier's work uses site-specific installations, drawings and objects to explore expectations while using subtle slippages to transgress propriety. Recent exhibitions and live events include: *Score: Mechanical Asynchronicity*, The Space at Nottingham Contemporary (with Martin Scheuregger), Nottingham; *Associated Thoughts on Line* as part of the *Convocation: On Expanded Language – Based Practices* within the Research Pavilion at the 58th Venice Biennale; *re-turning*, AirSpace Gallery, Stoke-on-Trent; *Bummock: Tennyson Research Centre* at the Collection Museum, Lincoln. Maier is associate professor in fine art at Nottingham Trent University, where she supervises Ph.D. candidates.

*****

ANGHARAD MCLAREN is a senior lecturer in textile design and Sustainable Clothing Research Coordinator at Nottingham Trent University, and a former lecturer and associate researcher at the School of Art and Design, University of Dundee, with prior textile industry and enterprise experience. Research interests include sustainable fashion and textiles, craft theory and practice, textile for performance, health and wellbeing, and relationships between community, place and heritage. www.angharadmclaren.co.uk/.

*****

VICTORIA MITCHELL is currently research fellow and was previously a senior lecturer at Norwich University of the Arts (1989–2016). In addition to contextual studies provision, mainly within fine art, she established the MA in Textile Culture (2001–10).

She has researched and published papers on various aspects of textile culture since 1995, pursuing an interdisciplinary approach which focuses on material, making, metaphor and meaning, most recently 'Judith Scott: capturing the texture of sensation' for *Textile* (2021). She was a co-investigator for the AHRC-funded *Beyond the Basket* (2009–11, University of East Anglia) and is co-editor (with Dr Stephanie Bunn, St. Andrews) of *The Material Culture of Basketry* (Bloomsbury 2020), for which she wrote on pattern in the context of braiding and dancing.

*****

ELEANOR MORGAN is an artist and writer working with video, printmaking and drawing to explore our entanglement with other species. This has included serenading a spider, making a diamond from the dead creatures of the River Thames and embracing a giant green sea anemone. Her book *Gossamer Days: Spiders, Humans and Their Threads* (Strange Attractor/The MIT Press 2016) examines the human uses of spider silk, from gun sights to sticky tunics and royal underwear. Her work has been funded by the Arts Council England, the Daiwa Anglo-Japanese Foundation, the Leverhulme Trust and the Royal Institute of British Architects. She has exhibited and performed with Eastside Projects, Ikon Gallery, RIBA London, Jerwood Space, the ICA and DOCUMENTA 13. She was awarded a Ph.D. from the Slade School of Art and is a lecturer in fine art at Loughborough University. www.eleanormorgan.com.

*****

DAKSHA PATEL's practice engages with scientific institutions researching themes connected to measurement, mapping and visualization. She has undertaken residencies in environmental modelling at Lancaster University, anatomy at King's College, London, applied mathematics at the University of Bristol, Parkinson's disease research at the University of Dundee, dermatology at Salford Royal National Health Service Trust and imaging science at the University of Manchester. Daksha's Ph.D. explored the concept of noise in medical imaging perceived through a contemporary drawing practice. She is a member of the research group The Cultural Negotiation of Science at Northumbria University. Her solo exhibitions include Dundee Contemporary Arts, Dundee; Paper2 Gallery, Manchester; Watermans Arts, London; Horniman Museum, London; and LifeSpace Gallery, Dundee.

*****

DR. BILL Prosser has been a Leverhulme Senior Research Fellow at the University of Reading, Andrew W. Mellon Research Fellow at the Harry Ransom Center,

University of Texas, and Centre Fellow at Regent's Park College, University of Oxford. He is now retired.

*****

JAMES QUINN is a lens-based practice-led researcher specializing in digital moving image and installation, drawing upon numerous theoretical discourses and fine art moving image practices (notably UK-based expanded cinema practices between the 1960s and 1970s). His doctoral thesis, 'A Critical Re-appraisal and Re-alignment of the Viewing Screen by Means of an Experimental Body of Fine Art Practice', interrogates the increasingly invisible, pervasive presence of contemporary screening technologies by employing video and digital installation artworks to disrupt habitual viewer engagement with viewing screens. This research has been exhibited internationally and has contributed to peer-reviewed articles published in 2020 and 2021, as well as featuring in various visual arts conferences in the United Kingdom and Europe.

*****

DORIS ROHR is a senior lecturer at Liverpool Hope University where she teaches theory and practice for fine art and design students based at Hope University's Creative Campus. Recent publications are mostly collaborative, with: Claire Penketh (2021) 'Drawingability' (editorial) in *Drawing: Research, Theory, Practice* Vol. 6.1; Charlie Gere (2018) 'Drawing on Text' in *Drawing: Research, Theory, Practice* Vol. 3.1; and Niamh Clarke (2020) Project Report: 'Collaborative Project: From Cradle to Parlour' in *Drawing: Research, Theory, Practice* Vol. 5.1. Rohr contributed a book chapter titled 'Art and Science in the Age of the Anthropocene' (2019) in Alannah Robins (ed) *From Dream to Dream: Where Art Meets Science* and a journal article 'Drawing in and out of Place' to *Environmental and Architectural Phenomenology* (2018/19, 30:1). Rohr exhibited with the Ulster-based collective The Drawing Journal in the exhibition 'Image of Thought' (Dec 2020–Jan 2021) archived at www.imageofthought.com.

*****

NICOLA SIMPSON is a research fellow at Norwich University of the Arts. Her interests are in discussing and performing the influence of Zen and Tantric Buddhisms on British counter-cultural art and writing. She is editor of *Notes from the Cosmic Typewriter: The Life and Work of Dom Sylvester Houédard* (Occasional Papers 2012) and co-editor of *Dom Sylvester Houédard* (Richard Saltoun, Riding House

2017). Other publications include 'polyphonicportraits & mantraportraits & mantrayantras', *NOTES ONE – onos & nonos: an introduction to the book of onomasticons by dsh* (Oxford: South Street 2016) and 'Attempts at repairing the universe – A Conversation between Nicholas Logsdail, Nicola Simpson and Charles Verey', *dsh* (Lisson Gallery 2018). Recent exhibitions include *Notes from the Cosmic Typewriter* (South London Gallery 2012), *Performing No Thingness: The Kinetic Poetry of Dom Sylvester Houédard, Ken Cox and Li Yuan-chia* (East Gallery, Norwich 2016) and *Dom Sylvester Houédard: tantric poetries* (Lisson Gallery 2020).

*****

JUDITH STEWART is an artist, writer and curator whose work prioritizes dialogue and collaboration. Her current research, *Hanse: Re-imagining the North Sea*, examines the role of affective responses to landscape and the role this plays in place-identity. Since 2017, she has worked collaboratively with James Quinn producing films, prints and artist books reflecting their encounters with specific places. Both her individual and collaborative works have been exhibited across the United Kingdom and Europe. As editor of *Place in Time: The Work of Tim Simmons* (Dewi Lewis Publishing 2020), Stewart used the work of the artist as a vehicle to explore aspects of contemporary responses to landscape. Other recent writings have arisen from her research with Lawrence Bradby, which critiques the hierarchical power structures of arts organizations and the role of the artist in participatory practices. Stewart is a lecturer at Sheffield Hallam University and a member of the Land2 research group. www.judith-stewart.com

*****

ANDREA STOKES is a fine artist who uses objects, imagery and strategies from 1980s feminism, working-class culture, new age spirituality and Utopian Humanism to interrogate meanings embedded within locations and objects. She studied Sculpture at Liverpool Polytechnic and Experimental Media at the Slade. She is associate professor in fine art and a member of the Contemporary Art Research Centre at Kingston University. Stokes is also a member of We Are Publication, a group of staff and students from the Kingston University Fine Art Ph.D. programme who conduct joint research and a speculative approach to publishing. www.andreastokes.com; www.uselesssplendour.wordpress.com/category/we-are-publication/.

*****

TOWNLEY AND BRADBY were a family artist practice and operated between 2005 and 2019. They made work in spaces that were already being used by someone else. This work included games, picnics, public conversations and site-specific drawings. The practice focused on the domestic, and on the emotional intensity of family life. Between 2017 and 2019, they were in residence with MK Gallery, Milton Keynes. They summarized the residency by making a book about a fictional family that plays and argues while exploring a city that is by turns colourful, magical and overwhelming. The book is called *How To Play With MK* (2020) and is published by MK Gallery.

*****

LUCY WARD is an artist and senior lecturer in drawing in the School of Art and Design at the University of the West of England (UWE), Bristol. Her teaching focuses on contemporary drawing theory and practices, with an emphasis on the performative and experimental. Ward's intricate drawings map the occurrence of anticipated and repeated events in ordered 2D space, intimating a friction between ideal pattern and repetitive behaviour in our everyday experience. Ward holds an MA in Drawing from Camberwell College of Art and a B.Sc. in Physics and Maths from the University of Bristol. She is a founding member of the research group, UWE Drawing Research.

*****

BRIAN WHALLEY is an Emeritus Professor at the University of Sheffield. He lives in Sheffield after spending most of his academic life at the Queen's University of Belfast where he was professor of geomorphology. His main interests in teaching and research in geomorphology are in glaciers and mountain landscapes and he is active in publishing in these fields and is an Honorary Life Fellow of the British Society for Geomorphology. He is a United Kingdom National Teaching Fellow and is part of the *Enhancing Fieldwork Learning* project where he has an interest and publishes in aspects of technology, pedagogy and active student participation in learning. He is involved in publishing 'geological poetry', particularly the poems of the Cumbrian poet Norman Nicholson. He is also actively involved in mountain walking, gliding and birdwatching.

*****

J. HARRY WHALLEY is a composer and academic whose interests lie in art–science collaboration. Recent works include a 'Plangency for Cello and Electronics' celebrating 850 years of Salisbury Cathedral, 'Reflections on Memory' with Alexander

McCall Smith and an extended work based on *Gödel Escher Bach* by Douglas Hofstadter and a film score for *Bonington*, which were premiered at the Banff Mountain Film Festival. His works have been performed by leading ensembles such as the Gildas Quartet, the Hebrides Ensemble and Ensemble Eunoia. Whalley is reader of sound and music at the University for the Creative Arts, where he established the Music Composition and Technology course and now runs the Audio Research Cluster collaborating across disciplines. He is a fellow of the London College of Music, Senior Fellow of the Higher Education Academy and composer-in-residence at St Vincent's Chapel, Edinburgh.

*****

KAROLINE WIESNER is professor of complexity science, Institute of Physics and Astronomy, Potsdam University, Germany; external faculty member at the Vienna Complexity Hub and the Centre for Philosophy of Science at the University of Bristol. She holds a Ph.D. in Physics, Uppsala University (2004) and was a postdoctoral fellow: Santa Fe Institute and University of California Davis (2004–07); assistant/associate professor of complexity sciences, University of Bristol (2007–21) and co-director of the Centre for Complexity Science at the University of Bristol (2007–18).

*****

CATHERINE YASS is a London-based artist and an associate lecturer at the Royal College of Arts, London. Major commissions include Flood Barrier (2023), commissioned by Create London; a Paddington Square permanent public art work commissioned by Sellar and Lacuna for St Mary's Hospital (2021–23), *Legacy*, The Supreme Court, to celebrate 100 years of women in the Law (2019); Royal London, Wellcome Foundation (2018); *Aeolian Piano*, to mark the departure of the BBC from the BBC TVCentre (2017); *Decommissioned*, The Jewish Community Centre, London (2013); Rambert Dance Company, London (2013); *Lighthouse*, De La Warr Pavilion (2011); *High Wire*, Artangel (2008); *Lock* (2006), a British Council residency at the Three Gorges Dam, China; and the set design for *Merce Split Sides* by Merce Cunningham (2003). Recent solo exhibitions and screenings include *Falling Away*, Ambika P3, London (2021), *Aeolian Piano* at RIBA (2018); *Lighthouse* at Alison Jacques Gallery, London (2012); a mid-career retrospective at De La Warr Pavilion, Bexhill-on-Sea (2011); *Flight*, The Phillips Collections, Washington D.C.; *The China Series*, Stedelijk-Hertogenbosch Museum, The Netherlands (2009); *Descent*, St. Louis Art Museum, St Louis, MO (2009). In 2002, Yass was shortlisted for The Turner Prize. She also represented the UK at the 10th Indian Triennial, 2001, and won the Glenn Dimplex Photography prize in 1999.

Bibliography

Abelian Sandpile Model (n.d.), 'Abelian Sandpile Model', Wikipedia (n.d) https://en.wikipedia.org/wiki/Abelian_sandpile_model. Accessed 20 October 2021.

Adamatzky, Andrew (ed.) (2010), *Game of Life Cellular Automata*, London: Springer.

Ades, Dawn (1981), 'Dada and Surrealism', in N. Stangos (ed.), *Concepts of Modern Art*, rev. ed., London: Thames and Hudson, pp. 110–37.

Ades, Thomas and Service, Tom (2012), *Thomas Ades: Full of Noises: Conversations with Tom Service*, London: Faber and Faber.

Agapakis, Christina (2014), 'Genes cannot be made from scratch', *Scientific American*, 14 June, Blog Network, https://blogs.scientificamerican.com/oscillator/genes-cannot-be-made-from-scratch/. Accessed 19 July 2021.

Alberro, Alexander and Stimson, Blake (1999), *Conceptual Art: A Critical Anthology*, Cambridge: MIT.

Anandagopu, Perumal, Suhanya, Sivakumar, Jayaraj, Veerasamy and Rajasekaran, Ekambaram (2008), 'Role of thymine in protein coding frames of mRNA sequences', *Bioinformation*, 2:7, pp. 304–07, https://doi.org/10.6026/97320630002304.

Anon. (2007), 'When in London ...', in V. Knoll and R. Gygax (eds), *The World of Interiors by Marc Camile Chaimowicz: The Artists' Book Arose from the Exhibition Marc Camille Chaimowicz*, Zürich: Mirgos Museum für Gegenwartskunst, p. 178.

Anon. (2012), 'Hillsborough report: Key findings', *BBC News*, 12 September, https://www.bbc.co.uk/news/uk-england-merseyside-19577033. Accessed 3 March 2021.

Anon. (2020), 'Coronavirus: Liverpool vs Atletico Madrid going ahead was "bad decision" says public health director', *The Independent*, 3 April, https://www.independent.co.uk/sport/football/european/coronavirus-liverpool-atletico-madrid-champions-league-uefa-gary-lineker-a9444916.html. Accessed 2 March 2021.

Arendt, Hannah (1958), *The Human Condition*, Chicago: University of Chicago Press.

Aristophanes (423 BCE), *The Clouds* (trans. I. Johnston), http://johnstoi.web.viu.ca//aristophanes/clouds.htm. Accessed 30 May 2021.

Aristotle (1970), *Poetics* (trans. G. E. Else), Ann Arbor: University of Michigan Press.

'Aurelian, adj.' (2005), *The Pocket Oxford Latin Dictionary* (ed. J. Morwood), Oxford, New York: Oxford University Press.

Bachelard, Gaston (1964), *The Psychoanalysis of Fire* (trans. C. M. Ross), London: Routledge & Kegan Paul.

Bak, Per (1996), *How Nature Works: The Science of Self-Organised Criticality*, New York: Copernicus.

Ball, Philip (2011), *Nature's Patterns: A Tapestry in Three Parts*, Oxford: Oxford University Press.

Ball, Philip (2016), *Patterns in Nature*, Chicago: University of Chicago Press.

Barad, Karen (2007), *Meeting the Universe Halfway: Quantum Physics and the Entanglement of Matter and Meaning*, Durham: Duke University Press.

Barnicoat, John (1972), *A Concise History of Posters*, London: Thames and Hudson.

Barrett Browning, Elizabeth (1870), *The Poetical Works of Elizabeth Barrett Browning*, New York: Crowell & Co., Digital Collections and Archives, Tufts University, https://dl.tufts.edu/concern/pdfs/1r66jc575. Accessed 1 September 2021.

Barthes, Roland (1974), *S/Z*, New York: Hill and Wang.

Barthes, Roland ([1973] 1975), *The Pleasure of the Text* (trans. R. Miller), New York: The Noonday Press.

Barthes, Roland (1977), 'The grain of the voice', in *Image-Music-Text* (trans. S. Heath), New York: Hill and Wang, pp. 179–89.

Bataille, Georges (1929a), '*Poussière*' ('Dust'), *Documents, 1*, 5:278, October, Rpt. in G. Bataille (1970), *Oeuvres Complètes*, I, 197, Paris: Gaillmard.

Bataille, Georges (1929b), 'L'informe', *Documents, 1*, Paris. Rpt. in G. Bataille (1970), *Oeuvres Complètes*, Paris: Gallimard.

Bateson, Gregory (1971), *Steps to an Ecology of Mind*, Chicago: University of Chicago Press.

Batty, Michael (2007), *Cities and Complexity*, Cambridge: The MIT Press.

Beckett, Samuel ([1950] 1959), *Molloy*, London: Calder.

Belmonte, Andrew (2007), 'The tangled web of self-tying knots', *Proceedings of the National Academy of Sciences of the United States of America*, 104:44, pp. 17243–244.

Benjamin, Walter (1999), *The Arcades Project* (trans. H. Eiland and K. McLaughlin), Cambridge: The Belknap Press of Harvard University Press.

Bennett, Jane (2010), *Vibrant Matter: A Political Ecology of Things*, Durham: Duke University Press.

Bennett, Phillippa (2021), 'Rewilding Morris: Wilderness and wildness in the last romances', in F. S. Boos (ed.), *The Routledge Companion to William Morris*, New York, Abingdon: Routledge.

Berg, Matthew D. and Brandl, Christopher J. (2021), 'Transfer RNAs: Diversity in form and function', in *RNA Biology*, vol. 18, 3rd ed., Columbia: Bellwether Publishing Ltd., pp. 316–39, https://doi.org/10.1080/15476286.2020.1809197.

Berger, John ([1960] 1979), *Permanent Red*, London: Writers and Readers.

Bergson, Henri ([1889] 1959), *Time and Free Will: An Essay on the Immediate Data of Consciousness* (trans. F. L. Pogson), New York: Macmillan.

Bernier, Ronald R. (2014), *Unspeakable Art of Bill Viola: A Visual Theology*, Eugene: Pickwick Publications.

Bhella, David (2016), 'Yes viruses are alive', *Microbiology Society*, Issue 'What is Life?', 10 May, https://microbiologysociety.org/resource_library/knowledge-search/are-viruses-alive-what-is-life.html. Accessed April 2021.

Binder, Timothy (1995), *In the Wave Lies the Secret of Creation*, Waynesboro: University of Science and Philosophy.

Blair, Sarah (2019), 'The ornament of grammar', *Journal of Illustration*, 6:1, pp. 137–60.

Blair, Sarah (2020), 'Pursuing the prepositions of drawing', *Drawing: Research, Theory, Practice*, 5:1, pp. 11–28.

Bois, Yves-Alain and Krauss, Rosalind (1997), *Formless: A User's Guide*, Paris, New York: Centre Georges Pompidou and Zone Books.

Borges, Jorge Luis (2000), *Labyrinths*, London: Penguin.

Bourdieu, Pierre (1984), *Distinction*, London: Routledge and Kegan Paul.

Bourgeois, Louise (2018), *Louise Bourgeois: Spirals*, New York: Cheim and Read.

Bragulla, Hermann H. and Homberger, Dominique G. (2009), 'Structure and functions of keratin proteins in simple, stratified, keratinized and cornified epithelia', *Journal of Anatomy*, 214:4, pp. 516–59, https://doi.org/10.1111/j.1469-7580.2009.01066.x. Accessed 19 July 2021.

Brandt, Erik (2019), 'Ficciones Typografika' workshop delivered at *In Progress*, AGI Open, Rotterdam, 24 September.

Brett, Guy ([1971] 2017), 'Introduction to Dom Sylvester Houédard: Visual poetries', in A. Hunt and N. Simpson (eds), *Dom Sylvester Houédard*, London: Richard Saltoun/Riding House.

Brown, Gill (2016), 'The paradigmatic evolution of scientific graphic design', *HARTS & Minds*, 3.1:8, pp. 2–18, https://docs.wixstatic.com/ugd/4b5f1a_f9364576db0d4d87a7e93c7348dd89a7.pdf. Accessed 26 May 2021.

Brown, Gill (2019), 'Visual elements and visual paradigm: Re-thinking scientific conceptual figures through graphic design', Ph.D. thesis, London: University of the Arts London, https://ethos.bl.uk/OrderDetails.do?did=3&uin=uk.bl.ethos.817782. Accessed 26 May 2021.

Brown, Nigel (2016), 'No viruses are not alive', *Microbiology Society*, Issue 'What is Life?', https://microbiologysociety.org/resource_library/knowledge-search/are-viruses-alive-what-is-life.html. Accessed April 2021.

Brown, Paul (2017), *Savage Enthusiasm: A History of Football Fans*, Durham: Goal Post.

Brownian Motion (n.d.), 'Brownian Motion', https://en.wikipedia.org/wiki/Brownian_motion. Accessed 12 August 2021.

Bucaille, Max (2019), 'Spiral and labyrinth', in E. Prestsæter (ed.), *These are Situationist Times! Deformations, Modifications, Derivations, and Transformations*, Oslo: Torpedo Press, pp. 146–47.

Buchloh, Benjamin H. D. (1982), 'Allegorical procedures: Appropriation and montage in contemporary art', *Artforum*, 21:1, pp. 43–56.

Buchmann, Sabeth (2016), 'Rehearsing in/with media: Some remarks on the relationship between dance, film, and painting', in I. Graw and E. Lajer-Burcharth (eds), *Painting Beyond Itself: The Medium in the Post-medium Condition*, Berlin: Sternberg Press.

Burke, Edmund ([1757] 2015), *A Philosophical Inquiry into the Origin of Our Ideas of the Sublime and Beautiful*, Oxford: Oxford University Press.

Burke, Janine (2006), *The Sphinx on the Table: Sigmund Freud's Art Collection and the Development of Psychoanalysis*, New York: Walker and Company.

Burleigh, Paula (2018), 'Ludic labyrinths: Strategies of disruption', *Stedelijk Studies*, 7, https://stedelijkstudies.com/journal/ludic-labyrinths-strategies-of-disruption/. Accessed 11 June 2021.

Burroughs, William S. (2005), *The Electronic Revolution*, Ubu Classics, http://www.ubu.com/historical/burroughs/electronic_revolution.pdf. Accessed 19 July 2021.

Busbea, Larry (2019), 'Geometries of thinking', in E. Prestsæter (ed.), *These are Situationist Times! Deformations, Modifications, Derivations, and Transformations*, Oslo: Torpedo Press, pp. 215–18.

Caillois, Roger ([1935] 1984), 'Mimicry and Legendary Psychasthenia' (trans. J. Shepley), *October*, vol. 31, Winter, London: The MIT Press, pp. 16–32.

C. D. (Dotremont, Christian) (1950), 'Le role de la spirale dans la cosmogonie du Capitaine Christeso', *Cobra*, 7, p. 21.

Caillois, Roger (2003), 'A new plea for diagonal science', in C. Frank (ed.), *The Edge of Surrealism: A Roger Caillois Reader* (trans. C. Frank and C. Naish), Durham, London: Duke University Press.

Cameron, Julia ([1994] 1995), *The Artist's Way*, London: Pan.

Carel, Havi ([2008] 2013), *Illness: The Cry of the Flesh*, London, New York: Routledge.

Carroll, Sean S. (2008), 'The cosmic origins of time's arrow', *Scientific American*, 298:6, pp. 48–57, http://www.jstor.org/stable/26000640. Accessed 3 May 2021.

Cartier-Bresson, Henri (2017), *Henri Cartier-Bresson: Interviews and Conversations 1951–1998*, New York: Aperture.

Cascella, Daniela (2015), *F.M.R.L: Footnotes, Mirages, Refrains and Leftovers of Writing Sound*, Winchester: Zero Books.

Castoriadis, Cornelius (1975), *The Imaginary Institution of Society*, Cambridge: The MIT Press.

Chan, Serena (2001), 'Complex adaptive systems', ESD.83 Research Seminar in Engineering Systems delivered at MIT, Cambridge, MA, 6 November, http://web.mit.edu/esd.83/www/notebook/Complex%20Adaptive%20Systems.pdf. Accessed 20 October 2021.

Clarke, Kenneth (1956), *The Nude*, London: John Murray.

Cocker, Emma (2013), 'Tactics for not knowing: Preparing for the unexpected', in E. Fisher and R. Fortnum (eds), *On Not Knowing How Artists Think*, London: Black Dog Press, pp. 126–35.

Cocker, Emma (2015), 'Sample of samplers: Repeating repeats, repeats', in D. Maier (ed.), *Grafting Propriety: From Stitch to the Drawn Line*, London: Black Dog Press, pp. 31–37.

Cohen, S. Robin, Mount, Balfour M., Tomas, Jon J. N. and Mount, Lauren F. (1996), 'Existential well-being is an important determinant of quality of life: Evidence from the McGill quality of life questionnaire', *Cancer*, 77, pp. 576–86.

Connor, Steven (n.d.), 'Pretty please: The pragmatics of petitionary practice', Steven Connor official website, https://stevenconnor.com/prettyplease.html. Accessed 12 June 2021.

Cook, Roger (2007), 'A sense of tact', in V. Knoll and R. Gygax (eds), *The World of Interiors by Marc Camile Chaimowicz: The Artists' Book Arose from the Exhibition Marc Camille Chaimowicz*, Zürich: Mirgos Museum für Gegenwartskunst, p. 37.

Cook, Theodore Andrea (1979), *The Curves of Life*, New York: Dover.

Coole, Diana and Frost, Samantha (eds) (2010), *Ontology, Agency and Politics*, Durham: Duke University Press.

Cosgrove, Denis E. (1984), *Social Formation and Symbolic Landscape*, Madison: The University of Winsconsin Press.

Cottingham, John (1997), *Descartes: Descartes' Philosophy of Mind*, London: Phoenix Books/Orion Publishing.

Coveney, Peter and Highfield, Roger (1995), *Frontiers of Complexity: The Search for Order in a Chaotic World*, London: Faber and Faber.

Crane, Walter (1902), *Line and Form*, London: George Bell and Sons.

Crawford, Dorothy H. (2011), *Viruses: A Very Short Introduction*, Oxford, New York: OUP Oxford.

Crick, Francis and Watson, James (1953), 'A structure for deoxyribose nucleic acid', *Nature*, 171, 737–38, https://www.nature.com/articles/171737a0. Accessed 19 July 2021.

Crimp, Douglas (2008), *On the Museum's Ruins*, Cambridge: The MIT Press.

Crutzen, Paul J. (2002), 'Geology of mankind', *Nature,* 415:51, p. 23.

cumulus n. (2022), *Collins Dictionary Online*, New York: HarperCollins Publishers, https://www.collinsdictionary.com/dictionary/english/cumulus. Accessed 6 September 2022.

Curd, Patricia (2019), 'Anaxagoras', *The Stanford Encyclopedia of Philosophy*, https://plato.stanford.edu/entries/anaxagoras/. Accessed 4 June 2021.

da Vinci, Leonardo ([c.1540] 1877), *Treatise on Painting* (trans. J. F. Rigaud), London: George Bell & Sons.

Dale, Henry (1934), 'Pharmacology and nerve-endings', *Proceedings of the Royal Society of Medicine*, XXVIII, pp. 319–32.

Damisch, Hubert (1995), *The Origins of Perspective* (trans. J. Goodman), Cambridge: The MIT Press.

Damisch, Hubert ([1972] 2002), *A Theory of /Cloud/ Toward a History of Painting*, California: Stanford University Press.

Darke, Colin and Mabb, David (2008), *Commodity Form: Colin Darke – The Capital Paintings, David Mabb – Rhythm 69*, Belfast: Golden Thread Gallery.

Daston, Lorraine and Galison, Peter (2007), *Objectivity*, New York: Zone Books.

Dawkins, Richard (1996), *Climbing Mount Improbable*, London: Viking.

Day, Brian J. (2005), 'The moral intuition of Ruskin's "Storm-Cloud"', *Studies in English Literature, 1500–1900*, 45:4, pp. 917–33, The Nineteenth Century.

de Certeau, Michel (1984), *The Practice of Everyday Life* (trans. S. Rendall), Berkeley: University of California Press.

De Vaan, Michiel (2008), *Etymological Dictionary of Latin and the Other Italic Languages*, Leiden: Koninklijke Brill NV.

Deleuze, Gilles (1968), *Différence et Répétition*, Paris: Presse Universitaires de France.

Deleuze, Gilles ([1968] 2014), *Difference and Repetition*, London: Bloomsbury.

Deleuze, Gilles (1997), 'Lewis Carroll', in G. Deleuze, *Essays Critical and Clinical* (trans. M. A. Greco and D. W. Smith), Minneapolis: University of Minnesota Press, pp. 21–22.

Deleuze, Gilles and Guattari, Félix (1994), *What is Philosophy?* (trans. H. Tomlinson and G. Burchell), New York: Columbia University Press.

Deleuze, Gilles and Guattari, Félix ([1988] 2020), *A Thousand Plateaus* (trans. B. Massumi), London: Bloomsbury.
Denny, Walter B. (2014), *How to Read Islamic Carpets*, New York: The Metropolitan Museum of Art.
Derrida, Jacques (2007), 'Derrida: "What comes before the question?"', YouTube, December, https://www.youtube.com/watch?v=Z2bPTs8fspk. Accessed 4 March 2021.
Descartes, René (1998), *Discourse on Method (1637) and Meditations on First Philosophy (1641)* (trans. D. A. Cress), 4th ed., Indianapolis: Hackett Publishing Company.
Dickens, Charles ([1861] 2008), *Great Expectations*, Oxford: Oxford University Press.
Dillon, Brian (2009), 'A dry black veil', *Cabinet*, 'Dust', issue 35, http://cabinetmagazine.org/issues/35/dillon/php. Accessed 31 August 2021.
Downs, Simon (2019), 'A drawing emerges: The drawing as a complex adaptive system', Loughborough: Loughborough University, https://hdl.handle.net/2134/11822946.v1. Accessed 8 August 2021.
Downs, Simon, Marshall, Russell, Sawdon, Phil, Selby, Andrew and Tormey, Jane (2007), *Drawing Now: Between the Lines of Contemporary Art*, London: I.B. Tauris.
Eagleton, Terry (2014), *Culture and the Death of God*, New Haven: Yale University Press.
Eastwood, Steven (2016), 'The interval and the instant: Inscribing death and dying', *Moving Image Review & Art Journal*, 5:1&2, pp. 26–42.
Eastwood, Steven (2017), 'The interval and the instant', Fabrica Gallery, Brighton, 7 October–26 November, https://www.fabrica.org.uk/the-interval-and-the-instant. Accessed 13 April 2021.
Eco, Umberto (1989), *The Open Work* (trans. A. Cancogni), Cambridge: Harvard University Press.
Edeline, Francis and Klinkenberg, Jean-Marie (2015), 'The appropriation of the work of art as a semiotic act', in P. F. Bundgaard and F. Stjernfelt (eds), *Investigations into the Phenomenology and the Ontology of the Work of Art: What are Artworks and How Do We Experience Them?*, Cham: Springer International Publishing, pp. 225–42, https://doi.org/10.1007/978-3-319-14090-2_13.
Ehn, Billy and Löfgren, Orvar (2009), 'Routines – Made and unmade', in E. Shove, F. Trentmann and R. Wilk (eds), *Time, Consumption and Everyday Life: Practice, Materiality and Culture*, Oxford, New York: Berg, pp. 99–114.
Eliade, Mircea (1969), *Yoga: Immortality and Freedom* (trans. W. R. Trask), Bollingen Series LVI, Princeton: Princeton University Press.
Engrenages (Spiral) (2005–present), France: Canal+.
Eversham, J. D. and Ruiz, Virginie (2011), 'Experimental analysis of the Reynolds flocking model', *Paladyn, Journal of Behavioral Robotics*, 2:3, pp. 145–55.
Featherstone, Mark (2016), 'Chaosmic spasm: Guattari, Stiegler, Berardi, and the digital apocalypse', *Communication and Media*, 11:38, pp. 243–68, https://www.researchgate.net/publication/314482391. Accessed 21 April 2022.
Ferreira, Ruiz and Umerez, Jon (2018), 'Dealing with the changeable and blurry edges of living things: A modified version of property-cluster kinds', *European Journal for Philosophy of Science*, 8, pp. 493–518, https://doi.org/10.1007/s13194-018-0210-z.
Fineberg, Jonathan (2006), *When We Were Young*, Berkeley: University of California Press.

Fish, Jonathan and Scrivener, Stephen (1990), 'Amplifying the mind's eye: Sketching and visual cognition', *Leonardo*, 23:1, pp. 117–26.

Fitch, Raymond E. (1982), *The Poison Sky: Myth and Apocalypse in Ruskin*, London: Ohio University Press.

Flache, Andres (n.d.), 'What is cellular automata modelling?', *Future Learn*, https://www.futurelearn.com/info/courses/complexity-and-uncertainty/0/steps/1902. Accessed 5 October 2021.

Flam, Jack D. (ed.) (1996), *Robert Smithson, The Collected Writings*, Berkeley: University of California Press.

Foa, Maryclare, Grisewood, Jane, Hosea, Birgitta and McCall, Carali (2000), *Performance Drawing: New Practices Since 1945*, London: Bloomsbury.

Ford, P. R. J. ([1981] 1997), *Oriental Carpet Design*, London: Thames and Hudson.

Forterre, Patrick (2010), 'Defining life: The virus viewpoint', *Origins of Life and Evolution of the Biosphere: The Journal of the International Society for the Study of the Origin of Life*, 40, pp. 151–60, https://doi.org/10.1007/s11084-010-9194-1.

Forterre, Patrick (2016), 'To be or not to be alive: How recent discoveries challenge the traditional definitions of viruses and life', *Studies in History and Philosophy of Biological and Biomedical Science*, 59, pp. 100–08, https://doi.org/10.1016/j.shpsc.2016.02.013.

Foucault, Michel (1977), *Discipline and Punish: The Birth of the Prison* (trans. A. Sheridan), Harmondsworth: Penguin.

Foucault, Michel ([1967] 1984), 'Of other spaces: Utopias and heterotopias' (trans. J. Miskowec), *Architecture/ Mouvement/ Continuité*, No. 5, October, pp. 1–9, http://web.mit.edu/allanmc/www/foucault1.pdf. Accessed 21 September 2021.

Foucault, Michel (2001), *The Order of Things*, London: Routledge.

Foucault, Michel ([1966] 2002), *The Order of Things*, London: Routledge.

Fractal Foundation (n.d.), 'What are fractals?', *Fractals are SMART: Science, Math and Art!*, https://fractalfoundation.org/resources/what-are-fractals/. Accessed 18 May 2021.

Freud, Sigmund ([1920] 1950), *Beyond the Pleasure Principle*, London: Hogarth Press.

Freud, Sigmund ([1933] 1968), 'Femininity' (trans. and ed. J. Strachey) in *New Introductory Lectures on Psychoanalysis. Standard Edition of the Complete Psychological Works of Sigmund Freud*, vol. 22, London: Hogarth Press, pp. 112–35.

Freud, Sigmund ([1901] 2001), 'On dreams' (trans. and ed. J. Strachey), in *The Standard Edition of the Complete Psychological Works*, vol. 5, London: Vintage.

Fry, Christopher (2018), 'Enchanting algorithms: How the reception of generative artworks is shaped by the audience's understanding of the experience', in *GA2018 XXI Generative Art Conference*, Verona: Italy.

Fry, Christopher (2019), 'Visuality and the haptic qualities of the line in generative art', in *GA2019 XXII Generative Art Conference*, Verona: Italy.

Fry, Roger ([1909] 1992), 'An essay in aesthetics', in C. Harrison and P. Wood (eds), *Art in Theory 1900–2000: An Anthology of Changing Ideas*, Oxford: Blackwell, pp. 78–86.

Fuss, Diane and Sanders, Joel (2004), 'Freud's ear', in D. Fuss (ed.), *The Sense of an Interior*, Abingdon: Routledge, pp. 71–107.

Gaiger, Jason (2008), *Aesthetics and Painting*, London: Continuum International Publishing Group.

Galafassi, Pierre (2009), 'Salon 129: Wool Dyeing History, with Focus on Dyeing of Rugs', Turkotek Salon, http://www.turkotek.com/salon_00129/salon.html. Accessed 21 September 2021.

Galanter, Philip (2003), 'What is generative art? Complexity theory as a context for art theory', in *GA2003 VI Generative Art Conference*, Milan: Italy.

Gell, Alfred (1994), 'The technology of enchantment and the enchantment of technology', in J. Coote and A. Shelton (eds), *Anthropology, Art and Aesthetics*, Oxford: Clarendon Press, pp. 40–66.

Gell, Alfred (1998), *Art and Agency: An Anthropological Theory*, Oxford: Clarendon Press.

Gidal, Peter (1978), *Structural Film Anthology*, London: BFI.

Giordano, Paolo (2020), *How Contagion Works: Science, Awareness and Community in Times of Global Crises – The Short Essay that Helped Change the COVID-19 Debate*, London: Weidenfeld and Nicolson.

Gleick, James (1987), *Chaos: Making a New Science*, New York: Viking Penguin.

Gold, Jeffrey (1985), 'Cartesian dualism and the current crisis in medicine – A plea for a philosophical approach: Discussion paper', *Journal of the Royal Society of Medicine*, 78, pp. 663–66.

Goldin, Amy (1975), 'Patterns, grids and painting', *Artforum*, 14:1, September, pp. 50–54.

Gombrich, Ernst (1979), *The Sense of Order*, Oxford: Phaidon.

Gordon, Beverly (2013), *Textiles: The Whole Story*, London: Thames and Hudson.

Górska, Elżbieta (2008), 'Four arguments for patterns of metaphorical thought', *Acta Philologica*, 35, pp. 15–31.

Grand View Research (2020), 'Viral vectors & plasmid DNA manufacturing market size, share & trends analysis report by vector (Lentivirus, Retrovirus), by workflow, by application, by end use, by disease, and segment forecasts, 2020–2027', Grand View Research, https://www.grandviewresearch.com/industry-analysis/viral-vector-manufacturing-market. Accessed 19 July 2021.

Graw, Isabelle (2004), 'Dedication replacing appropriation: Fascination, subversion, and disposition', in P. Kaiser (ed.), *Louise Lawler and Others*, Ostfildern-Ruit: Hatje Cantz.

Graw, Isabelle and Lajer-Burcharth, Ewa (eds) (2016), *Painting Beyond Itself: The Medium in the Post-Medium Condition*, Berlin: Sternberg Press.

Gray, Laura (2017), *Contemporary British Ceramics and the Influence of Sculpture: Monuments, Multiples, Destruction and Display*, New York: Routledge, https://doi.org/10.4324/9781315114132.

Griffin, David (2016), '*Ut pictura poesis*: Drawing into space', *Project Anywhere*, Centre of Visual Art, University of Melbourne and Parsons School of Art, Media and Technology, The New School, http://www.projectanywhere.net/ut-pictura-poesis-drawing-into-space-david-griffin/. Accessed 5 April 2021.

Griffiths, Frances and Lynch, Christine (2009), *Reflections on the Cottingley Fairies in Her Own Words*, Sheffield: JMJ Publications.

Grossman, Wendy (2014), 'Squaring the circle: The math of art', in W. Grossman and E. Sebline (eds), *Man Ray: Human Equations. A Journey from Mathematics to Shakespeare*, Ostfildern: Hatje Cantz, pp. 184–207.

Grosz, Elizabeth (2008), *Chaos, Territory, Art: Deleuze and the Framing of the Earth*, New York: Columbia University Press.

Groys, Boris (2016), *In the Flow*, London: Verso.

Guarnieri, Michael and Balmes, John R. (2014), 'Outdoor pollution and asthma', *Lancet*, 383:9928, pp. 1581–92, https://www.ncbi.nlm.nih.gov/pmc/articles/PMC4465283/. Accessed 31 August 2021.

Guattari, Félix (1995), *Chaosmosis: An Ethico-Aesthetic Paradigm*, Sydney: Power Press.

Guillery, Ray W. (2005), 'Observations of synaptic structures: Origins of the neuron doctrine and its current status', *Philosophical Transactions of the Royal Society of London, Series B, Biological Sciences*, 360, pp. 1281–307.

H. D. (Doolittle, Hilda) ([1956] 2012), *Tribute to Freud*, New York: New Directions.

Hamblyn, Richard (2017), *Clouds: Nature and Culture*, London: Reaktion Books.

Hansell, Mike (2007), *Built by Animals: The Natural History of Animal Architecture*, Oxford: Oxford University Press.

Harraway, Donna (2016), *Staying with the Trouble: Making Kin in the Chthulucene*, Durham: Duke University Press.

Harris, Dale B. (1963), *Children's Drawing as Measures of Intellectual Maturity*, New York: Harcourt, Brace & World.

Haraway, Donna (2020), 'Situated knowledges: The science question in feminism and the privilege of partial perspective', *Feminist Theory Reader*, London: Routledge, pp. 303–10.

Hartl, F. Ulrich and Hayer-Hartl, Manajit (2009), 'Converging concepts of protein folding in vitro and in vivo', *Nature Structural & Molecular Biology*, 16:6, pp. 574–81, https://doi.org/10.1038/nsmb.1591.

Harvey, Doug (2020), 'Memorials grown from clay and ink', in Ward, Cathy, Harvey, Doug and Wallis, Robert J. (2020), *Liberty Realm*, London: Strange Attractor Press, pp. 9–12.

Heald, Rebecca (2016), 'Thinking Tantra', *Thinking Tantra Research Papers*, London: The Drawing Room, https://drawingroom.org.uk/uploads/resources/Tantric_Research_Papers_web.pdf. Accessed 7 September 2021.

Hearne, Thomas (1885), *Remarks and Collections*, vol. 1, Oxford: Clarendon Press.

Heidegger, Martin (1962), *Being and Time* (trans. J. Macquarrie and E. Robinson), Oxford: Blackwell.

Herzfeld, Chris and Lestel, Dominique (2005), 'Knot tying in great apes: Etho-ethnology of an unusual tool behavior', *Social Science Information*, 44:4, pp. 621–53.

Hesiod (1953), *Theogony* (trans. N. O. Brown), New York: The Liberal Arts Press.

Hewison, Robert (2016), 'If walls could talk ... The ethics of dust at Westminster Hall', *Apollo Magazine*, 29 June, https://www.apollo-magazine.com/if-walls-could-talk-the-ethics-of-dust-at-westminster-hall/. Accessed 14 April 2021.

Highmore, Ben (2016), 'Daily habits: Pattern at the edge of meaning', YouTube, 17 October, https://www.youtube.com/watch?v=ezClcFecoXI. Accessed 13 October 2021.

Hill, John T. and Liesbrock, Heinz (eds) (2015), *Walker Evans: Depth of Field*, Munich: Prestel.

Hillier, Robert (2012), 'The design and development of the Sylexiad typeface', in N. Brunswick (ed.), *Supporting Dyslexic Adults in the Workplace*, Chichester: Wiley-Blackwell, pp. 185–96.

Hillier, Robert (2016), 'Design practice as palimpsest', in L. Imbesi and L. Di Lucchio (eds), *The International Journal of Visual Design*, vol. 10, 3rd ed., Champaign: Common Ground, pp. 1–18.

Hirst, Damien (1996), *No Sense of Absolute Corruption*, New York: Gagosian Gallery.

Hecker, Ewald ([1899] 2008), 'Karl Ludwig Kahlbaum', *History of Psychiatry*, 19:1, pp. 80–85, https://doi.org/10.1177/0957154X080190010602.

Hitchcock, Alfred (1958), *Vertigo*, Los Angeles: Paramount Pictures.

Hitchcock, Alfred (1960), *Psycho*, Los Angeles: Paramount Pictures.

Hoelscher, Jason (2013), 'Pattern and deregulation: Beauty and non-order in contemporary painting', *ArtPulse Magazine*, 5:17, pp. 22–25.

Holden, Kate and Crooks, Daniel (2016), 'Futurist Daniel Crooks', *The Saturday Paper*, https://annaschwartzgallery.com/news/futurist-daniel-crooks. Accessed 20 October 2021.

Houédard, Dom Sylvester (1960), 'Heathen holiness', *The Aylesford*, Review, 3.3, Autumn, London: Lisson Gallery.

Howell, Anthony (1999), *The Analysis of Performance Art: A Guide to its Theory and Practice*, London, New York: Routledge.

Howell, Anthony and Templeton, Fiona (1977), *Elements of Performance Art*, London: The Ting.

Hughes, Joe (2009), *Deleuze's Difference and Repetition: A Reader's Guide*, London: Continuum.

Hunt, Kevin (2018), 'Review of *Flow and Friction*', *Eye Magazine*, 6 April, http://www.eyemagazine.com/blog/post/follow-the-glitch. Accessed March 2021.

Hutchinson, Thomas (ed.) (1907), *The Complete Works of Percy Bysshe Shelley*, London, New York, Toronto: Oxford University Press.

Hyman, John (2006), *The Objective Eye*, Chicago: The University of Chicago Press.

Igoe, Elaine (2010), 'The tacit-turn: Textile design in design research', *Duck Journal for Research in Textiles and Textile Design*, 1, pp. 1–11.

'Image, n.' (1963), *The Concise Oxford English Dictionary of Current English* (eds H. W. Fowler and F. G. Fowler), London: Oxford University Press.

'Image, n.' (2005), *The Pocket Oxford Latin Dictionary* (ed. J. Morwood), Oxford, New York: Oxford University Press.

Ingold, Tim (2010), 'The textility of making', *Cambridge Journal of Economics*, 34:1, pp. 91–102, https://doi.org/10.1093/cje/bep042.

Ingold, Tim (2017), 'Surface visions', *Theory, Culture and Society*, 34:7&8, pp. 99–108, https://doi.org/10.1177/0263276417730601.

Invisible Committee (2014), 'Fuck off Google', *Anonymous France*, www.anonymous-france.info. Accessed 26 August 2020.

Jacobs, Jane ([1961] 1965), *The Death and Life of Great American Cities*, New York: Penguin.

Jaffray, Sarah (2019), 'Gained in translation: Drawing art history', in *TRACEY: Drawing and Visualisation Research*, 14:1, pp. 1–13, https://ojs.lboro.ac.uk/TRACEY/article/view/2547. Accessed 26 August 2020.

James, Henry ([1916] 2013), *The Figure in the Carpet*, Project Gutenberg eBook, Transcribed from the 1916 Martin Secker edition by David Price, https://www.gutenberg.org/files/645/645-h/645-h.htm. Accessed 21 September 2021.

James, William ([1890] 2000), *The Principles of Psychology*, pp. 50–63, http://www.public-library.uk/ebooks/50/61.pdf. Accessed 30 May 2021.

Jennifer, Noémie (2016), 'Should perishable art be saved? I Conservation Lab', *Vice*, Vice Media Group, https://www.vice.com/en/article/53wvzk/how-to-save-perishable-unruly-art-conservation-lab. Accessed 31 August 2021.

Jenny, Hans ([1967–72] 2001), *Cymatics: The Study of Wave Phenomena*, Eliot: MACROmedia Publishing.

Jorn, Asger (2019), 'Mind and sense: On the principle of ambivalence in Nordic *husdrapa* and mind singing', in E. Prestsæter (ed.), *These are Situationist Times! Deformations, Modifications, Derivations, and Transformations*, Oslo: Torpedo Press, pp. 3–5.

Joyce, James ([1914] 1968), 'Eveline', *The Dubliners*, Harmondsworth: Penguin, pp. 34–39.

Jullien, François ([1991] 2004), *In Praise of Blandness* (trans. P. M. Varsano), New York: Zone.

Junji, Ito (2013), *Uzumaki*, deluxe edition, San Francisco: Viz.

Kale, Sirin (2020), 'Richie Mawson was a beloved dad and Liverpool fan. Did a late lockdown cost him his life?', *The Guardian*, 20 August, https://www.theguardian.com/society/2020/aug/11/richie-mawson-was-a-beloved-dad-and-liverpool-fan-did-a-late-lockdown-cost-him-his-life. Accessed 2 March 2021.

Kay, Lily E. (2000), *Who Wrote the Book of Life?: A History of the Genetic Code*, Stanford: Stanford University Press, http://www.sup.org/books/title/?id=377. Accessed 19 July 2021.

Kemp, Martin (2000), *Visualizations: The Nature Book of Art and Science*, Oxford: Oxford University Press.

Kempske, Bonnie (2021), *Kintsugi: The Poetic Mend*, London: Herbert Press, Bloomsbury Publishing.

Kentridge, William (2014), *Six Drawing Lessons*, Cambridge: Harvard University Press.

Kiesler, Frederick (1996), 'Art and architecture: Notes on the spiral-theme in recent architecture', in S. Gohr and G. Luyken (eds), *Frederick J. Kiesler: Selected Writings*, Ostfildern: Verlag Gerd Hatje, pp. 47–52.

Klausmeyer, Bryan (2021), 'Spirale (Spiral)', *Goethe-Lexicon of Philosophical Concepts*, 1:1, n.pag., https://goethe-lexicon.pitt.edu/GL/article/view/22. Accessed 11 June 2021.

Klee, Paul ([1956] 1973), *Pedagogical Sketchbook* (trans. S. Moholy-Nagy), London: Faber & Faber.

Koller, Christian and Brändle, Fabian (2015), *Goal! A Cultural and Social History of Modern Football*, Washington: The Catholic University of America Press.

Korn, H. and Faure, P. (2003), 'Is there chaos in the brain? II. Experimental evidence and related models', *Comptes Rendus Biologies*, 326:9, pp. 787–840, https://www.sciencedirect.com/science/article/pii/S1631069103002002?via%3Dihub. Accessed 17 March 2022.

Kostelanetz, Richard (ed.) (1988), *Conversing with Cage*, New York: Limelight Editions.

Kraft, Kerstin (2004), 'Textile patterns and their epistemological functions', *TEXTILE*, 2:3, pp. 274–89.

Krips, Henry (2010), 'The politics of the gaze: Foucault, Lacan and Žižek', *Culture Unbound*, Vol. 2, pp. 91–102, Linköping University Electronic Press, http://www.cultureunbound.ep.liu.se. Accessed 20 April 2022.

Kristeva, J. (1984), *Revolution in Poetic Language* (trans. M. Waller), New York: Columbia University Press.

Kucharski, Adam (2020), *The Rules of Contagion: Why Things Spread – And Why They Stop*, London: Wellcome Collection.

Küchler, Suzanne and Were, Graeme (2005), *Pacific Patterns*, London: Thames and Hudson.

Lacan, Jacques ([1973] 1978), 'The seminar of Jacques Lacan', in *Book 11: The Four Fundamental Concepts of Psychoanalysis* (trans. A. Sheridan), New York: W. W. Norton.

Ladyman, James and Wiesner, Karoline (2020), *What is a Complex System?*, London: Yale University Press.

Land, Christopher (2005), 'Apomorphine silence: Cutting-up Burrough's theory of language and control', *Ephemera: Theory and Politics in Organisation*, 5:3, pp. 450–71.

De Landa, Manuel (1997), *A Thousand Years of Nonlinear History*, New York: Zone Books.

Lander, E. S. et al. (2001), 'Initial sequencing and analysis of the human genome', *Nature*, 409:6822, pp. 860–921, https://doi.org/10.1038/35057062.

Lange-Berndt, Petra (2015), *Materiality*, Cambridge: The MIT Press.

Latour, Bruno (2004), *The Politics of Nature*, Cambridge: Harvard University Press.

Lawton, S. and Miles, G. (2019), 'Occupational skin and lung disease in coalfield communities', *Nursing Times*, 115:7, pp. 58–60.

Lefebvre, Henri ([1992] 2004), *Rhythmanalysis: Space, Time and Everyday Life* (trans. S. Elden), London, New York: Continuum.

Leslie, Esther (2014), 'Animation and history', in K. Beckman (ed.), *Animating Film Theory*, Durham: Duke University Press.

Levi, Primo (1985), *The Periodic Table* (trans. R. Rosenthal), New York: Schocken Books.

Lissitsky, El (2014), *About Two Squares: A Suprematist Tale of Two Squares in Six Constructions* (trans. O. Belkeddar), London: Tate Publications.

Lissitzky-Kuppers, Sophie (1968), *El Lissitzky: Life, Letters, Texts*, London: Thames and Hudson.

Litherland, Geoff Diego and McLaren, Angharad (2017), 'Woven within the surface: Interconnected explorations of fine art linen canvas production', in *Proceedings of Intersections: Collaborations in Textile Design Research Conference*, 13 September, Loughborough: Loughborough University London, https://www.lboro.ac.uk/media/wwwlboroacuk/external/content/schoolsanddepartments/aed/downloads/Litherland_INTERSECTIONS2017.pdf. Accessed 18 October 2021.

Loermans, Helena (2021), 'Helena Loermans Handwoven Textiles, Research and reconstruction of historic canvases at Lab O', https://helenaloermans.blogspot.com/. Accessed 1 June 2021.

Lomas, David (2006), 'Labyrinth and vertigo: On some motifs in André Masson and their meaning', in E. Adamowicz (ed.), *Surrealism: Crossings/Frontiers*, Oxford, Bern: Peter Lang, pp. 83–108.

Lomibao, Laila and Ombay, Santos (2017), 'Does repetition with variation improve students' mathematics conceptual understanding and retention?', *International Journal of Science and Research*, 6:6, pp. 2131–37, https://www.ijsr.net/search_index_results_paperid.php?id=ART20174479. Accessed 4 April 2012.

London School of Hygiene and Tropical Medicine (2013), 'Chris Grundy and Anne Eggebert – John Snow legacy, art and how mapping is used in research today', *SoundCloud*, https://soundcloud.com/lshtm/chris-grundy-and-anne-eggebert. Accessed 26 August 2020.

Lopez-Perez, Daniel (2019), *R. Buckminster Fuller: Pattern-Thinking*, Switzerland: Lars Muller.

Love, Johanna (2020), 'Drawing dust', *Drawing: Research, Theory, Practice*, 5:2, pp. 193–208.

Lowe, Nicholas (2006), 'Corruption contraption: The *gidouille* as a transubstantiating device', *Pataphysica*, 3, pp. 130–49.

Lowndes, Sarah (2021), *One Immediate Huge Voice: Raymond Williams and Community*, London: Raymond Williams Foundation, https://www.raymondwilliamsfoundation.org.uk/centenary. Accessed 16 July 2021.

Lowndes, Sarah (2022), *Crowds, Community and Contagion in Contemporary Britain*, London, New York: Routledge.

Lubbock, Tom (2012), *English Graphic*, London: Frances Lincoln.

Luquet, Georges-Henry ([1927] 2001), *Children's Drawings* (trans. A. Costall), London: Free Association.

Mabille, Pierre (1936), *La Construction de l'Homme*, Paris: Jean Flory.

Mabb, David (2006), 'Art into everyday life', in *Mabb*, Vilnius: Siuolaikinio Meno Centras, pp. 28–33.

Mackenzie, Louise (2018), *Tentacular Resonances*, Dundee: Generator Projects.

Maharaj, Sarat (2002), 'Xeno-Epistemics: Makeshift kit for sounding visual art as knowledge production and the retinal regimes', *Documenta 11*, Platform 5, Hantje Cantz, pp. 71–84.

Maharaj, Sarat (2009), 'Know-how and No-How: Stopgap notes on "method" in visual art as knowledge production', *Art & Research: A Journal of Ideas Contexts and Methods*, 2:2, pp. 1–11, http://www.artandresearch.org.uk/v2n2/maharaj.html. Accessed 6 September 2022.

Mainberger, Sabine (2010), '"In the vortex of the spiral tendency": Questions of aesthetics, literature and natural sciences in the work of Goethe', *Estudos Avançados*, 24:69, pp. 203–18.

Mall, Amar (2007), 'Structure, innovation and agency in pattern construction: The Kolam of Southern India', in E. Hallam and T. Ingold (eds), *Creativity and Cultural Improvisation*, Oxford: Berg, pp. 55–78.

Mandler, Jean M. (2010), 'The spatial foundations of the conceptual system', *Language and Cognition*, 2:1, pp. 21–44.

Man Ray ([1920] 1967), *Dust Breeding*, New York: The Metropolitan Museum of Art, https://www.metmuseum.org/art/collection/search/271420. Accessed 31 August 2021.

Margulis, Elizabeth (2014), *On Repeat: How Music Plays the Mind*, New York: Oxford University Press.

Marinelli, Lydia (2006), 'Vorstellungen eines möbels', in L. Marinelli (ed.), *Die Couch: Vom Denken im Liegen*, Munich: Prestel, pp. 7–31.

Marks, Laura U. (2004), 'Haptic visuality: Touching with the eyes', *Framework: The Finnish Art Review*, 2, pp. 79–82.

Marks, Laura U. (2009), 'Information, secrets, and enigmas: An enfolding–unfolding aesthetics for cinema', *Screen,* 50:1, pp. 86–98.

Marks, Laura U. (2010), *Enfoldment and Infinity: An Islamic Genealogy of New Media Art*, Cambridge: The MIT Press.

Marks, Laura U. and Dana, Polan (2000), *The Skin of the Film: Intercultural Cinema, Embodiment, and the Senses*, Durham: Duke University Press.

Marta de Menezes Official Website (2003), 'Inner Cloud', Marta de Menezes, https://martademenezes.com/art/portraits/inner-cloud/. Accessed 19 July 2021.

Massey, Doreen (2005), *For Space*, London: Sage Publications.

Mayer, Andreas (2013), *Sites of the Unconscious: Hypnosis and the Emergence of the Psychoanalytic Setting* (trans. C. Barber), Chicago: University of Chicago Press.

McCosker, Anthony and Wilken, Rowan (2014), 'Rethinking "big data" as visual knowledge: The sublime and the diagrammatic in data visualisation', *Visual Studies*, 29:2, pp. 155–64, https://doi.org/10.1080/1472586X.2014.887268.

McQuillan, Patrick (2008), 'Small-school reform through the lens of Complexity Theory: It's good to think with', *Teachers College Record*, 110:9, pp. 1772–801.

McRobie, Allan (2017), *The Seduction of Curves: The Lines of Beauty that Connect Mathematics, Art, and the Nude*, Princeton, Oxford: Princeton University Press.

Mega, M. S., Cummings, J. L., Salloway, S. and Malloy, P. (1997), 'The limbic system: An anatomic, phylogenetic, and clinical perspective', *The Journal of Neuropsychiatry and Clinical Neurosciences*, 9:3, pp. 315–30.

Meggs, Philip B. (1992), *A History of Graphic Design*, 2nd ed., New York: Van Rostrand Reinhold.

Meillassoux, Quentin (2009), *After Finitude: An Essay on the Necessity of Contingency* (trans. R. Brassier), London: Bloomsbury.

Mershon, Ella (2016), 'Ruskin's dust', *Victorian Studies*, 58:3, pp. 464–92.

Mietzsch, Mario and Agbandje-McKenna, Mavis (2017), 'The good that viruses do', *Annual Review of Virology*, 4:1, pp. iii–v.

Mijuskovic, Slobodan (2009), 'Discourse in the indefinite person', in D. Evans (ed.), *Appropriation*, London: Whitechapel, pp. 142–48.

Ministry of Housing, Communities & Local Government (2021), 'Celebrating Religious Festivals During Coronavirus (COVID-19)', GOV.UK, 16 June, https://www.gov.uk/government/publications/covid-19-guidance-for-the-safe-use-of-places-of-worship-during-the-pandemic-from-4-july/celebrating-religious-festivals-during-coronavirus-covid-19. Accessed 16 July 2021.

Monod-Herzen, Édouard (1927), *Science et esthétique. Principes de morphologie générale*, Paris: Gauthier-Villars.

Mookerjee, Agit (1971), *Tantra Art, Its Philosophy & Physics*, New Delhi: Ravi Kumar.

Mookerjee, Agit (1975), *YOGA ART*, London: Thames & Hudson.

Moran, Joe (2019), *First You Write a Sentence: The Elements of Reading, Writing ... and Life*, London: Penguin Random House UK.

Morris, William (1972), *The Wood Beyond the World*, New York: Dover.

Mulhall, Stephen ([1996] 2005), *Heidegger and Being and Time*, London, New York: Routledge.

Mulvey, Laura (2006), *Death 24x a Second: Stillness and the Moving Image*, London: Reaktion.

Mundry, Isabel (2012), 'Regular irregular – On the fleeting quality of pattern in contemporary music', in A. Gleiniger and G. Vrachliotis (eds), *Pattern: Ornament, Structure, and Behavior*, Basel: Birkhäuser, pp. 85–96.

Murdoch, Iris ([1959] 1997), 'The sublime and the good', in P. Conradi (ed.), *Existentialists and Mystics: Writings on Philosophy and Literature*, London: Chatto and Windus, pp. 205–20.

Müri, René and Göbel, Nicole (2020). 'See faces in the clouds? It might be a sign of your creativity', in *Psyche*, London: Aeon Media Group Ltd., 2012–21, https://psyche.co/ideas/see-faces-in-the-clouds-it-might-be-a-sign-of-your-creativity. Accessed 20 October 2021.

Nancy, Jean-Luc ([2009] 2013), *The Pleasure of Drawing* (trans. P. Armstrong), New York: Fordham University Press.

Necipoğlu, Gülru (1995), *The Topkapı Scroll: Geometry and Ornament in Islamic Architecture*, Santa Monica: Getty Center for the History of Art and the Humanities.

Necipoğlu, Gülru (2015), 'The scrutinizing gaze in the aesthetics of Islamic visual cultures: Sight, insight, and desire', *Muqarnas*, 32, pp. 23–61.

Nelson, Robert S. and Shiff, Richard (eds) (2003), *Critical Terms for Art History*, Chicago: University of Chicago Press.

Neuroscience & Graphic Design (2022), 'A collaboration between neuroscientists and a graphic design researcher', https://neuroscience-graphicdesign.com/. Accessed 2 March 2022.

Newhall, Beaumont (1982), *The History of Photography*, London: Secker & Warburg.

Newman, Avis and de Zegher, Catherine (2003), *The Stage of Drawing, Gesture and Act*, London: Tate Publishing.

Newman, Eric A., Araque, Alfonso and Dubinsky, Janet M. (eds) (2017), *The Beautiful Brain: The Drawings of Santiago Ramón y Caja*, New York: Abrams.

O'Donoghue, Diane (2019), *On Dangerous Ground: Freud's Visual Cultures of the Unconscious*, London: Bloomsbury.

Olalquiaga, Celeste and Blackmore, Lisa (eds) (2018), *Downward Spiral: El Helicoide's Descent from Mall to Prison*, New York: Urban Research.

Ortero-Pailos, Jorge (2016), *The Ethics of Dust at Westminster Hall*, London: Palace of Westminster, Studio Otero-Pailos, http://www.oteropailos.com/the-ethics-of-dust-series. Accessed 31 August 2021.

Owens, Craig (1980), 'The allegorical impulse: Toward a theory of postmodernism', *October*, Vol. 12, Spring, London: The MIT Press, pp. 67–86.

Pacheco, Francisco ([1649] 1990), *El Arte de la Pintura* (ed. B. Bassegoda i Hugas), Madrid: Catedra.

'pareidolia', n. (n.d.), *Merriam-Webster Dictionary Online*, Springfield: Merriam-Webster, https://www.merriam-webster.com/dictionary/pareidolia. Accessed 24 June 2021.

Parsons, Talcott (1951), *The Social System*, New York: The Free Press.

Passalacqua, Karla, John, Purdy and Christiane, Wobus (2019), 'The Inert Meets the Living: The expanding view of metabolic alterations during viral pathogenesis', *PLoS Pathogens*, 15:7, p. e1007830, https://doi.org/10.1371/journal.ppat.1007830.

Peacock, J. (1995), *Book Production*, 2nd ed., London: Blueprint.

Pearson, Helen (2008), 'Virophage suggests viruses are alive', *Nature*, 454:7, p. 677, August.

Perec, Georges ([1985] 2011), *Thoughts of Sorts* (trans. D. Bellos), London: Notting Hill Editions.

Perret, J. L., Plush, B., Lachapelle, P., Hinks, T. S., Walter C., Clarke, P., Irving, L., Brady, P., Dharmage, S. C. and Stewart, A. (2017), 'Coal mine dust lung disease in the modern era', *Respirology*, 22:4, pp. 662–70.

Perry, Grayson (2014), *Playing to the Gallery*, London: Particular Books.

Phillips, Stephen J. (2017), *Elastic Architecture: Frederick Kiesler and Design Research in the First Age of Robotic Culture*, Cambridge: The MIT Press.

Plato (1971), *Timaeus and Critias* (trans. D. Lee), Harmondsworth: Penguin Books.

Plato ([*c*.375 BCE] 1974), *The Republic* (trans. D. Lee), London: Penguin.

Poe, Edgar Allan (1840), 'Philosophy of furniture', *Burton's Gentleman's Magazine and Monthly American Review*, 6, pp. 243–45.

Powers, Melvin (1956), *Advanced Techniques of Hypnosis*, London: Thorsons.

Prigogine, Ilya (1997), *The End of Certainty: Time, Chaos, and the New Laws of Nature*, Cambridge: The Free Press.

Prosser, Bill (2010), 'Nothings in particular', in D. Caselli (ed.), *Beckett and Nothing*, Manchester: Manchester University Press, pp. 84–106.

Proust, Marcel ([1954] 2017), *Chardin and Rembrandt* (trans. J. Feldman), New York: David Zwirner.

Purce, Jill (1974), *The Mystic Spiral*, London: Thames & Hudson.

Rampley, Matthew (2005), 'Art history and cultural difference: Alfred Gell's anthropology of art', *Art History*, 28:4, Autumn, pp. 524–51.

Rancière, Jacques ([2009] 2011), *The Emancipated Spectator*, London: Verso.

Rauschenberg, Robert (*c*.1950), *Dirt Painting for John Cage*, New York: Robert Rauschenberg Foundation, https://www.rauschenbergfoundation.org/art/artwork/dirt-painting-john-cage. Accessed 20 October 2021.

Rauschenberg, Robert (*c*.1953), *Untitled (Gold Painting)*, New York: Robert Rauschenberg Foundation, https://www.rauschenbergfoundation.org/art/artwork/untitled-gold-painting. Accessed 31 August 2021.

Rawson, Philip (1969), *Drawing*, Oxford: Oxford University Press.

Reddy, Michael J. (1993), 'The conduit metaphor: A case of frame conflict in our language about language', in A. Ortony (ed.), *Metaphor and Thought*, Cambridge: Cambridge University Press, pp. 164–201.

Rickard, Scott (2011), *The beautiful math behind the world's ugliest music*, Ted Talks, www.ted.com/talks/scott_rickard_the_beautiful_math_behind_the_world_s_ugliest_music/. Accessed 6 July 2021.

Rickles, Dean, Penelope, Hawe and Shiell, Alan (2007), 'A simple guide to chaos and complexity', *Journal of Epidemiology and Community Health*, 61:11, pp. 933–37.

Riley, N. D. (1938), *The Observer's Book of Butterflies*, London: Frederick Warne.

Rodés Sarrablo, Thaïs (2012), 'El Soporte de tela en la pintura Europea de los siglos XVI, XVII y XVIII', Bachelor thesis, Catalonia: Universitat de Lleida, http://hdl.handle.net/10459.1/46448. Accessed 21 August 2021.

Rohr, Doris (2016), 'Drawing on nature: The legacy of Ruskin's moral cosmos', Unpublished Ph.D., thesis Lancaster: Lancaster University, http://ethos.bl.uk/OrderDetails.do?did=1&uin=uk.bl.ethos.695894. Accessed 31 August 21.

Roossinck, Marilyn J. (2011), 'The good viruses: Viral mutualistic symbioses', *Nature Reviews Microbiology*, 9:2, pp. 99–108.

Ruskin, John ([1860] 1905a), *Modern Painters*, London: George Allen.

Ruskin, John ([1866] 1905b), 'The ethics of the dust', in E. T. Cook and A. Wedderburn (eds), *The Library Edition of the Works of John Ruskin*, Vol. 18, London: George Allen, pp. 189–368.

Ruskin, John ([1884] 2012), *The Storm-Cloud of the Nineteenth Century*, London: Pallas Athene.

Russell, Dave (1997), *Football and the English: A Social History of Association Football in England, 1863–1995*, Preston: Carnegie Publishing.

Saint Mary's University (2009), *Sand Vibrations Pattern – Chladni Plate*, Halifax: St. Mary's University, You Tube, 23 June, https://youtube/YedgubRZva8. Accessed 12 August 2021.

Sampson, Tony (2004), 'A virus in Info-Space: The open network and its enemies', *M/C Journal*, 7:3, (n. pag.), https://doi.org/10.5204/mcj.2368.

Sanderson, Grant (2020), 'Simulating an epidemic', 3Blue1Brown, 27 March, https://www.3blue1brown.com/lessons/epidemic-simulations. Accessed 5 October 2021.

Sanghvi, Yogesh S. (2005), 'A roadmap to the assembly of synthetic DNA from raw materials', in M. S. Garfinkel, D. Endy, G. L. Epstein and R. M. Friedman (eds), *Working Papers for Synthetic Genomics: Risks and Benefits for Science and Society*, MIT Libraries: Dpace@MIT, vol. 2007, pp. 15–33, https://dspace.mit.edu/handle/1721.1/39657. Accessed 19 July 2021.

Scandinavian Institute for Computational Vandalism (2019), *The Situationist Times*, issues 1–5 and presentation, http://vandal.ist/thesituationisttimes/. Accessed 14 June 2021.

Schade, Virtus (1964), 'Forgotten knowledge of the universe in the [*sic*] children's hopscotch', *The Situationist Times*, 5, pp. 36–39.

Schaeffer, Pierre (1985), *Faber et Sapiens: Histoire de Deux Complices*, Paris: Belfond, http://catalogue.bnf.fr/ark:/12148/cb348788048. Accessed 5 April 2021.

Schwenk, Theodor (1965), *Sensitive Chaos*, London: Rudolf Steiner Press.

Schwenk, Theodor (1996), *Sensitive Chaos: The Creation of Flowing Forms in Water and Air*, Forest Row, Sussex: Sophia Books.

Serres, Michel (1982), *The Parasite* (trans. L. R. Schehr), Baltimore: The Johns Hopkins University Press.

Service, Tom (2016), 'Stuck on repeat: Why we love repetition in music', *The Guardian*, 29 April, http://www.theguardian.com/music/2016/apr/29/why-we-love-repetition-in-music-tom-service. Accessed 3 April 2021.

Shakespeare, William ([1609] 1982), *Hamlet*, London: Methuen & Co.

Shaykh, Hannah (1983), 'The Persian carpet', in *Arabic Short Stories* (trans. D. Johnson-Davies), Berkeley: University of California Press, pp. 106–10.

Shearer, Rhonda Roland (n.d.), 'Marcel Duchamp's impossible bed and other "not" readymade objects: A possible route of influence from art to science', Part II, Marcel Duchamp.net, http://www.marcelduchamp.net/marcel-duchamp-impossible-bed/. Accessed 12 June 2021.

Sherrington, C. S. (1942), *Man on His Nature*, Cambridge: Cambridge University Press.

Sheldrake, Merlin (2020), *Entangled Life: How Fungi Make Our Worlds*, New York: Penguin Random House.

Shepherd, Walter (1971), 'Principles of approximation', in W. Shepherd (ed.), *Glossary of Graphic Signs and Symbols*, London: J.M. Dent and Sons, pp. 30–36.

Simmer, Stephen (1981), 'The academy of the dead: On boredom, writer's block, footnotes and deadlines', *Spring 1981*, pp.89–106.

Simpson, Nicola (ed.) (2012), *Notes from the Cosmic Typewriter: The Life and Work of Dom Sylvester Houédard*, London: Occasional Papers.

Skafsgaard Lowery, Rebecca (2019), 'Criss-cross and the gender of pattern painting', in A. Katz (ed.), *With Pleasure: Pattern and Decoration in American Art, 1972–1985*, New Haven: Yale University Press, pp. 105–21.

Snapes, Laura (2020), 'Captain Tom Moore becomes oldest artist to claim UK No 1 single', *The Guardian*, 24 April, https://www.theguardian.com/music/2020/apr/24/captain-tom-moore-becomes-oldest-artist-to-claim-uk-no-1-single. Accessed 2 March 2021.

Stevens, Peter S. (1974), *Patterns in Nature*, Harmondsworth: Penguin Books.

Stevens, Wallace (1971), *The Collected Poems of Wallace Stevens*, New York: Alfred A. Knopf.

Steyerl, Hito (2011), 'In free fall: A thought experiment on vertical perspective', *e-flux Journal*, 24 April, https://www.e-flux.com/journal/24/67860/in-free-fall-a-thought-experiment-on-vertical-perspective/. Accessed 22 October 2019.

Steyerl, Hito (2017), 'A sea of data: Apophenia and pattern (mis-recognition)', *Duty Free Art*, London, New York: Verso, pp. 47–63.

Stoppani, Teresa (2007), 'Dust revolutions. Dust, *informe*, architecture (notes for a reading of Dust in Bataille)', *The Journal of Architecture*, 12:4, pp. 437–47.

Sutton, Denys (1963), *Nocturne: The Art of James McNeill Whistler*, London: Country Life.

Tarkovsky, Andrei (1979), *Stalker*, Soviet Union: Mosfilm.

The Twilight Zone (1960–64, USA: CBS).

Thompson, D'Arcy Wentworth (2014), *On Growth and Form*, Cambridge: Cambridge University Press.

Thompson, Edward Palmer (2011), *William Morris, Romantic to Revolutionary*, Oakland: PM Press and Merlin Press.

Thompson, Jon (1983), *Carpets: From the Tents, Cottages and Workshops of Asia*, London: Barrie and Jenkins.

Thorp, Liam (2020), 'Liverpool coronavirus deaths soared after Atletico Madrid match', *Liverpool Echo*, 3 June, https://www.liverpoolecho.co.uk/news/liverpool-news/liverpool-coronavirus-deaths-soared-after-18354705. Accessed 3 March 2021.

Toombs, S. Kay (1988), 'Illness and the paradigm of the lived body', *Theoretical Medicine*, 9:2, pp. 201–26.

Toombs, S. Kay (1993), *The Meaning of Illness: A Phenomenological Account of the Different Perspectives of Physician and Patient*, vol. 42, Dordrecht: Kluwer Academic Publishers.

Tremblay, Caroline (2004), 'Le Carnivalesque dans Ubu d'Alfred Jarry', MA thesis, Quebec City: University of Quebec.

Tversky, Barbara (2002), 'What do sketches say about thinking?', in *AAAI Spring Symposium, Sketch Understanding Workshop*, Stanford University, AAAI Technical Report SS-02-08 2002, https://www.tc.columbia.edu/faculty/bt2158/faculty-profile/files/2002_Tversky_Whatdosketches-sayaboutthinking. Accessed 5 March 2021.

Tversky, Barbara (2011), 'Visualizing thought', *Topics in Cognitive Science*, 3, pp. 499–535.

Typography Deconstructed Website (2010), 'Tail', http://www.typographydeconstructed.com/tail/. Accessed 2 April 2016.

Tzara, Tristan ([1920] 1981), 'To make a Dadaist poem', in *Dada Manifesto on Feeble Love and Bitter Love*, reprinted in *Seven Dada Manifestos and Lampisteries* (trans. Barbara Wright), London: John Calder.

Van Regenmortel, Marc H. V. (2016), 'The metaphor that viruses are living is alive and well, but it is no more than a metaphor', *Studies in History and Philosophy of Science Part C: Studies in History and Philosophy of Biological and Biomedical Sciences*, 59, pp. 117–24, https://www.sciencedirect.com/science/article/abs/pii/S1369848616300115. Accessed April 2021.

Van Regenmortel, Marc H. V. and Mahy, Brian (2004), 'Emerging issues in virus taxonomy', *Emerging Infectious Diseases*, 10:1, pp. 8–13, https://www.ncbi.nlm.nih.gov/pmc/articles/PMC3322749/. Accessed April 2021.

Vasold, Georg (2015), 'Woven thinking: Textile art and the advent of modern art scholarship' in H. von Rike Frank and G. Watson (eds), *Textiles: Open Letter*, Berlin: Sternberg Press, Vienna: Generali Foundation, pp. 166–79.

Venturi, Robert (1966), *Complexity & Contradiction in Architecture*, New York: The Museum of Modern Art.

Villarreal, Luis (2004), 'Are viruses alive?', *Scientific American*, 291:6, pp. 100–05, http://www.jstor.org/stable/26060805. Accessed April 2021.

Virilio, Paul (2009), *The Aesthetics of Disappearance* (trans. P. Beitchman), Cambridge: The MIT Press.

'virus, n'. (2021), *Etymonline*, Tupelo: Douglas Harper, https://www.etymonline.com/word/virus. Accessed 8 August 2021.

Walker, Michelle Boulous (2017), *Slow Philosophy: Reading Against the Institution*, London: Bloomsbury Publishing.

Ward, Richard (1963), *Brooke Bond Picture Card British Butterflies*, London: Brooke Bond.

Warner, Marina (2011), 'Freud's couch: A case history', *Raritan*, 31:2, pp. 146–63.

Watson, Lani (2015), 'What is inquisitiveness', *American Philosophical Quarterly*, 52:3, pp. 273–88.

Weil, Simone (1952), *Gravity and Grace* (trans. A. Wills), Lincoln: University of Nebraska Press.

Wellcome Collection (1980), *RNA Tie Club*, https://wellcomecollection.org/works/j76x7bk6. Accessed 19 July 2021.

Whalley, J. Harry (2016), 'Before Alpha-Go: Interpretation though improvisation and LISP,' in *UK Music Informatics and Cognition (MIC)*, 20 July, Informatics Forum, Edinburgh: University of Edinburgh, https://research.uca.ac.uk/3403/. Accessed 19 October 2021.

Whalley, W. Brian (2019), 'Drawing as pattern information extraction: Linking geomorphology and art', *Drawing: Research, Theory, Practice*, 4:1, pp. 29–53.

Wheeler, Michael (ed.) (1995), *Ruskin and Environment: The Storm-Cloud of the Nineteenth Century*, Manchester, New York: Manchester University Press.

Williams, Raymond (1976), *Keywords*, London: Fontana Press.

Williams, Raymond ([1977] 1978), *Marxism and Literature*, Oxford: Oxford University Press.

Williams, Raymond ([1958] 1989), *Resources of Hope: Culture, Democracy, Socialism*, London: Verso.

Williams, Raymond ([1961] 2011), *The Long Revolution*, Cardigan: Parthian.

Williams, Raymond ([1973] 2016), *The Country and the City*, London: Vintage.

Williams, Raymond ([1958] 2017), *Culture and Society*, London: Vintage.

Winnicott, Donald ([1971] 1991), *Playing and Reality*, London: Routledge.

Winsby, Mira Bai (2006), 'Into the spiral: A conversation with Japanese horror maestro Junji Ito', *78 Magazine*, February–March, http://www.78magazine.com/issues/03-01/arts/junji.shtml. Accessed 18 June 2021.

Wollheim, Richard (1987), *Painting as an Art*, London: Thames and Hudson.

Woolf, Virginia (1982), *The Diary of Virginia Woolf, Volume 3, 1925–30* (ed. A. O. Bell), London: Penguin.

Yanagi, Sōetsu ([1972] 2013), *The Unknown Craftsman; A Japanese Insight into Beauty*, New York: Kodansha USA, Inc.

Yau, John (2018), *Philip Taaffe (Contemporary Painters)*, London: Lund Humphries.

Zuboff, Shoshana (2019), *The Age of Surveillance Capitalism: The Fight for a Human Future at the New Frontier of Power*, London: Profile Books.

Zuccari, Frank, Véliz, Zahira and Fiedler, Inge (2005), 'Saint John in the wilderness: Observations on technique, style, and authorship', *Art Institute of Chicago Museum Studies*, 31:2, pp. 30–45, see also pp. 105–08.

Index